Cancer on Trial

Cancer on Trial

Oncology as a New Style of Practice

PETER KEATING AND
ALBERTO CAMBROSIO

THE UNIVERSITY OF CHICAGO PRESS   CHICAGO AND LONDON

PETER KEATING is professor of history at the Université du Québec à Montréal. ALBERTO CAMBROSIO is professor in the Department of Social Studies of Medicine at McGill University. Together, they are the authors of *Exquisite Specificity: The Monoclonal Antibody Revolution* and *Biomedical Platforms: Realigning the Normal and the Pathological in Late Twentieth-Century Medicine*.

The University of Chicago Press, Chicago 60637
The University of Chicago Press, Ltd., London
© 2012 by The University of Chicago
All rights reserved. Published 2012.
Printed in the United States of America
21 20 19 18 17 16 15 14 13 12   1 2 3 4 5

ISBN-13: 978-0-226-42891-8 (cloth)
ISBN-10: 0-226-42891-5 (cloth)

Library of Congress Cataloging-in-Publication Data

Keating, Peter, 1953–
   Cancer on trial : oncology as a new style of practice / Peter Keating and Alberto Cambrosio.
      p. cm.
   Includes bibliographical references and index.
   ISBN-13: 978-0-226-42891-8 (cloth : alk. paper)
   ISBN-10: 0-226-42891-5 (cloth : alk. paper)   1. Oncology—Research.   2. Cancer—Research.
   I. Cambrosio, Alberto, 1950–   II. Title.
   RC254.5.K43 2012
   616.99'40072—dc23
                                                            2011019445

To my father, Dr. Douglas Keating (1924–2009)
PK

to pascale
a

Contents

Foreword

The publication of this book, *Cancer on Trial: Oncology as a New Style of Practice*, is most welcome, and the authors, Peter Keating and Alberto Cambrosio, are to be congratulated for such an achievement.

Cancer clinical trials are essential to develop more effective therapeutic strategies for cancer patients. Facing a tremendous increase in the incidence of cancer, both in the Northern and Southern Hemispheres, the medical community is facing a major challenge. And yet cancer clinical research is a rather new discipline. Its history and progress during the past sixty years are described very well in this book. The first clinical trials were conceived and conducted in the late 1950s. However, in Europe it is now the practice to celebrate Clinical Trials Day on 20 May, which is the anniversary of James Lind's observation (20 May 1747) on the prevention of scurvy. Although a young discipline, cancer clinical trials have brought to clinical investigators the need for cooperation, even international cooperation, and also that of multidisciplinarity.

This book provides an unparalleled look at the development of cancer clinical trials, the cooperative group systems in the United States and Europe, and the emergence of biostatistics as a key discipline in the development of scientifically rigorous studies. It describes in detail the transatlantic cooperation that took place early on: most of the European pioneers have indeed spent several years in the United States while at the same time Canadian investigators have spent time in Europe. The various chapters of this book illustrate the evolution of clinical trial methodology during this period as well as the current complexity of this discipline. Such knowledge is essential for the new young generation of clinical investigators to move forward.

Taking into account that 150 years ago surgery was conducted without anesthetic or anti-infective agents, we are indeed observing, in the current era of molecular biology and targeted therapy, significant progress in the understanding of cancer. As a result, numerous new possibilities to improve the survival and the quality of life of patients with this life-threatening disease have emerged and are being welcomed with cautious optimism. There are currently various opportunities to implement the results of basic science and to translate these discoveries into changes in clinical practice, all of which require sophisticated clinical trials. Of course these trials are expensive, and increased support to all partners for accelerating the development of optimal treatment is essential to maintain the capacity for medical excellence.

In 1909, Bleriot crossed the channel and only sixty years later in 1969, Armstrong went to the moon; it is hard to dream what oncology could be sixty years from now. We sincerely believe this book will significantly contribute to generate enthusiasm and commitment among the new generation of clinical trialists.

Françoise Meunier, MD, PhD, FRCP
Director General, European Organization
for Research and Treatment of Cancer (EORTC)

Richard Sylvester, ScD
Assistant Director and Head of Biostatistics, EORTC

The cooperative groups sponsored by the US National Cancer Institute (NCI) were formed beginning in the mid-1950s and have provided a publicly funded standing infrastructure to conduct cancer clinical trials ever since. The accomplishments of the US cooperative groups, often in collaboration with similar organizations in Europe and Canada, are notable for the development of curative therapies for most pediatric cancers; the introduction of effective combined modality treatments for many diseases, including head and neck cancer, cervix cancer, esophageal cancer, non-small-cell lung cancer, and anal cancer; the development of adjuvant chemotherapy as a standard of care for many solid tumors; and the introduction of chemoprevention strategies for patients at high risk of developing breast and prostate cancer. The groups have also worked collaboratively with the pharmaceutical industry to introduce new cancer drugs into medical practice and to develop new indications for already

approved drugs. Cooperative group studies have also provided important insights into optimal scheduling and sequencing of chemotherapy
drugs and the preferred route of drug administration that have established new standards of care. Importantly, cooperative groups have also
rigorously evaluated perceived advances in oncology care and, in some
cases such as high-dose chemotherapy and autologous stem cell trans-
plant for breast cancer, proven them ineffective. Beyond their scientific
contributions, the cooperative groups developed many of the commonly
used methods to conduct multicenter clinical trials, extended clinical trials to community oncologists, and provided a fertile training ground for
generations of clinical investigators.

In the current era of molecular medicine, the cooperative group bio-
specimen repositories, derived from patients enrolled in clinical trials
with uniform treatments and known outcomes, have become invaluable
resources for the development of new prognostic and predictive biomarkers. And with the recent emphasis on comparative effectiveness
and cost, the cooperative groups, as publicly funded entities free from
commercial bias, are ideally positioned to directly compare therapeutic
approaches or drugs from different sponsors, study drug combinations,
develop treatments for rare diseases, and rigorously examine the cost-
effectiveness of new treatments and their impact on patient quality of
life.

Despite their achievements, the cooperative groups have been subjected to scrutiny and criticism for nearly their entire existence. In the
last fifteen years alone, no fewer than five committees have been convened by NCI or at the request of NCI to evaluate the cooperative group
program. Most recently the Institute of Medicine (IOM) of the US National Academy of Sciences produced a report titled "A National Cancer Clinical Trials System for the 21st Century: Reinvigorating the NCI
Cooperative Group Program" that identified many challenges faced by
the cooperative groups at the present time, including stagnant funding,
inefficient processes, excessive and complex government oversight, and
increasing competition with the pharmaceutical industry for patient accrual to cancer clinical trials. The time to activate a phase III trial in the
cooperative groups now averages nearly two years, and the trials that
take the longest to launch are at high risk of failing to complete enrollment, often because the science has moved on and enthusiasm for the
trial design has waned during its long gestation. The IOM report calls

for strengthening the publicly funded cancer clinical trials in the United States and also for modifying the system so that it is scientifically nimble, efficient in trial launch and completion, and accessible to all members of the US population.

As we look to the future of cancer clinical trials in general, and the cooperative groups in particular, this volume by Peter Keating and Alberto Cambrosio provides a comprehensive and fascinating overview of the history of cancer clinical trials that will help us chart a course for the future. Clinical trials remain the fundamental tools by which we translate novel laboratory findings into products that benefit patients, and a robust and global approach to cancer clinical trials is vital to introduce safe and effective new treatments into the population, to target those treatments to patients most likely to benefit, and to thereby improve patient outcomes and control the cost of cancer care.

Richard L. Schilsky, MD
Professor of Medicine
Chief, Section of Hematology-Oncology
University of Chicago
Chairman, Cancer and Leukemia Group B 1995–2010

Acknowledgments

The photograph on the cover of this book provides a glimpse of *Le bois sacré* (1993), a work by French artist and friend Alain Diot (http://www.alain-diot.com/). Alain died of cancer in March 2004. But this is not why we are using his work. As one unfolds the rolls of *Le bois sacré*, a tight mesh of bodies gradually appears, their silhouettes carved out less by their own bodily boundaries than by the interstices interconnecting them. They emerge from a common magma, from their "fragilities and solidarities," as Alain would have said. The silhouettes seem to fall, or maybe they float; in any event, they move in a space without obvious coordinates or clear territory. They remind us of the hundreds of thousands of past, present, and future patients enrolled in clinical trials, all the more so in that the rolls of *Le bois sacré* simultaneously evoke the long sheets of paper on which the sequential steps of clinical trials protocols are inscribed. We would like to thank Alain's wife, Janine Ledu, and the photographer, Patrick Box, for permission to reproduce this picture.

As the adage goes, "money is the sinew of war." We therefore gratefully acknowledge financial support from the following agencies: the Social Sciences and Humanities Research Council of Canada (grants 410–2002-1453, 410–2005-1350, and 410–2008-1833), the Canadian Institutes of Health Research (grants MOP-64372 and MOP-93553), and the Fonds Québécois de Recherche sur la Société et la Culture (grants 01-ER-70743, SE-95786, and SE-124896).

Intellectual connections are also essential. We were lucky enough to be able to spend sabbatical years and short-term stays in stimulating environments. PK had the luxury of a full year at the National Humanities Center in North Carolina in 2000–2001 funded by the Burroughs Wellcome Fund to get the ball rolling. Fellow fellows know the value of the experience and the deep sense of gratitude we feel toward the center. Fred Tauber, UQÀM (Université du Québec à Montréal), and the Department of Philosophy at Boston University made possible a subsequent year at the Boston University Center for the Philosophy and History of Science (across the street from the ECOG Data Center). Finally, Harry Marks and the Department of the History of Medicine at Johns Hopkins welcomed PK for a half-sabbatical in 2010, a leave of absence once again made financially possible by UQÀM. AC spent a sabbatical year in 2004–5 at Centre de Recherche Médecine, Science, Santé et Société (CERMES, INSERM Unit 502) in Paris, with support from a Canadian Institutes of Health Research/INSERM International Exchange Agreement. Previous short-term stays at the Max Planck Institute for the History of Science in Berlin, and at the now defunct History of Modern Biomedical Science Unit of the Wellcome Institute for the History of Medicine in London (both in 1999), as well as a full year spent in 2001 at the INSERM Unit 379 (Épidémiologie et sciences sociales appliquées à l'innovation médicale) in Marseille, coincided with the beginnings of the project on which this book is based. Additional short stays at the ESRC Centre for Social and Economic Aspects of Genomics (CESA-Gen), University of Cardiff (UK) in 2005, at the Centre de Sociologie de l'Innovation of the École Nationale Supérieure des Mines de Paris in 2007, and at the UMR 912 SE4S (Sciences Économiques et Sociales, Systèmes de Santé, Sociétés) in Marseille in 2008 and 2009 provided opportunities for further fieldwork in Europe.

A series of workshops on the history of cancer gave us a chance to test our ideas on a number of experts: we would like to mention, in particular, the meetings organized in Manchester by John Pickstone and Carsten Timmermann in March 2007 and October 2005; in Bethesda (Maryland) by David Cantor in November 2004, and in Paris by Bernardino Fantini in May and October 2006. Several other meetings, workshops, and invited talks that for brevity's sake we will refrain from listing provided ample occasion to reformulate our initial understanding of the development of cancer clinical trials. Our thanks to all the colleagues who took time to listen to, read, and criticize our presentations, among whom we would like to mention, once again, the late Harry Marks (1947–2011) whose unfailing and critical support made a world of difference.

Throughout the footnotes readers will encounter the names of many of the clinicians and scientists who agreed to be interviewed, answer our often-repeated questions, and give us access to personal and institutional documents and archives and without whom this book would not have been possible. Time, especially for clinicians who have to take care of patients, is a rare commodity; we are therefore especially grateful for their availability and welcome. In particular, we would like to thank Dr. Françoise Meunier (director general, EORTC) who welcomed us at the organization's headquarters in Brussels and gave us free access to the EORTC archives.

The network maps used in the book, especially in chapter 8, have been produced in collaboration with Andrei Mogoutov, who developed the software program ReseauLu X2, taught us the basic tricks of sciento-metric and network analysis, and carried out most of the computer data processing. AC has a particular debt vis-à-vis the faculty, staff, and graduate students of McGill University's Department of Social Studies of Medicine, who have created an ideal environment in which to pursue interdisciplinary work on the development of biomedicine; a special thank-you to Lana Povitz for her truly careful formatting of footnotes and references; Heike Faerber, who graciously performed the mind-boggling task of obtaining copyright permissions for the book's illustrations; and Jean-Paul Acco of the Neuro Media Services of the Montreal Neurological Institute & Hospital for skillfully reworking the figures. PK similarly thanks members of the Département d'histoire of UQÀM and the Centre interuniversitaire de recherche sur la science et la technologie, also housed at UQÀM, for their support and for the convivial academic climate created by the researchers and students.

Early versions of some of the material used in this book have been published in a number of articles:

Keating, P. & Cambrosio, A. (2009). Who's minding the data? Data managers and data monitoring committees in clinical trials. *Sociology of Health & Illness* 31: 325–42.

Keating, P. & Cambrosio, A. (2007). Cancer clinical trials: The emergence and development of a new style of practice. *Bulletin of the History of Medicine* 81: 197–223.

Keating, P. & Cambrosio, A. (2006). Le criblage des médicaments: Les modèles animaux en recherche thérapeutique au National Cancer Institute (1955–2000). In G. Gachelin, ed., *Les organismes modèles dans la recherche médicale*. Paris: Presses Universitaires de France, 181–208.

Cambrosio, A. & Keating, P. (2006). Cancer clinical trials: New style of research, new forms of risk. In F. Walter, B. Fantini & P. Delvaux, eds., *Les cultures du risque (XVIe–XXIe siècle)*. Geneva: Presses d'Histoire Suisse, 169–86.

Keating, P. & Cambrosio, A. (2005). Risk on trial: The interaction of innovation and risk factors in clinical trials. In T. Schlich & U. Tröhler, eds., *The Risks of Medical Innovation: Risk Perception and Assessment in Historical Context*. London: Routledge, 225–41.

Keating, P. & Cambrosio, A. (2003). Beyond bad news: The diagnosis, prognosis, and classification of lymphomas and lymphoma patients in the era of biomedicine (1945–1995). *Medical History* 47: 291–312.

Keating, P. & Cambrosio, A. (2002). From screening to clinical research: The cure of leukemia and the early development of the cooperative oncology groups, 1955–1966. *Bulletin of the History of Medicine* 76: 299–334.

Introduction

From "Nonentity" to Global Network

The Rise of a New Style of Biomedical Practice

C'est toujours au confluent de rencontres, de hasards, au fil d'une histoire fragile, précaire, que ce sont formées les choses qui nous donnent l'impression d'être les plus évidentes. (Foucault, 1994b: 449)

An Early Career in Clinical Oncology

In the early 1960s a young Canadian physician nearing the end of his medical degree, Pierre Band, developed an interest in cancer treatment. In order to explore his options for further study in the field, he contacted the American Medical Association (AMA). The reply was somewhat disheartening: the AMA representative was "not sure that [he could] provide [Dr. Band] with exactly the sort of information" he sought. The AMA had discontinued approval of cancer residency-training programs because the few that existed had "varied so widely in nature and duration." He suggested instead that Dr. Band contact the American Cancer Society (ACS) and the US National Cancer Institute (NCI).[1] Like the AMA, the ACS replied that "there really [wasn't] any recommended special training for oncology" and pointedly added that a "cancer specialist" was a "non-entity."[2]

---

1. American Medical Association (AMA) to Dr. Pierre Band, 26 October 1961 (Dr. Band's personal documents).

2. Dr. Ronald N. Grant (ACS) to Dr. Band, 20 November 1961 (Dr. Band's personal documents).

While there were no "cancer specialists" at the beginning of the 1960s, the ACS did admit the existence of two "outstanding" cancer institutions—namely, Buffalo's Roswell Park Memorial Institute and New York City's Memorial Cancer Center. In other words, the lack of cancer specialists did not signify a deficiency of treatments or treating institutions but rather that the traditional specialties such as internal medicine, surgery, or radiology continued to control cancer treatment. Doctors could become a "so-called clinical cancerologist" only by specializing in those fields.[3] In response to Band's inquiry, the NCI thus counseled that he first train in one of the established medical specialties and then join a clinical center devoted the study and treatment of malignant diseases.[4]

In spite of the discouraging news, Dr. Band continued his quest for training in cancer therapy. Funded by a National Institute of Cancer of Canada fellowship in 1964–65, he spent eighteen months in Georges Mathé's unit at the Gustave-Roussy Institute in Villejuif on the outskirts of Paris, arguably then, as today, the leading French cancer hospital. Subsequently, from 1966 to 1969, having availed himself of a National Institutes of Health (NIH) fellowship, he worked at the Roswell Park Memorial Institute.[5] The two fellowships not only familiarized him with the routines of two premier cancer institutions, they also introduced him to a new organization for clinical research that embodied a new "style of practice," cooperative oncology groups.

Unlike previous cancer treatment and research programs that had functioned in single institutions, cooperative groups treated and studied patients across multiple cancer treatment centers using common protocols. In Europe, Band's mentor, Georges Mathé, was one of the originators of Europe's first cooperative oncology group: the Groupe Européen de Chimiothérapie Anticancéreuse (GECA), founded in 1961 and the forerunner of the European Organization for Research and Treatment of Cancer (EORTC). In America, Roswell Park acted as one of the principal sites of the Acute Leukemia Group B and of the Eastern Cooperative Oncology Group (ECOG), established in 1955 by members of the NCI under the name of Eastern Solid Tumor Group (figure 1.1). Dr. Band

3. Dr. Grant to Dr. Band.

4. The NCI nonetheless offered special fellowships in cancer research. Dr. Norah DuV. Tapley (NCI) to Dr. Band, 14 March 1962 (Dr. Band's personal documents).

5. "Band, Pierre Robert," *American Men & Women of Science*, 23rd ed., p. 326. Interview with Dr. Pierre Band, Montreal, 18 October 2007.

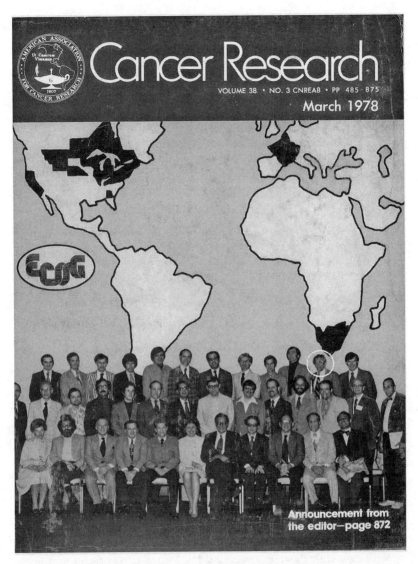

FIGURE I.I. Front cover of the March 1978 issue of *Cancer Research* showing a picture (undated, but probably taken that same year) of ECOG principal investigators sitting in front of a map of their locations. Dr. Band is the second from the right on the third row (circle). Adapted and reprinted by permission from the American Association for Cancer Research.

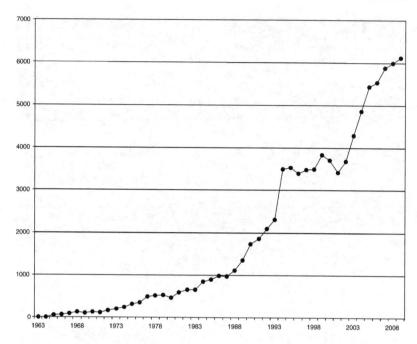

FIGURE I.2. Number of articles related to cancer clinical trials listed in PubMed.

attended his first ECOG meeting in 1966 and when he later joined the University of Alberta in Edmonton, the university became the first Canadian ECOG affiliate. In the early 1970s Dr. Band was elected chairman of ECOG's breast cancer committee and designed a landmark clinical trial protocol that will be discussed later in this book. Through these organizations, Dr. Band ultimately pursued a successful career as a chemotherapist and a cancer clinical trial specialist, both as a member of Canadian clinical and academic centers and as a principal investigator for ECOG and GECA/EORTC.

We begin with this vignette about Dr. Band's career because it highlights the remarkable rise of cancer treatment research. Consider the following figures that contrast sharply with the idea of cancer treatment as a marginal pursuit. In the United States at the end of the first decade of the twenty-first century, the US National Cancer Institute, the largest funding source for clinical cancer trials, sponsored 14,000 investigators at over 3,000 institutions, enrolling 25,000 patients in clinical trials each year. In Europe, 300 hospitals and cancer centers in more

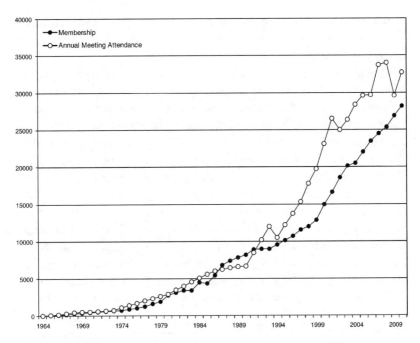

FIGURE 1.3. Number of ASCO members and of people attending the ASCO annual meeting. Data sources: ASCO (2004: 15, 21, 27, and 37) and ASCO membership office.

than thirty countries and almost 3,000 scientists and clinicians participated in EORTC activities by 2010. The number of yearly articles related to cancer clinical trials grew from fifty-six in 1965 to over 6,000 in 2009 (figure 1.2). Similarly, the American Society of Clinical Oncology (ASCO) expanded from sixty-six members in 1964 to more than 27,000 in 2010, when it attracted 32,700 attendees (26,600 professional and 4,370 exhibitors) to its annual meeting (figure 1.3). Founded in 1975, the European Society for Medical Oncology (ESMO), the continental equivalent of ASCO, counted some 5,000 members in 2010 and drew 16,000 participants to its annual meeting. Last but not least, anticancer drugs have become a major pharmaceutical market, and cancer research is now pursued on a global basis (Eckhouse, Lewison & Sullivan, 2008; Chabner & Roberts, 2005) with an estimated global anticancer market of $70 billion in 2008.[6] The exponential growth of clinical cancer research has pro-

6. At http://knol.google.com/k/global-cancer-market-review-2008-world-top-ten-cancer-drugs#.

duced concomitant changes in routine cancer treatment: the vast majority of (Western) cancer patients are treated according to treatment protocols established via multicenter clinical trials.

In short, while "cancer specialists" or "clinical oncologists" have become common figures in the medical landscape, they did not exist a few
decades ago. In the early 1960s a cancer specialist was not only a nonentity, many practitioners also regarded the treatment of terminally ill cancer patients with heroic courses of chemotherapy as highly questionable.
The randomized clinical trials that today form the basis of medical oncology were relatively rare and aroused outright opposition from physicians loath to assign patients randomly to competing treatments. Barth
Hoogstraten, a Dutch-born physician who immigrated to the United
States after the war to become one of the pioneers of cancer chemotherapy, has recalled how in the mid-1950s patients admitted to the NCI's
Department of Medicine were treated by physicians "who had no formal training in oncology for the simple reason that the medical schools
did not yet offer such programs, and the word *oncology* did not even exist" (Hoogstraten, 2005: 82).[7] Emil Frei III, another founding father of
US chemotherapy, recalled that the "greatest underlying controversy"
of those early years was whether one should treat cancer patients with
chemotherapy since the treatment was often worse than the disease. In
developing cancer chemotherapy, he and his colleagues in Europe thus
confronted a "cynical" or "skeptical" old guard.[8] Maurice Tubiana, a
leading postwar French oncologist, has described how, during the introduction of clinical trials at the Gustave-Roussy Institute in the early
1960s, "several physicians were scandalized and reacted with hostility. . . .
The controversy soon turned into a quasi-religious war; some physicians,
including the head of the surgery department, a very distinguished and
prestigious person, decided to resign from a hospital where the fate of
patients was decided by tossing a coin" (1995: 301–2, our translation).
How then did such a spectacular change in the medical landscape occur?

7. According to the *OED*, "oncology" is listed in an 1860 lexicon of medical terminology as an 1857 term "for the doctrines of boils and tumours." It appears again in 1915 in a
statistical analysis of "cancer, or what is, perhaps, more appropriately termed the science
of oncology," and a 1935 issue of *Science* mentions a professorship of oncology at Cornell
University's Medical College.

8. E. Frei III, statement made for the DVD that is included with the ASCO (2004)
monograph.

How did medical oncology move from a nonentity and in some regards a reviled practice to the central position it now occupies in modern medicine? This book attempts to answer these questions.

About This Book

Writing the history of a laboratory or organizations such as the Roswell Park Memorial Institute, the NCI, ASCO, or ECOG is a relatively straightforward task; writing the biography of a few key individuals in cancer research or an intellectual overview of the development of the field would also be a relatively clear-cut undertaking. Examples of these historical genres abound, and they are valuable additions to the literature. None of these forms of historical inquiry seemed to capture the research we wished to pursue. We had decided from the outset not to write a stand-alone *institutional* history of cancer organizations, a *conceptual* history of the evolving understanding of cancer, or a *political* or *economic* account of the governmental and industrial shaping of the cancer market. Rather, our research sought to capture all those aspects of cancer research and treatment within a single frame. We hoped to avoid the snare of writing a "total" history of the meteoric rise of cancer treatment and research. Rather than the elusive global picture, we sought more modestly to describe the evolving configurations of institutional, organizational, conceptual, human, and material resources that, through a remarkably dense network, have produced the now-taken-for-granted pillars of modern oncology: cancer clinical trials, treatment protocols, and the like.

The organizing principle of this book is simple: in addition to the present introductory section, there are three main parts, each corresponding to a major period in the development of cancer clinical trials. Each part begins with the description of a clinical trial that is paradigmatic of the corresponding period. Subsequent chapters focus on the novel aspects of the trial, such as the kinds of drugs used; the material, organizational, and statistical infrastructure of the trial; and its goal(s) and nature. Of course, no single trial captures all the subtleties and complexities of any given period, and each part thus describes other trials and related events. In spite of its shortcomings, introducing each part with a concrete undertaking—a specific clinical trial—has a clear and distinct advantage: it allows us to discuss *practices*—to "approach our topic by analyzing what people did" (Foucault, 1994b: 634–35, our translation)—rather than focusing on structures such as professions, specialties, or organizations.[9]

Practices do involve such institutions (macroactors, some sociologists would say) and we discuss them in the course of this book. But—*pace* the social sciences with their neat disciplinary borders—practices, especially in emerging areas such as clinical oncology in the post–World War II era, do not respect the settled boundaries between social, economic, cultural, and organizational domains; they reconfigure existing structures in often unexpected ways and they mobilize resources assigned to other tasks. In short, they generate fluid arrangements that sometimes congeal into stable innovations (at least temporarily) and sometimes do not. Given that our interest lies in the establishment of cancer clinical trials and related practices, we have opted for an empirical analysis of the processes that led to this outcome.

By focusing on practices and their outcomes, we have resisted the temptation, felt by many social scientists, to use the past to explain the present and, in particular, to account for innovations by reducing them to preexisting practices or structures: while novelty necessarily emerges from existing structures—fully de novo creations do not exist—such novelty cannot be reduced to elements of those structures or to any deterministic interaction thereof.[10] According to Paul Rabinow (2000: 44), a social science bent on explaining away the emergence of new forms by condensing them into a set of historical or social determinants is unable to acknowledge singularities—that is, new assemblages that "make things work in a different manner."[11] Accounts of innovation should similarly avoid privileging, a priori, single factors or structures, whether social, technical, cultural or economic. As in our previous work on biomedical platforms (Keating & Cambrosio, 2003), we consider technological determinism—the explanation of history through technical or scientific breakthroughs—and sociological reductionism—positing social or

9. In addition to being central to Foucault's genealogical approach, a "practice turn" characterizes recent developments in the field of science and technology studies. Key contributions include Bruno Latour's description of the main tenets of "actor-network-theory" (ANT) (1987; 2005) and Andrew Pickering (1992). Joseph Rouse (2002) provides a general philosophical argument about the role of scientific practices.

10. For a criticism of the notion of explanation in the social sciences, see Latour (1988: 155–76).

11. For another contribution that explicitly follows Rabinow's Foucauldian advice, see Landecker (2007).

political or cultural factors as the ultimate explanation of medical and techno-scientific developments—as mirror images of the same mistake. In short, we look for singularities rather than regularities, and we do so by recognizing that "there are no fundamental phenomena. There are only reciprocal relations, and perpetual gaps between them" (Foucault, 1994b: 277, our translation).[12]

One of the consequences of our approach is that we do not deploy one of the standard sociohistorical tricks of the trade—namely, comparisons: between countries, patterns of research and education, different kinds of cancer, and different kinds of hospitals. To understand why, recall that in our initial vignette we saw how Dr. Band's early training spanned two continents and how clinicians and researchers established institutions such as ECOG and the EORTC on both sides of the Atlantic. While this would normally prompt comparisons between America and Europe in order to uncover sociocultural differences between them, our take here is quite different: the vignette foregrounds a network that links European and North American oncologists. In other words, the emergence and development of this network attracts our attention. Comparative analysis, in contrast, reworks the recurrent theme of the constraints and determinants of practice by showing how perceptions and actions are molded through vaguely defined local "institutions." We have instead tried to understand how practitioners produced increasingly similar perceptions, actions, and practices through the constitution of clinical research networks that span countries and continents or, to use a more abstract terminology, through the establishment of a new style of practice.

The establishment of oncology networks linking North America, Europe, and Japan has been the explicit goal of organizations such as the NCI. Most of the founders of clinical oncology in Europe transited through US institutions, and the NCI established a European outpost (a liaison office) in Brussels in 1972 to foster the integration of Western oncology and to gain greater access to European anticancer agents. Thus instead of looking for differences corresponding to preset categories (nations, hospitals, professions, insurance schemes, etc.), we suggest that one may profitably investigate the work implicated in the production of similarities without overlooking manifest differences. More often than not the constitution of a common practice is antecedent to the production of

12. For a discussion of this issue and of Foucault's criticism of causal approaches, see Veyne (2008).

differences in unexpected quarters (e.g., larger differences within, rather than between nations). The history of cancer clinical trials amounts to an attempt to create a network spanning Europe and North America, articulated through a few major hubs (e.g., the NCI) and through which all sorts of entities (oncologists, protocols, drugs, etc.) circulate.

Contrary to the traditional medical specialties that developed locally and nationally before joining transnational initiatives, medical oncology and cancer clinical trials have pursued an international or, at least, a multicenter project (maps such as the one on the background of figure 1.1 are often used in the trialists' reports) from the outset. The alignment of clinical and research interfaces across numerous national and international sites has constantly preoccupied practitioners in this field, as has the production of clinically meaningful differences across clinical sites. During its 1966 congress in Tokyo, the Union for International Cancer Control (UICC) launched a Project on Controlled Therapeutic Trials to list ongoing and unpublished trials *and* to "keep under review" the methodology of the trials. In 1974 the UICC redefined the scheme to focus on the difficulties ("whether methodological, ethical or practical") raised by clinical trials. Reoriented yet again in 1978, the project now concentrated on the methodology of trials "with a view to rendering them easier to conduct and to have them yield more valid conclusions" (Flamant & Fohanno, 1982: 1–3). As part of the project, members published international surveys of randomization practices in cancer trials. The surveys had an explicitly normative edge ("to make recommendations regarding how randomization should be carried out") (Pocock & Lagakos, 1982: 5). Similar studies investigated alternative trial designs and the rationale for designing large trials (Grage & Zelen, 1982; Peto, 1982). Explicit recommendations for large trials included "Make complexity optional," "'Flag' all patients," "Keep a log," and "Telephone randomization" (no e-mail back then). As in the field of international health where American bilateral agreements competed with international projects, the NCI sponsored collaborative programs with Europe and Japan that promoted the free flow of drugs, scientists, and information and that led therefore to the harmonization of practices (Wittes & Yoder, 1998; Newell & Sugano, 1977; Goldin, Muggia & Rozencweig, 1979a).

At this stage another component of our opening vignette becomes relevant. In his quest to become a bona fide clinical oncologist, Dr. Band underwent training in elite cancer centers. But as we also saw, he also became an active member of cooperative oncology groups. Band's parallel pursuits highlight the dual nature of cancer treatment and research and the fact that although both activities take place within local institutions they cannot be reduced to local practices. Treatment and research in cancer both marshal multiple centers and, as we will see, evolve in a virtual world of statistical analysis and multi-institution databases. Moreover, not only do cancer trials occur at a local and a global level, but also the "location" of the cooperative groups that conduct these trials is less individual institutions than the abstract entity, the "protocol" (more on this below) that defines a given trial. The names of the cooperative groups similarly defy territorialization either through vague boundary definitions (i.e., *Eastern* Cooperative Oncology Group, *Southwest* Oncology Group, *North Central* Cancer Treatment Group), or by calling upon organizational principles that transcend any single institution by reference to historical contingency (Cancer and Leukemia *Group B*), an organ system or a closely related specialty (*Gynecologic* Oncology Group, National Surgical Adjuvant *Breast* and *Bowel* Project), a medical specialty (American College of *Surgeons* Oncology Group, *Radiation Therapy* Oncology Group), or an age group (*Children's* Oncology Group).

The cooperative groups are both distributed and centralized. Historically, the management of multi-institution clinical trials soon fell under the responsibility of group "statistical centers" or "data centers" that dealt with protocol approval, quality control, data auditing, and such, in short, with the definition of trial procedures. The centralized detection of "outlier centers"—of participating clinics that submitted data of questionable value—became an ongoing concern of the central units and exemplified the key regulatory role they played (e.g., Sylvester et al., 1981). The reasons for disciplining participating centers were not simply practical or organizational: they struck at the epistemic heart of the undertaking. By developing an activity predicated upon the collation of data from multiple sites, trialists not only had to ensure that the participating centers followed the protocol—without which a trial would amount to little more than a set of unrelated observations—but also, and more interestingly, that the differences in outcomes (if any) that the trial was designed to measure could be ascribed to natural (biological, pharmacological) factors rather than institutional or practice-based differences. Cancer clinical trials, in other words, generated their own regulatory tools beyond those offered by more traditional medical institutions such as hospitals and specialties, and, in so doing, became self-regulating entities.

A Capsule History of Cancer Clinical Trials

It is now time to briefly introduce the three periods into which we have divided the history of cancer clinical trials. The boundaries of each period are permeable, not only because there are no clear-cut borders (there is always some continuity in the discontinuity and vice versa) but also because our focus on the emergence of a new style of practice (we discuss this term below) rather than on national institutions must accommodate minor disparities between North America and Europe. Far from emerging fully armed from Minerva's head, oncology is the product of tinkering and ad hoc adjustments. In other words, contrary to the impression fostered by methodology handbooks, clinical cancer trials were not the simple and direct application of a few principles or methods; rather, they grew out of a complex history of innovations and solutions to emerging problems. Implementation of the first multicenter trials and of the institutions for their conduct (the cooperative groups) generated problems whose solution modified established practice, a process that in turn generated new problems, and so on.

As we will see in the Prelude, prior to World War II cancer therapy rested primarily on two modalities: surgery and radiation therapy. Chemotherapy emerged as a third modality (in parallel with endocrine cancer therapy) in the 1950s. Access to these different modalities varied not only in time but also according to the type of cancer, the country, and the treating hospital (Pickstone, 2007). Rather than focusing on these differences, in this section we briefly examine the emergence and development of chemotherapy (and, correlatively, of medical oncology) and clinical experimentation as embodied in clinical trials. Once established, clinical trials went on to comprise radiotherapy and surgery in multimodal treatment protocols. In other words, the experimental turn and the new style of practice engendered by chemotherapy created ripple effects that touched on all aspects of cancer treatment.

Initially devoted to testing the chemical compounds provided by an NCI screening program established in the 1950s (see below), by the 1970s the NCI cooperative group system had spread to community hospitals from their original sites in research organizations like the NCI or the Roswell Park Institute through programs designed to increase accrual to clinical trials. As a consequence, clinical investigators—"trialists" and their advanced research protocols—came into direct contact with community oncologists (most of whom had initially trained with the trialists).

At the same time, as distributed research organizations, by the mid-1960s the cooperative groups no longer simply tested drugs: they tested hypotheses concerning therapy, examined pathophysiological mechanisms, and sought means to prevent cancer. These studies were integrated into the clinical trial protocols and became such a significant part of clinical cancer research that by the 1990s, half of all Phase III clinical trials conducted by the cooperative groups incorporated ancillary laboratory and correlative studies.

As in the United States, transnational cooperative groups in Europe under the aegis of the EORTC performed clinical trials by specializing in a given type of cancer or in a given type of trial or task, and many discussions targeted the distinctive features of a cooperative group as opposed to working parties or task forces and how an EORTC cooperative group should properly function. In this respect the establishment of the data center in Brussels in 1974, in the wake of similar centralizing moves in the United States, constituted an important institutional milestone. The creation of the EORTC data center coincided with the definition of common statistical approaches and procedural guidelines that came to regulate clinical trials and thus went hand in hand with the certification of clinical cancer trials as authentic products of the new style of practice.

Following successes in the mid-1960s and early 1970s with the leukemias and the lymphomas, clinical trials moved from relatively short-termed studies, using single agents and patients in advanced-stage disease, to much longer-term, Phase III studies entailing the treatment of earlier phases of disease. While the initial wave of immunotherapies produced relatively disappointing results in the 1980s, the 1990s saw the emergence of a panoply of biologicals and, subsequently, of so-called targeted-therapy drugs. In spite of an apparently continuous development, we can distinguish three periods in the development of cancer clinical trials:

The first, stretching from the 1950s to the mid-1960s, corresponds to the emergence of cancer chemotherapy treatment and research. This period saw the establishment of major programs to find anticancer drugs, such as the screening program sponsored by the NCI's Cancer Chemotherapy National Service Center and the emergence of the first chemotherapeutic regimens and the cooperative oncology groups, whose stated goal was the production of those regimens. This initial period also produced several prominent characteristics of the cancer clinical trial system, such as the division of trials into different kinds or "phases."

The second, running from the mid-1960s to the end of the 1980s, was dominated by large-scale clinical trials that compared potentially limitless combinations of substances and modalities as part of what recent critics have characterized, somewhat unfairly, as "Coke vs. Pepsi" investigations. Work carried out during this period shared assumptions concerning the fractional efficacy of chemotherapeutic drugs in killing tumor cells (the "cell kill hypothesis") and the growth behavior of tumor cells ("cell kinetics"), both of which provided a rationale for the development of "combination" and "adjuvant" chemotherapies and thus a justification for what trialists, branded as blatant empiricists, described as a "rational empiricist" approach (Carter, 1995). At the end of the 1980s, however, a sense of crisis pervaded the system. Fueled by poor results (the lack of a "cure" for cancer), researchers concluded that chemotherapy had reached a "plateau in cytotoxic drug discovery" (Hollander & Gordon, 1990).[13]

The third, beginning in the 1990s, can be characterized as a "molecular turn" insofar as it is based on the re-alignment of the relations between (molecular) biology, the clinic and the new biotech industry. Almost paradoxically, the molecular turn also gave rise to the perception of a growing gap between basic and clinical research. Under the heading of *translational research*, numerous research and funding programs sought to recapture the lost, though largely "mythical" (Barthes, 1987), unity of treatment and research implied by the term *biomedicine*. While not limited to cancer, it is noteworthy that the expression *translational research* made its debut in 1993 with the discovery of cancer genes, "which suggested immediate applications in early detection and treatment of cancer" (Butler, 2008: 841). Massive investments in molecular biology resulted in the development of "targeted therapies," of which Imatinib (Gleevec) and Herceptin have become paradigms (Vasella, 2003; Bazell, 1998). Without eliminating large Phase III clinical trials, the proliferation of targeted compounds has refocused the attention of researchers and investors (looking for lucrative new drugs) on exploratory Phase I/II trials.

To sum up, medical oncology is built around the performance of clinical trials, the principal vector for the development of chemotherapy. Clinical

13. See the debate between John Cairns, Peter Boyle, and Emil Frei (Cairns & Boyle, 1983) about the number of cancer patients cured by chemotherapy.

trials also envelop the two other anticancer therapies, radiotherapy and surgical therapy, and combinations of the three modalities. In addition, clinical trials function as an autonomous platform for the investigation of the biology and pathogenesis of cancer. But what does it mean to say that they amount to a new style of practice?

Cancer Clinical Trials as a New Style of Practice

Given what we have said so far, it should be obvious that cancer clinical trials are not unique events or even a series of related events: they participate in a veritable system of biomedical research. Subsequent chapters will provide a sense of the complexity of the system. For present purposes we refer readers to figures 1.4 and 1.5 that display the attempts of the NCI and GECA in the late 1950s and early 1960s to develop new chemotherapeutic agents. Figure 1.4 offers a simplified version of the process through which different substances (synthetic compounds, plant extracts, etc.) provided by different institutions (government labs, industry, universities, and research organizations) inched their way through the Cancer Chemotherapy National Service Center, where they underwent tests in animal models such as the Leukemia L1210 mouse, tissue cultures, hormone assays, microbiological assays, among others. If a substance passed the screen created by the tests, researchers then examined the compounds in clinical studies that were themselves subjected to further assessment. Only a tiny fraction of the substances that survived these tests entered the clinic. Interestingly enough, the figure also shows the continuity between testing locations: animal models and human trials are part of the same platform.

In the mid-1960s, administrators of the NCI's chemotherapy program described their activities as a series of decisions in a process that regulated the flow of drugs, regimens, and disease concepts through rodent models, then into Phase I, II, and III clinical trials, and finally, into the hands of practicing physicians. While the definition of phases and stages differed somewhat in Europe, both GECA and the EORTC adopted a very similar approach. Figure 1.5 shows a set of sequential procedures (the names in parenthesis are those of the European clinical researchers in charge of a particular procedure) that correspond to those in the US illustration, albeit with some additional "technical details." Notice, in particular, the decisive roles played by statistical analysis and group

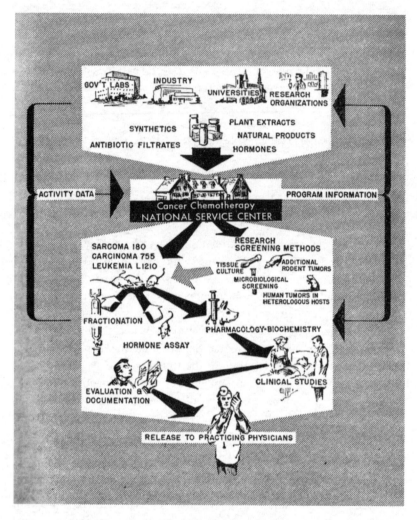

FIGURE 1.4. A 1957 diagram showing the early US cancer clinical trial system. Reprinted by permission of Oxford University Press from K. M. Endicott, "The Chemotherapy Program," *Journal of the National Cancer Institute* 19 (1957): 279.

discussions at the end of several stages. Both figures illustrate how, even at this early juncture, the clinical trial system mobilized distinctive sets of practices that were all articulated by a "common good."

The two figures hide almost as much as they show. A more complete picture of the systems at play would include different kinds of disease

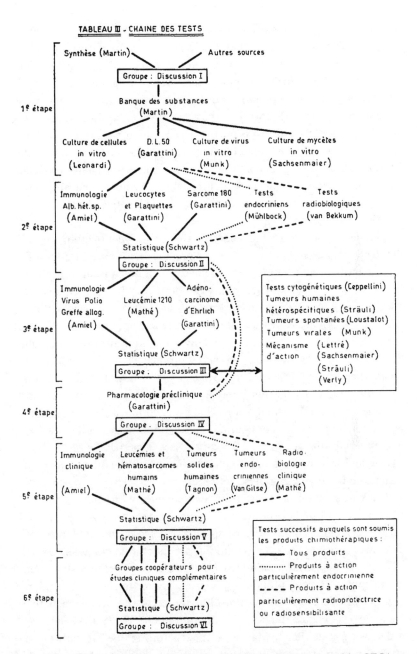

FIGURE 1.5. An early 1960s diagram showing the "testing sequence" used by GECA to bring a chemotherapeutic drug from the initial, screening phase to its clinical acceptance and use. Source: Undated promotional brochure (EORTC Archives, Brussels).

(cancer is not a single disease); all sorts of in vivo and in vitro models; a motley of different practitioners (the emerging clinical oncologists, internists, hematologists, surgeons, radiotherapists, statisticians, and laboratory researchers including pharmacologists, biochemists, etc.); public and private, commercial, and academic institutions; and, of course, the patients themselves who, with the passage of time, became increasingly active participants. This list is far from complete: it has evolved over the years in a process punctuated by numerous public debates and reforms. Through these reforms and additional ad hoc tinkering, trialists have designed an expedient to manage and regulate the connections among the different components of a trial and thus provide it with a single, common identity: the *protocol*.

The protocol lies at the heart of this self-authenticating (or self-vindicating) new style of practice. As previously proposed by Joan Fujimura and Danny Chou (1994), we have drawn this notion from the related notion of a *style of reasoning* developed by Ian Hacking (1992a; 1992b; 2004). According to Hacking, observational sciences, field sciences, laboratory sciences, and statistical sciences are distinctive styles of reasoning insofar as they consist not only of a unique group of techniques but also deploy new domains of objects, produce new kinds of facts, and create their own criteria of truth (hence, their "self-authentication" that, it should be noted, does not imply a concession to relativism). In short, and adding a sociological flavor to Hacking's notion, a style of reasoning calls upon a distinctive configuration of institutions, scientific practices, and materials that generates new entities as well as specific ways of identifying and investigating research questions, of producing and assessing results, and of regulating these activities.

Hacking's notion can be usefully supplemented with Michel Foucault's idea of a *dispositif*, a term that roughly translates as *apparatus*. According to its canonical definition, a dispositif refers to the "decidedly heterogeneous set of discourses, institutions, architectural arrangements, regulatory decisions, laws, administrative decisions, scientific statements, philosophical, moral, philanthropic propositions, in short, the said and the unsaid. . . . The 'dispositif' itself is the network that connects these elements" (Foucault 1994a: 299, our translation). As Paul Veyne notes, a dispositif is neither a Marxist infrastructure (a sociostructural constraint) nor an ideology (2008: 45–58). Rather, and like Hacking's *style*, it generates new classes of objects, new types of sentences, new forms of evidence, new kinds of explanations, new criteria for assessing results, new forms of truth, as well as new forms of intersubjectivity. In other words, in addition to introducing its own form of objectivity—in the dual sense of producing distinctive objects and distinctive epistemic values (Daston & Galison, 2007; Kusch, 2009)—a dispositif also generates the agents that perform these activities: clinical cancer trials, for instance, produced medical oncologists and the stylized relations they entertain with one another and other specialists. Once again, this notion does not presuppose any form of relativism. Far from denying the possibility of observing empirical facts and of assigning a truth-value to them, it uses this possibility as a starting point for investigating the procedures that allow clinicians and scientists (or any other epistemic subject) to do so.

Why use the term *styles of practice* or, in other words, why abandon "reasoning"? Beyond the linguistic fact that "practice" has a less "cognitivist" connotation than "reasoning," there are two additional motives.[14] First, Hacking's examples draw mainly on long-haul history and thus have a somewhat more epochal referent than that denoted by our investigation. We wanted to capture changes that took place within a few decades and that thus obey a less "historical," more pragmatic, temporality (Dodier, 1995). Second, cancer clinical trials articulate different styles of reasoning, for instance, a clinical, a statistical, and (increasingly) a laboratory style. Articulation does not necessarily entail smooth sailing. Trialists commonly evoke conflicts between statisticians and clinicians that in some cases have acquired epic proportions (Marks, 1997: 197–228). Nonetheless, despite occasional frictions, trials do work as a collective endeavor and routinely produce results that, in turn, inform clinical practices.

Clinicians have their own tools, skills, objects, and organizational structures, as do statisticians and laboratory practitioners. To participate in a trial they must combine these distinctive resources. Their participation is thus no mere "collaboration" between different specialties (Shrum, Genuth & Chompalov, 2007). Rather, as this book will show, trials have generated a whole new class of sui generis objects that, in turn, have redefined the practices of clinicians, statisticians, and biologists giving rise, for instance, to a specific breed of *biostatisticians*. Nor are trials merely a testing device, able to produce empirical facts of no theoretical import: clinical trials have become full-fledged experiments.

14. For a critique of cognitivism, see Descombes (2001). For a criticism of Hacking's notion that partially overlaps with ours, see Kusch (2010).

The trial system also contains its own reflexive machinery for establishing facts as well as how those facts should be integrated into the evolving network of concepts. At the same time trials are not experimental systems that can be controlled by a single researcher working within the boundaries of a specific line of work (Rheinberger, 1997). A trial is a collective, distributed activity whose performance lies beyond the grasp of the individual epistemic subject. The judgment about a trial's outcome is a collective judgment (Bourret, 2005) that cannot be reduced to either an individual, statistical, clinical, or laboratory style of reasoning.

At this point, critical readers will no doubt exclaim: "Oh well, all this talk about a new style of practice seems to be little more than an attempt to create yet another buzzword; can't you give us a better sense of what you mean?" We believe we can. Using figure 1.6, for instance, we can allow the reader to witness (virtually) a session of the 2007 San Antonio Breast Cancer Symposium (SABCS). For the subspecialty of breast cancer the SABCS plays a role similar to ASCO's annual Grand Mass for oncology as a whole. The first SABC symposium took place in 1978 as a one-day regional conference attended by 141 US physicians and surgeons from a local five-state area. It has since become a major international event: in 1990, 526 participants from twenty-four countries attended the symposium and by 2007 their number had risen to 8,507 from eighty-six countries. As figure 1.6 shows, to enable attendees to see the speakers and their slides, large screens are installed at regular intervals in the audience. While such a massive congregation is surely a sign that a

FIGURE 1.6. A view of the 2007 San Antonio Breast Cancer Symposium. Photo Courtesy © SABCS/Todd Buchanan 2007.

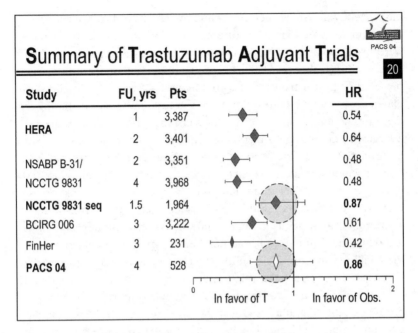

FIGURE 1.7. Comparing clinical trials during a meeting (see text for explanations). Courtesy of Dr. Marc Spielmann, Institut Gustave-Roussy (France).

critical mass of some sort has been reached, numbers alone do not demonstrate a distinctive style of practice. So: what are these oncologists doing when they watch countless PowerPoint presentations?

Figure 1.7 shows one of the slides presented by researchers at the SABCS meeting. It has been selected almost at random and is representative of the summary slides often used toward the end of a talk. This particular presentation focused on a clinical trial testing a new, targeted substance (trastuzumab is the generic name of Herceptin) in an "adjuvant chemotherapy" regimen (i.e., chemotherapy treatment after surgical removal of the tumor). The column on the left (under Study) contains a list of acronyms such as NCCTG (North Central Cancer Treatment Group) and NSABP (National Surgical Adjuvant Breast and Bowel Project), generally followed by a number that designates the trial carried out by that group. Columns to the right show the median follow-up (in years) and the number of patients. The purpose of the slide is to compare the results of a trial (PACS 04, carried out by investigators at eighty-two sites) with the results of similar trials by other groups. Without going into specifics, it can be seen that while the results of PACS 04 show that trastuzumab failed to make a difference (and in this respect those results were similar to those of NCCTG 9831seq), they also contradicted the results of all the other trials. This kind of comparison is a common way of presenting clinical trial results: arguments take the form of slides displaying a systematic confrontation of the results of several studies that therefore no longer appear as isolated instances but, rather, as pieces of a larger, communal puzzle-solving activity. A puzzle, moreover, whose pieces and contours are continuously redefined by related experiments.

To account for discrepancies or even outright contradictions, trialists engage in a critical examination of the strengths and weaknesses of each trial. For instance, could the diverging results have been caused by a different trial design (e.g., sequential versus concomitant use of a given drug)? Or by differences in the number or type of patients enrolled in the trial? Or, maybe, more subtle interactions between factors such as time and efficacy played a role . . . and so on. This is not rank empiricism. Each of the quantitative variables in play is underwritten by years of research into the qualities those variables represent: the numbers are not stand-alone variables but have acquired their sense and meaning through an array of clinical trials and evolving notions of the right way to conduct a clinical trial. In other words, the variables are conceptually informed. To say that a trial lacked sufficient participants, for instance, refers implicitly to an entire subfield of biostatistics specializing in clinical trials, and to question the possible molecular heterogeneity of the type of patients enrolled references the accumulation of data on genomic parameters that are *coproduced* at the clinical-laboratory interface. Finally, it is also essential to understand that debates do not stop at this level of critical inquiry: they are performative insofar as hypotheses about plausible causes of differential results are used to design new trials.

Charts such as the one in figure 1.7 illustrate the new style of practice "in action." More importantly, they clearly show that individual trials are not isolated events but rather, elements of an integrated, open-ended set of activities that stretch back and forth in time. We insist on "open-ended" because practitioners act as the bearers of the new style that continues to evolve as novel biomedical objects, such as genetic mutations, biochemical pathways, and targeted anticancer substances, enter the fray, generating unanticipated problems, innovative responses, and new interfaces. In other words, cancer clinical trials cannot be dismissed as a "science of the particular"; that is, they are not merely data collecting and collating activities "subjected to regulatory 'scientific' norms articulated, administered, and arguably enforced by government and corporate organizations" (Lynch et al., 2008: 8 n. 6). Rather, they are full-fledged experiments that continually redefine the new style of practice and, as such, produce new objects of study, substances, and treatment regimes.

The Protocol

Let us begin, once again, with a picture that, in this case, is truly worth a thousand words. Figure 1.8 shows the cover of a recent book on cancer clinical trials. While we will not comment on the organizations represented in the illustration (they will be discussed in subsequent chapters), readers will quickly note the presence of the cooperative groups. This illustration provides a glimpse into the intricate institutional structure of trials, whereby multiple interactions among manifold institutional and biological entities ensure appropriate checks and balances. The development of a clinical trial lies at the center of the illustration, and the

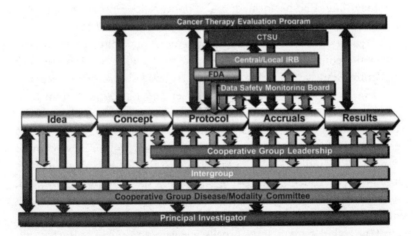

FIGURE 1.8. The front cover illustration of a cancer clinical trials publication, showing the central place occupied by the protocol. Reprinted from S. P. L. Leong, *Cancer Clinical Trials: Proactive Strategies* (New York: Springer, 2007), with kind permission of Springer Science+Business Media, and the Coalition of Cancer Cooperative Groups © 2002.

protocol lies in the middle of this linear flow. According to the diagram, the initial idea for a trial, translated into a concept, leads to the definition of a protocol, which results in the recruitment of patients ("accruals" in the trialists' slang) and, hopefully, to the production of evidence for or against the intervention under study. Such is the centrality of the role played by protocols in clinical trials that the latter are often equated with the former; trialists, for instance, refer to a clinical trial XY as Protocol XY.

The term *protocol* originated in Medieval Latin where it referred to the first (proto) paper sheet glued (kolla) to the top of the minutes of public transactions and that outlined the contents of the resulting volume. As chronicled in the *OED*, the semantic field of the term has since undergone considerable extension, leading to the present-day scientific and clinical meanings that include the list of the successive steps of an experiment, the outline of a planned examination, or the agreed-upon schedule of chemotherapeutic drugs and dosages. The term thus has multiple meanings referring simultaneously to a legal authority, since its content is binding on the participating parties; a convention, for it is the result of a transaction between participants; a public, because as a communal document it is open to inspection by interested individuals or, at least, overseeing agencies and organizations; a prescription, since it dictates both the activities that have to be undertaken by participants and how they should be performed; and, finally, a description, insofar as it acts as a record of what has been done (see also Lynch, 2002).

As readers understand by now, cooperative oncology groups are the principal medium of protocol development and administration in both the United States and Europe. Figure 1.9 shows how protocols were produced in a cooperative group (the Cancer and Leukemia Group B or CALGB) and traveled to the NCI in the 1970s. Since then protocol development has become considerably more complex. While in the 1970s a typical protocol occupied seven pages, today's protocols are always more than one hundred in length, a testimony to the increasing complexity of human experimentation. Using once again the example of CALGB, "study activation" circa 2006 required 370 distinct processes that included "317 work steps, 42 decision points and 29 processing loops." Since the complete map of the process measured "243.5 × 41 [inches, approx. 6.18 m × 1.06 m] in 8-point font" it could not be printed in a scientific journal (Dilts et al., 2006). As can be seen in figure 1.9, even in the

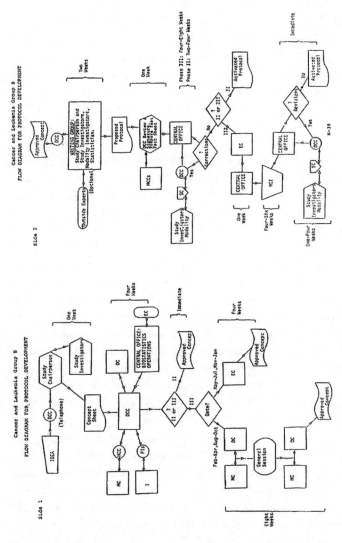

FIGURE 1.9. Flow diagram of a CALGB protocol (ca. 1977; unpublished document; James Holland, personal papers). MC = Modality Committee, MCC = Modality Committee Chairperson, DC = Disease Committee, DCC = Disease Committee Chairperson, SC = Study Committee, EC = Executive Chairperson, NCI = National Cancer Institute, PI = Principal Investigator(s), and I = Investigators.

early days the process of setting up a protocol was almost as complex as the protocol itself.

The tortuous nature of protocol development reflects, in part, a long series of administrative initiatives undertaken by the NCI and the office that managed the groups, the Cancer Therapy Evaluation Program (CTEP) and their European equivalents, to overcome problems such as protocol redundancy, patient recruitment, and the timely introduction of novel techniques. These initiatives have structured protocol development in ways that are not immediately obvious. For example, all Phase III protocols in the United States have to be submitted to CTEP for final approval. CTEP defines funding priorities according to protocol significance (as opposed to trivial)—but CTEP's notion of "important" draws in part on constant negotiations and interactions with the cooperative groups as well as with its contacts with outside experts and policymakers. Practitioners refer to this process of priority setting as "oncopolitics," a term adopted by the *Journal of Clinical Oncology* as one of the keywords used to classify domains of expertise in the cancer field.[15]

Let us return to the process laid out in figure 1.9, keeping in mind two things. First, although the chart dates from the 1970s, the considerably more complex process used today adopts a similar pattern. The discussion that follows is thus of more than historical interest. Second, this is merely the tip of an iceberg of corridor conversations, telephone calls, and telexes (subsequently faxes and more recently e-mails). Now, to grasp the process described in this chart, we must first understand that protocols embody what US trialists call "concepts" and European trialists "rationale"—by which they mean the therapeutic and/or biological hypothesis to be tested by a given protocol as well as the information to be provided by the ancillary studies. More than simple rules, in other words, concepts concern the information that a successful protocol produces: novelty, not routine. In practice, trialists distinguish concepts and protocols. As the flow chart shows, concepts precede protocols in the sense that during protocol development, the initial starting point is the "idea." Typically, somebody, usually a principal investigator, communicates ideas verbally to the disease committee chairperson, who then speaks to the study chairperson. The latter, in collaboration with other investigators, produces a "concept sheet" that serves as the basis for further discussions with biostatisticians and members of the disease

15. Interview with James Holland, 24 July 2003, New York, NY.

committee, the modality committee, and the executive committee. The entire process takes several weeks, and once everybody agrees, an "approved concept" starts down the road to becoming an "approved protocol." In this sense one could say that concepts drive protocols.

But since one protocol leads to another, concepts also follow protocols. This is quite obvious when we look at the development of a concept according to the chart. Following their first—verbal—formulation, ideas have to be reduced to a concept sheet that in turn has to be submitted to a checklist specifying what concerns a concept must address and how the concept should be presented. The "concept sheet checklist" tells us, for example, that a concept answers the question, "What needs to be done and why?"[16] Thus, like a research proposal, the concept describes the history and the nature of the problem and the novelty of the proposed solution. The concept thus becomes the introduction to the protocol and has to be distinguished from concepts animating previous protocols. Once a concept is approved, a writing group composed of the study chairperson, the study investigators, the modality investigators, and the statisticians takes about two weeks to transform the concept into a protocol. The protocol then moves back through the disease committee, the modality committee, and the executive committee before (if Phase III) going off to the NCI for final approval. Members of CTEP often participate in these discussions; not only do they relate CTEP priorities to group members (who, in turn, relate group priorities to CTEP), but they also keep group members abreast of developments in other cooperative oncology groups so as to reduce redundancies (Dilts et al., 2006).

Based on this description of the production of a cancer trial protocol (see also Löwy, 1995b), how should we understand the nature and role of this device? To illustrate our understanding of a protocol, it is useful to compare it to Marc Berg's (1998) attempt—based on a description of a multi-institutional research protocol in the field of cancer chemotherapy—to develop a synthetic view of protocols in general. Whereas many authors have suggested that the point of a clinical trial protocol is to impose order on some previous disorder, Berg's analysis concludes that the relevant division is not between order and disorder, but between an old order and a new order, both of which contain intrinsic elements of disorder. He nonetheless agrees that despite its negotiated nature, some

16. CALGB, Minutes of Group Meeting, Biltmore Hotel, New York, 8–11 October 1980, p. 37, James Holland personal papers.

degree of imposition follows from the acceptance of a protocol by participants. The imposition arises from the rule-bound nature of the operation. According to Berg, "a protocol is a formalism: it operates using a collection of specific, explicit rules, which turn input data into output" (1998: 233). In virtue of the rules they express, protocols coordinate activities in different times and places. In the case of clinical medicine, as a device for the normalization of practices, "the protocol makes the administration of highly complex treatment schedules *possible* in the first place" (1998: 232, emphasis in the original). Aligning practices and getting others (both persons and things) to follow rules clearly requires a good deal of work and negotiation, and Berg rightly observes that the process of setting up and successfully running a clinical trial protocol is a highly political process.

While we agree with Berg about many aspects of his description, we find his definition and his approach somewhat restrictive. First, Berg tends to see the protocol as an organizational device or tool, and he consequently downplays the creative nature of the process. Instituting a protocol entails more than just the imposition of a new order (although we do not deny that this may often be the case): the new order implies the creation of new things or entities (disease categories, patient categories, etc.) and new ways of acting; it is not just the reorganization of already existing entities. Second, Berg's definition of a protocol tends to reduce it to one of its parts—namely, the schema, a highly stylized diagram that accompanies protocols and that shows what substances will be given to patients on what schedule and under what conditions. Figure 1.10 shows a schema excerpted from a mid-1970s ECOG protocol illustrating the schedule according to which substances were to be administered to the patients following two different regimens that treated metastatic breast cancer—namely, a three-drug combination and a single-drug treatment (see chapter 6). As we have seen, more than a plan or a schema, a protocol comprises a number of other components that link it both to the past and to the future and that articulate the protocol with other scientific practices. To focus exclusively on the schema—an "immutable mobile" (Latour, 1987) that circulates unscathed from one location to the other—is to frame the protocol as a mere tool and not as a program of situated action that encompasses its own performance.

Finally, while we agree that protocols do not necessarily transform disorder into order but create a new order out of an older order, the older order is not "a nonprotocolized situation" where "medical personnel can

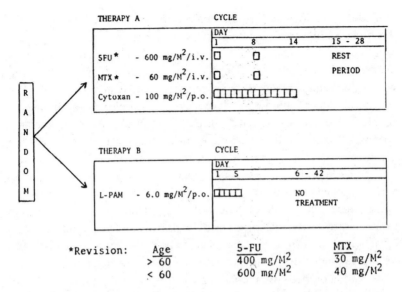

FIGURE 1.10. Randomization scheme of the ECOG Protocol 0971 against metastatic breast cancer: patients received either a multidrug CMF regimen or a single drug L-PAM regimen. Reprinted with kind permission from G. P. Canellos et al., "Combination Chemotherapy for Metastatic Breast Carcinoma: Prospective Comparison of Multi-Drug Therapy with l-phenylalanine mustard," *Cancer* 38 (1976): 1883. © 1976, John Wiley and Sons.

allow a myriad of more or less precise laboratory tests, historical data, psychosocial circumstances, and so forth to shape the course of action undertaken" (Berg, 1998: 241). We contend that prior to the imposition of any given protocol, the situation is *already* highly "protocolized"—in the sense that an archaeology of medical judgments would show that much of common practice flows from deeply embedded protocols whose "rules," so to speak, have sunk out of sight. In medical oncology, in particular, it is not so much the case that "protocols discipline practices," as that the constitution and administration of protocols is itself a practice. In fact, it amounts to what we have called a new style of practice.

Conclusion

This chapter has provided an initial account not simply of how we have approached the subject matter of this book but also of the subject matter itself. Readers may wish to treat this book as a history of the emergence of modern oncology and its defining practice, cancer clinical trials. Yet this book is not a traditional history in the sense that it does not consist of a fine-grained analysis of the institutional development of oncology in a particular country. Nor does it provide a detailed comparative overview of the establishment of the specialty in the Western world. Rather, as indicated in the subtitle, we have tackled our topic by examining cancer clinical trials as a new style of practice. This, in turn, has dictated how we selected and analyzed the empirical material that supports our argument.

It could be objected that clinical trials (in any field) would be more perfectly understood as a technology. From this point of view the history of the emergence and development of clinical trials could be reduced to the history of "protocols" as a way of organizing and managing biomedical activities. In turn, such a history could be understood as the history of a powerful undercurrent of twentieth-century medicine aiming at the rationalization of medical practices and leading to present-day "evidence-based medicine" in which randomized clinical trials are the gold standard (Timmermans & Berg, 2003; Berg, 1997; Marks, 1997). While there is no denying the existence and import of the trend toward rationalization in biomedicine, our goal in this book is to provide a different view of cancer clinical trials—one that, as already mentioned, examines them as a new style of practice, viewing them as platforms rather than "mere" technology (Keating & Cambrosio, 2003). While protocols are central to clinical trials, they are but the central organ of the new style of practice and do not subsume it in its entirety. The new style cannot, moreover, be explained solely in terms of the expression of a largely external trend toward rationalization. To understand the style, we must examine the content and the emergent properties of the components of clinical trials. And in so doing we must attend to the temporal, institutional, and sociotechnical specificities of the domain (in the present case: cancer) in which clinical trials have been deployed. This is why we have centered our analysis on developments since the 1950s, a period during which cancer clinical trials emerged and developed their distinctive style. The following Prelude sets the stage for our analysis of the emergence of this style by reviewing the practices and institutions that postwar oncologists inherited from their prewar predecessors. The Prelude will also provide brief technical descriptions of practices of diagnosis and prognosis that guide oncologists' work.

Before There Were Trials

The cooperative oncology groups discussed in this book inherited a network of institutions devoted to cancer treatment and research, a substantial body of clinical and biological knowledge, and deeply entrenched therapeutic practices. It is the intent of this prelude to provide readers with some background knowledge of the field of cancer therapy prior to the emergence of clinical cancer research. Our overview will be selective, focusing on topics that are relevant for subsequent chapters wherein we examine how, against this institutional and epistemic backdrop, the cooperative groups combined practices to create a new bioclinical collective; that is, one that cut across traditional lines of communication and transformed how clinicians and researchers interact, the standards of evidence, and the relations between the different therapeutic modalities and the purveyors of those modalities.

Cancer Hospitals and Cancer Institutes

Cancer therapy in the immediate postwar era consisted primarily of radiotherapy and surgery accompanied by an occasional foray into experimental chemotherapy. Practitioners pursued all three therapies in both general and specialized cancer hospitals that conducted research and therapy together. Such institutions were relatively recent creations. Although hospitals specializing in the treatment of cancer patients had existed since the eighteenth century, with little to offer in terms of treatment and generally regarded by more orthodox practitioners as a haven for quacks, most failed to outlast their founders. A second wave of cancer hospitals funded mainly by charities rolled over Europe in the second half of the nineteenth century. More successful as institutions, some such as the London Cancer Hospital (1851; renamed the Royal Marsden after its founder), the Christie Hospital in Manchester (1892), and the New York Cancer Hospital (1884; now the Memorial Sloan-Kettering Cancer Center [MSK]) are still with us (Cantor, 1993). The first cancer hospitals in the Americas were founded toward the end of the nineteenth century, and in keeping with the times the new institutions sought to combine both treatment and research, thus creating the modern "cancer institute." The oldest such institute in the world remains the Roswell Park Cancer Institute founded in 1898 and funded by the New York State legislature.

Prior to World War I, groups of physicians and charities around the world created similar public and private sources of funds for treatment and research, forming such well-known associations as the German Committee for Cancer Research (1900), the Imperial Cancer Research Fund (1902), the American Association for Cancer Research (1907; subsequently the American Cancer Society), the Japanese Foundation for Cancer Research (1908), and the Cancer Society in Stockholm (1910). Specialist cancer journals appeared at about the same time, beginning with the French *Revue des maladies cancéreuses* (1896) and continuing with the German *Zeitschrift fur Krebsforschung* (1904), the Japanese *Gann* (1907), the Italian *Tumori* (1911), and the *American Journal of Cancer Research* (1916) (Cantor, 1993).

The interwar years saw the emergence of a number of national and international, social and scientific groups devoted to the control of cancer, ranging from the League of Nations Cancer Commission (1925) to what is today the largest cancer research institute, the US National Cancer Institute. Established in 1937 "to promote research in the cause, prevention and methods of diagnosis and treatment of cancer," the NCI had been preceded by a decade of ineffectual legislative efforts beginning with a wildly optimistic $5 million "reward for the discovery of a successful cure for cancer" offered in 1927 (Faguet, 2005: 95). Located on the new Bethesda campus of the National Institutes of Health, the NCI was housed in its own building and by 1939 counted twenty researchers and a small extramural funding program (Nass & Stillman, 2003: 233). As an experimental enterprise, the NCI also purchased almost half of the inbred mice produced by the premier experimental mice breeders in

America, the Jackson laboratories (Rader, 2004: 160). Clinical research did not begin in earnest until after the war with the establishment of the Laboratory of Experimental Oncology. A joint project of the NCI and the University of California Medical School at San Francisco, the laboratory was a fifteen-bed hospital. Opened in 1947, it acted as an outpost of the NCI, and during its six years of operation laboratory researchers tested 605 treatments on 467 end-stage cancer patients (Shimkin, 1978). The construction of a five-hundred-bed NIH Clinical Center at the Bethesda campus resulted in the termination of the San Francisco operation and the beginning of a new era in clinical cancer research. By the end of World War II, in other words, cancer had an extensive institutional infrastructure that spanned treatment, research, and prevention (Pinell, 2000).

Diagnosing Cancer

By the mid-1950s, the study of cancer had also generated a rich vocabulary of description, much of which is still in use. Based on the accumulation of more than two centuries of clinical, anatomic, and microscopic observations, these expressions may sometimes appear obscure and/or confusing to today's average reader. To begin with, while the swellings and lumps typical of cancer are commonly known as tumors, there are other kinds of tumors not caused by cancer, such as those produced by bacterial or viral infection. To separate cancerous from noncancerous tumors, therefore, oncologists frequently use the word *neoplasia* or "new growth" to signify cancer. Even though all cancers are classified as neoplasia, not all neoplasia are malignant or dangerous: some are benign and hence no cause for concern.

Additional terminological problems arise once a tumor has been diagnosed as cancerous and malignant (Majno & Joris, 2004: 735–46). Since the 1860s, for example, tumors have been classified according to their tissue of origin and the cells that compose the tissues. Pathologists still recognize the four original tissue types—epithelial, connective, muscular, and nervous—and thus four principal categories of neoplasia. Experience since the nineteenth century has shown that cancers are not equally divided among the tissue types: approximately 80% of all tumors derive from the epithelial tissue since this tissue lines the esophagus, stomach, bowels, bladder, and lungs and is thus most commonly exposed to environmental irritants and mutagens such as tobacco and alcohol. Tumors that originate here are called carcinomas. Because the vast majority of cancers are carcinomas, some writers conflate cancer and carcinoma even though the second is a proper subset of the first. As for the remaining 20% of tumors, pathologists refer to those of the connective and muscular tissues as sarcomas. The exceptions to the rule—and these are hardly minor exceptions—are the leukemias and the lymphomas. Although both are cancers of the connective tissues, they are not considered sarcomas. Finally, tumors of neural origin have the simple suffix "oma" appended to the name of the cell type involved (as in Schwannoma for a cancer of the Schwann cells that line the nerves).

The cells that compose cancer tissues escaped investigation until the 1840s when major improvements in microscopy made them visible and the rise of Theodor Schwann's (1810–1882) cell theory conceptualized cells as the ultimate functional unit of living beings. The cells then under study included not only plants and animals but also the equally lively, yet deadly, cancerous neoplasia. At present, the two hundred different human cell types give rise to almost six hundred different types of cancer. And while a cell-based classification has long been accepted as the most rational classification, it does not exhaust current nomenclature: some present-day cancers stood out as distinct diseases long before they were classified as cancers. Such is the case of Hodgkin's disease described by the nineteenth-century British physician Thomas Hodgkin (1798–1866) on the basis of anatomical dissections and clinical observations well before pathologists described the cells at the source of the complaint. Although there have been attempts to change the name of Hodgkin's disease to Hodgkin's lymphoma in order to reflect the renewed understanding of the cellular basis of the disease, Hodgkin's disease remains firmly ensconced in the contemporary medical lexicon.

By the beginning of the twentieth century, pathologists had attributed a cellular basis to most common cancers. Diagnosis consequently depended upon the discerning eye of the pathologist. The division of the cancers into kinds, however, did not end there as individual cancers can be further categorized in terms of their developmental status or prognosis. The first concerns the determination of the degree of spread or stage of a given case of cancer. The second grades tumor tissues into different categories of aggressiveness based upon the appearance of the tissue under the microscope. Staging and grading are subsets of diagnosis: first one determines what kind of cancer is present in terms of the tissue type and cell type, then one assigns a stage and a grade to the cancer in order to understand how advanced and dangerous the disease has become. Before further examining the staging and grading systems created during the first half of the twentieth century, we must briefly review the cancer treatments developed in that period and that served as a platform for the aforementioned systems. For although the logical sequence of medical interventions often appears to move ineluctably from diagnosis to therapy, in both clinical practice and in historical terms this sequence is often inverted when treatment results provide diagnostic information and create a need for diagnostic refinements and/or calls for diagnostic standardization.

Treating Cancer

Surgery and Radiotherapy

As noted above, prior to World War II, cancer therapy rested primarily on two modalities, neither of which was devoted exclusively to the treatment of cancer: radiation therapy and surgery. Although radiation therapy was relatively new, extirpation of the most accessible tumors such as those of the breast and lip had been practiced since antiquity (Olsen, 2002). Prior to antisepsis, cauterization served as the principal weapon against the infection of surgical wounds. This, as can be imagined, severely limited the practice of cancer surgery. Since surgery involving less visible, internal tumors requires both anesthesia and antisepsis, procedures for these cancers did not develop until the latter half of the nineteenth century when surgeons such as the celebrated Austrian surgeon Theodor Billroth (1829–1894) and his disciple William Halsted (1852–1922), often described as the father of American surgery, devised a number of cancer surgery protocols such as Halsted's radical mastectomy for breast cancer (see chapter 6) and Billroth's procedure for gastric cancer (Ellis, 2002). Even with the knowledge of modern bacteriology, however, such procedures remained risky, and the early survival rates registered were strikingly low, giving rise to vigorous debate as to when exactly such drastic measures were justified. The two-year survival rate for gastric surgery in 1900, for example, ran around 10% with operative mortality itself running at 37% (Martensen, 1997).

Despite these gloomy numbers, procedures improved incrementally to the point where, by 1906, William James Mayo (1861–1939), of Mayo

Clinic fame, had reduced operative mortality for gastric cancer surgery to 14% (Martensen, 1997). Similarly, by the mid-1920s, the British public health official and epidemiologist Janet Lane-Claypon (1877–1967) was able to report that cancer patients who underwent breast surgery, for example, survived 60% longer (5.7 years) than patients who did not (3.6 years) (Aronowitz, 2001). While such statistics might not pass muster under modern scrutiny, they inspired enough confidence in both patients and practitioners at the time to generate a sizeable clientele. If we consider only the United States, although we lack specific numbers, the cancer surgery population was sufficiently large prior to World War II to engender three cancer surgery subspecialties divided among anatomic sites: breast, head and neck, and colorectal. Since surgeons have long resisted the division of their domain, the degree to which the subspecialties held sway over their respective practices was far from complete. Even today the American colorectal cancer patient has the choice (or confusion) among a general surgeon, a colorectal surgeon, and what has recently become known as a surgical oncologist (O'Shea, 2004).

Specialists in radiation therapy followed a path similar to that outlined above for cancer surgeons and remained closely identified with their specialty of origin (radiology) long before the development of a practice devoted exclusively to the treatment of cancer (radiation oncology). Nevertheless, although deprived of their own specialty, practitioners of radiotherapy often worked within their own institutions. Following the discovery of x-rays by Wilhelm Roentgen (1845–1923) in 1895 and radium in 1903 by Marie Curie (1867–1934), the first decades of the twentieth century saw widespread experimentation with the new therapeutic modalities in dedicated institutions in Europe such as the Laboratoire Biologique du Radium (Paris, 1906), the Radiumhemmet (Stockholm, 1910) and the Radium Institute (London, 1911). Given the expense of and the increasing demand for radium, beginning in the 1920s a growing number of governments on the continent and in Canada funded radium supplies for clinical use (Hayter, 1998: 670). At the same time, centers initially devoted to radium therapy alone began to incorporate x-ray therapy equipment so that by the 1930s it became possible to speak of radiotherapy centers.

Outside of the radium institutes, however, radiologists specializing in radiotherapy initially had an even slighter hold on their therapeutic modality than cancer surgeons. Before 1945 both diagnostic and therapeutic radiologists usually deployed the same energy source for producing radiation and thus shared facilities and equipment, meaning that no firm distinction could be drawn among practitioners. This was less the case in Europe, where the proliferation of radium hospitals and institutes allowed for an earlier development of therapeutic radiology as a distinct practice during the 1930s and 1940s (del Regato, 1993; Holsti, 1995). Researchers at the Radium Institute in Paris, for example, established the current principles of radiotherapy dose fractionation whereby the total amount of radiation given in the course of a treatment is distributed over the course of a number of days (Lenz, 1973).

Despite advances such as these, early x-rays were relatively weak and penetrated poorly. Outside the radium and radiotherapy centers, therapeutic use of x-rays was thus largely restricted to topical applications such as the treatment of fungal diseases (del Regato, 1996). The situation in the United States did not change dramatically until the 1950s when radiotherapy became a default option for applicants in general radiology who had failed the diagnostic part of the exam (del Regato, 1984). In France the public financing of radiotherapy equipment led directly to the formation of the Centres Anticancereux that remain the backbone of cancer treatment in France (Pinell, 2002). The treatment centers that grew up around the publicly financed purchases of radium often encountered opposing and confounding forces, including a jealous medical profession intent on combating so-called state medicine and a public hungry for the curative rays for any and all purposes.

In the United States, the original National Cancer Institute mandate of 1937 stipulated that the NCI provide radium to hospitals throughout the country for routine cancer treatment. Confronted with the option of providing cancer care for the poor or creating an institute for cancer research, Congress had opted for the latter. Free radium addressed the national shortage of radium without taking the fatal step down the road to state medicine. While Britain, France, Sweden, and Canada offered clear examples of the alternative with their national networks of radium treatment centers, the NCI lobbied vigorously to divest itself of the radium program shortly after its founding even though it continued to supply radium to hospitals into the 1960s (Cantor, 2008).

Despite their polyvalent treatment practices, radiotherapy centers outside the United States often served as platforms for more comprehensive cancer programs that combined preclinical research and treatment. Since the state funding of a medical treatment, like the state funding of vaccinations a century before, demanded justification, the mere existence of radiotherapy centers stimulated the collection of treatment statistics that in turn generated a series of ripple effects: they called attention to the dire prognosis that was cancer and they created a thirst for more systematic data collection and the formation of tumor registries. The cancer statistics produced within the new radiotherapy clinics also laid bare discrepancies between treatment patterns in areas directly served by clinics and outlying regions, creating what has remained a central tension within the oncology community between cancer center treatment and community-based treatment (Hayter, 2005).

Chemotherapy and Drug Screening before 1955

Before 1955 chemotherapy contributed little to cancer therapy even though Paul Ehrlich (1854–1915), one of the founders of immunology, had developed the bases for the modern understanding of chemotherapy with his notion of cell surface receptors (Silverstein, 2002). On the eve of the inauguration of the NCI chemotherapy program in the 1950s, there were three classes of compounds in use: nitrogen mustards, antimetabolites, and hormones. The first two classes of drugs were not developed until the 1940s, although the first, the nitrogen mustards, had been introduced to the world in 1917 as the British response to the use of chlorine gas by German troops. Unlike chlorine gas, mustard gas could be absorbed through the skin thus bypassing the gas masks that had quickly sprung up as a defense in gas warfare. Despite knowledge of its use and consequences, understanding of its mechanism eluded biochemists until 1941 when the US Army hired Louis Goodman (1906–2000) and Alfred Gilman (1908–1984) of Yale University to inquire into the pharmacologic effects of the weaponized gas.

In the course of their studies, Goodman and Gilman noticed that the compound destroyed the lymph nodes and the bone marrow in mice, and so they tested the compound on mice with cancer of the lymph nodes. Having obtained a reduction in the mouse tumor, they moved on to humans the following year, injecting the compound into a patient suffering from terminal lymphoma. The result was a qualified success. Although the patient died, the tumor shrank, leading to a dramatic, albeit short-lived, remission (Goodman, 1963). Throughout the late 1940s and early 1950s, research on both sides of the Atlantic uncovered a number of other analogous compounds with similar effects on the leukemias and the lymphomas.

The study of antimetabolites, the second class of anticancer substances, grew out of the study of antibiotics. Work on the ability of sulphonamides to inhibit bacterial growth had led, in 1940, to the suggestion that the substance blocked bacterial growth by metabolizing compounds essential to the growth of bacteria. Richard Lewisohn (1875–1961) at the Mount Sinai Hospital in New York transferred the concept to the field of cancer when he showed in 1944 how folic acid caused the regression of mammary cancer tumors in mice (MacGregor, 1966). Work then shifted to humans when Sidney Farber (1903–1973) at Harvard University treated leukemia patients with folic acid. Unlike the tumor reduction in mice, the folic acid seemed to have the opposite effect on humans. Hoping that a folic acid *inhibitor* would have therapeutic consequences, Farber turned to his former colleague in biochemistry, Yellapragada SubbaRow (1895–1948), and convinced him to develop analogues of folic acid as potential inhibitors. SubbaRow did so in his newly acquired position as research director at Lederle Laboratories (then a division of American Cyanamid). Among the analogues was amethopterin (methotrexate), which SubbaRow produced shortly before his premature death (Diamond, 1985). Used once again with leukemia patients in 1948, methotrexate produced significant symptom palliation. Following these initial breakthroughs, however, enthusiasm for chemotherapy gradually turned to outright hostility as repeated use produced repeated failure despite brief remissions (DeVita & Chu, 2008).

The systematic study of hormone therapy in the United States began in 1947 when the American Medical Association's therapeutic trials committee organized a group of clinicians to study the use of hormones in advanced breast cancer. In doing so the AMA sought to advance a practice that was by then almost fifty years old. Hormonal therapy for breast cancer in the form of removal or, later on, irradiation of the ovaries had first been suggested for inoperable cases of breast cancer at the end of the nineteenth century by the Scottish surgeon G. T. Beatson (1848–1933) (Beatson, 1896; Paul, 1981). By midcentury, most observers agreed that while this therapy reduced symptoms, it did not improve survival, a consensus finally confirmed in a cooperative group study carried out in the 1960s by Bernard Fisher's National Surgical Adjuvant Project (Radvin et al., 1970; see chapter 6). The scope of hormone therapy widened further with the discovery of the sex hormones in the first half of the century as hormone therapy opened up an entirely new field of research of considerable complexity (Oudshoorn, 1994). Initially suspected to be one of the causes of breast cancer, estrogens, for example, had by the late 1940s come to be seen as possible inhibitors of the same disease; hence the interest in hormones and their manipulation (Aub & Nathanson, 1951). While not a clinical trial, the AMA's clinical evaluation of hormone therapy showed that there was indeed some value in the use of androgens and estrogens in breast cancer. The study failed nonetheless to offer criteria for the selection of patients for their use or to quantify the results of such use (AMA Therapeutic Trials Committee, 1960).

While the discovery of the nitrogen mustards and the antimetabolites was largely the product of serendipity, the search for anticancer substances was also the subject of more systematic efforts. Prior to the Second World War, laboratories in Germany, Britain, and the United States screened a variety of substances for anticancer activity (Löwy, 1998; Bud, 1978: 445–46). Other small-scale private and public screening programs emerged in the postwar era at the Chester Beatty Institute in Great Britain and the University of Tokyo in Japan (History of the Cancer Chemotherapy Program, 1966). Similarly, in its 1946 Annual Report the NCI described a program "to test the chemotherapeutic value of various synthetic chemicals on transplantable, spontaneous, and induced animal cancers." By 1947 screeners had accumulated 1,200 compounds and screened over 500. The number of interesting compounds, however, had fallen from 10% to around 1%.[1]

Screening accelerated after the war when private research institutions in the United States such as the Sloan-Kettering Institute joined the fray and initiated their own screening programs for anticancer agents in the cancer-bearing mice that had been developed at the Roswell Park Institute during the First World War (Bud, 1978; Löwy & Gaudillière, 1998). The Sloan-Kettering program, then the largest in the United States, brought together "twenty-eight major pharmaceutical, industrial and university laboratories."[2] In 1951 MSK inaugurated a clinical testing program and opened a ward of forty-seven beds for "otherwise untreatable" patients. The following year they reported that although "no chemical cure [had] been found . . . nevertheless, the palliative usefulness of agents already at hand [had] been clearly established."[3] Sometimes cited

1. Annual Report, Federal Security Agency, 1947, p. 303, NCI Archives.

2. The Sloan-Kettering Institute: Report of the Chairman and President of the Board, 1951–1952, 110 (MSK Cancer Center, Record Group 101.1, Box 9, Annual Report).

3. Ibid., 113.

as a precursor to the NCI's screening program, the Sloan-Kettering program differed from NCI's in a number of respects. As we will see in the next chapter, unlike the NCI, MSK adopted an industrial approach. Moreover, the Sloan-Kettering program failed to develop clinical research that went beyond the simple testing carried out in the forty-seven-bed ward.

Prognosticating Cancer

The plurality of treatments, their often public funding, and the rich clinical vocabulary that provided many different designations for the "same" diseases in the field of cancer created a demand for common categories to evaluate the severity of the disease. The interwar period laid the groundwork for a number of nomenclature, staging, and grading schemes that have persisted until the present day and that are crucial to the conduct of clinical cancer trials. Now pursued as autonomous enterprises, classifications have been established for most cancer sites. We will not, of course, consider all of them here. A brief glance at the earliest schemes is instructive with regard to how statistical and clinical demands brought such systems into existence.

In cervical cancer, for example, the desire for uniform statistics on the results of the increasingly popular use of radiotherapy prompted the Radiological Sub-Commission of the Cancer Commission of the Health Organization of the League of Nations to form an expert subgroup in 1928 charged with devising a means to this end. Reporting back the following year, the experts proposed the subdivision of the development of uterine cancer into four separate stages based on the anatomic extent of growth. The scheme became known as the League of Nations Classification for Cervical Cancer. Subsequently adopted by the International Federation of Gynecology and Obstetrics, the federation continues to revise and update the widely used scheme.[4] In the early 1940s researchers developed a similar classification for breast cancer for approximately the same reasons (Portmann, 1943).

Cancer surgeons had preceded radiologists and radiotherapists into this classificatory domain. By the beginning of the twentieth century, the growing number of cases that went under the knife had given rise to the

4. At http://www.figo.org/default.asp?id=27.

hypothesis that cancer progressed through a series of continuous stages. The aforementioned William Halsted was among the first to argue that solid cancers such as breast cancer spread in an orderly manner from the initial tumor site in the breast, to the adjacent lymphatic system, and then on to distant organs, with each stage conferring an increasingly grim outlook on patient survival. Clinicians and pathologists subsequently translated this surgical insight into a clinical staging system in the interwar years when the rising numbers of patients treated for cancer through radiotherapy and surgery and competing applications of these treatments created the need for distinguishing subtypes of patients. As we will see in chapter 6, breast cancer chemotherapists came to dispute this local understanding of cancer *etiology* and promote a more systemic understanding of the disease. This more systemic view did not lead to a rejection of the utility of *staging systems*. In other words, staging and etiology have followed parallel tracks since the 1950s.

The earliest staging scheme in surgery was the Dukes's staging classification of rectal cancer first formulated by Sir Cuthbert Dukes (1890–1977) at St. Mark's Hospital in London in the 1920s. As chief pathologist at St. Mark's, Dukes had a unique view of the field of colorectal cancer, and his staging system first materialized as a response to a problem raised by the hospital's surgeons during a sort of "naturally occurring" clinical trial. Throughout the 1920s surgeons at St. Mark's had pursued two different surgical techniques in the treatment of rectal cancer, and using survival statistics each group claimed superior results. To sort out the claims, Dukes compared patient statistics and the corresponding biopsy results and concluded that the previous comparisons of survival rates were biased: unknown to the surgeons, patients with ostensibly the same disease had, in fact, formed two distinct groups in terms of extent of disease at the time of surgery as revealed through the initial biopsies. By treating the two groups as a single entity, comparisons of the procedures had thus masked this fundamental difference and led the competing surgeons into a maze of spurious claims. So in 1932 Dukes reorganized the fifty-two cases that had accumulated at St. Mark's between 1927 and 1930 into three categories based, like the cervical cancer cases, on the anatomic extent of disease at the time of the operation.

Dukes did not publish the final results of his twenty-five years of research until after World War II, whereupon his system found widespread acceptance in both the United States and Britain (Morson, 1985; Weiss, 2000a). By then Dukes and his colleagues at St. Mark's Hospital had accumulated over 3,500 cases of rectal cancer that offered a substantial empirical basis for their classification (Dukes & Bussey, 1958). In cervical cancer and rectal cancer, then, the same impetus—to determine the therapeutic efficacy of a single treatment or to compare treatments—and the same strategy prevailed: the standardization of the clinical phenomena under treatment, that is, staging. In tandem with clinical trials, staging schemes proliferated in the postwar period, and with the exception of cancers of the brain, spinal cord, and blood, all cancer-staging schemes follow a plan similar to that described above, one based on the anatomic origin and spread of the disease.

The most widespread staging scheme presently used in the field of cancer is the TNM (tumor, node, metastasis) system (see Sobin et al., 2010 for the latest edition) and it, too, is largely a postwar phenomenon with roots in the 1940s. As the acronym suggests, the classification follows the basic pattern of staging systems and describes the spread of the disease from the initial tumor (T), to the lymph nodes (N), and finally to other organs through the process of metastasis (M). TNM delineates the stages of disease spread according to a mix of qualitative and quantitative "grades" (not to be confused with pathological "grading": see below), whereby each of the three principal stages of cancerous growth are further subcategorized according to varying degrees of severity. Tumor size (T) is usually graded from 1 to 4 based on size. N and M have, minimally, two major subdivisions (0 and 1) corresponding to the presence or absence of cancer and, if higher grades are given, to the number of lymph nodes or distant sites involved. A typical designation might therefore be: cT4N1M0 (large tumor, present in the lymph nodes, no metastasis).

Not originally intended to direct treatment, the TNM system grew out of a French wartime effort to develop a common nomenclature for the Permanent Cancer Survey. Organized by the French National Hygiene Institute, the survey set out in 1943 to create a cancer registry that would include all cases of cancer treated in French cancer centers. Initial results showed that the categorization of the cancers was consistently stymied by the fact that the same "histological" cancer could be registered under a variety of names depending upon the anatomical extent of the disease. To overcome the proliferation of entities based on anatomical extent, Pierre Denoix (1912–1990), the head of the project, proposed standardized reporting procedures (Ménoret, 2002). Denoix developed his system throughout the 1950s, at the end of which it was adopted by the Union for International Cancer Control (UICC, 1958; Sobin, 2001).

In the United States, the American Joint Committee on Cancer created its own system at about the same time. For many years, although the two systems evolved in parallel, they contained significant differences. In 1987 an international committee fused the systems, thus endowing TNM with "worldwide" recognition (Hutter, 1987).

Just as surgeons had preceded clinicians, pathologists had preceded surgeons in the categorization of subdivisions of cancers within specific disease entities through the development of a second technique for determining the prognosis of a cancerous lesion known as *grading*. Grading emerged at the end of the nineteenth century largely under the impetus of researchers like David von Hansemann (1858–1920), a one-time research assistant of the German physician Rudolf Virchow (1821–1902), the celebrated founder of cellular pathology (Hansemann, 1892; Wagner, 1999; Weiss, 2000b; Gonzalez-Crussi, 1999). Hansemann based his cancer grades on a theory of the development of cancer that hypothesized that normal cells were transformed into cancer cells through a process he termed *anaplasia*, which represented the reverse process of normal cell development from an undifferentiated germ cell to a differentiated member of a specific tissue. This reversion to a more primitive type, also called *dedifferentiation*, signified that the less normal cells resembled their tissue of origin (when stained and viewed under a microscope), the more cancerous they had become. In addition, in Hansemann's view the greater a cancerous tissue displayed anaplasia, the more likely it was to metastasize (Hansemann, 1903).

On this basis Hansemann proposed that cancerous tissues could be divided into low, intermediate, and high grades depending upon the degree of anaplasia. In practice, however, Hansemann's grades often referred to different cancers rather than to degrees of danger within a specific cancer. Pathologists consequently attribute the beginnings of the modern system of grading of specific cancers to work carried out by the American pathologist Albert C. Broders (1885–1964) at the Mayo Clinic in the 1920s and 1930s (Nascimento, 1999; Kilpatrick, 1999; Oliveira & Nascimento, 2001). Starting with cancer of the lip, Broders went on to develop criteria for the grading of a number of other cancers. The most durable of his grading techniques have been those established for rectal cancer whose resultant classification has often been used in tandem with the Dukes staging system (Shampo, 2001). As can be imagined, grading requires the services of an experienced surgical pathologist. During staging, on the other hand, scans and x-rays can sometimes determine the anatomic extent of a given cancer without recourse to the services of a surgeon.

Testing Therapies

In addition to diagnosing, grading, staging, and treating cancer, clinicians and researchers in the first half of the twentieth century also tested cancer therapies. Given its novelty, radiation therapy was clearly fertile ground for testing. To gain a better grasp of how tests for therapy functioned prior to the advent of clinical trials, we will briefly review some of the difficulties encountered in the evaluation of the efficacy of radiotherapy in a specific cancer: Hodgkin's disease. As we will see, the test of a therapy had entirely different connotations in this period, particularly with regard to the kinds of skill required by a physician to undertake radiotherapy.

The problems of therapy testing often have their roots in diagnosis. Tests of therapy—with or without recourse to clinical trials—presuppose cases of specific disease: without specific entities to treat, tests of therapy amount to little more than tests of panaceas. In the case of Hodgkin's disease, Hodgkin had originally described the disease in 1832 on the basis of a series of seven patients, all of whom presented similar symptoms, a shared natural history that resulted in death, and common postmortem findings. The cellular basis of Hodgkin's disease escaped Hodgkin himself, who worked without a microscope in the so-called anatomo-clinical tradition of medicine that correlated autopsy findings and clinical findings in the form of *clinical pictures* (Faber, 1923; Temkin, 1963). Despite the perspicacity of his observations, the prevailing technology limited their accuracy. Subsequent reviews of Hodgkin's original biopsy specimens have shown that only two of his seven patients would today be diagnosed with Hodgkin's disease (Postan, 1999).

The observation of an abnormally large cell with two nuclei—named the Reed-Sternberg cell after its codiscoverers—that accompanied most cases of Hodgkin's disease led to the definition of the cellular basis of Hodgkin's disease at the beginning of the twentieth century and is now considered a hallmark of the disease. Understanding of the disease remained nonetheless clouded in many respects well into the postwar period. In the late 1940s, for example, pathologists recognized at least three different types of Hodgkin's based on the appearances of cells in stained

tissue samples (Jackson & Parker, 1947). These types were not the subject of hard and fast distinctions; in an extended series of biopsies conducted at the US Army Institute of Pathology just after World War II,
the pathologists noted that the vast majority of cases could be placed in
at least two different categories, an observation that extended to lymphomas as a whole (Custer & Bernhard, 1948).

Despite the cellular basis, in the first half of the twentieth century, clinicians continued to rely heavily on clinical pictures to manage Hodgkin's, and by the eve of World War II, clinical work had further subdivided these pictures (Ziegler, 1911; Karnofsky, 1966). The difficulty in
nailing down the specific nature of the disease was reflected in the fact
that Hodgkin's continued to be described under numerous designations
well into the twentieth century. In a review in the early 1930s, Wallhauser
enumerated over fifty different names for Hodgkin's and lamented that
"perhaps no other disease has been encumbered with a more surprising array of names than has Hodgkin's disease" (Wallhauser, 1933: 522).
Therapists, however, embraced this luxuriant variety as it enabled them
to set out a manifold of treatment strategies and to tailor them to the
individual.

The Swiss radiotherapist René Gilbert (1892–1962) is a case in point.
A firm believer in the infectious and inflammatory nature of Hodgkin's
disease—consensus over the cancerous nature of the disease did not
emerge until the 1950s—Gilbert made significant advances in the treatment of what was an invariably fatal affliction (Kaplan, 1962). Present-
day Hodgkin's specialists generally credit Gilbert with having almost
single-handedly developed the radiotherapy of Hodgkin's disease. In
turn, Gilbert (1939) himself believed that much of his success depended
upon his clinical acumen and, in particular, his ability to recognize and
navigate his way through the four evolutionary paths that he believed
Hodgkin's could take. Gilbert achieved, moreover, substantial success:
he claimed survival times double or even triple those of untreated patients even though he generally qualified these periods of apparent recovery as "remissions" that gave "the illusion of recovered health."

Despite his success, research in the postwar era cast a shadow over
Gilbert's accomplishments. As he did not use controls in his therapeutic excursions, many of the younger generation of radiotherapists had
"a tendency to dismiss his apparently superior results as being due to
case selection" (Kaplan, 1962: 553). Equally as problematic as the lack
of controls—which was problematic only in retrospect—was the fact that

the treatment required considerable clinical expertise and, given the variety of forms assumed by the protean disease, had to be adapted to individual cases. As Gilbert described his approach on the eve of the Second World War after almost twenty years of experience:

> In this very special systemic disease, the therapeutic radiologist must act as a general physician; he should know well the manifestations of the disease and the regions involved by it, its modes of extension, its types and evolutionary caprices, and naturally the mode of action of roentgen therapy as well as the reactions of the patient during treatment. . . . It is on this basis that he should establish a plan of treatment the execution of which will remain subordinate to the evolution of each case. (1939: 234)

Gilbert's last counsel—that therapists should establish a "plan of treatment the execution of which will remain subordinate to the evolution of each case"—bears repetition. It is the exact opposite of the strategy used in clinical trials where the individual case is subordinate to the plan of treatment or protocol. As can be imagined, in Gilbert's mind, the physician's expertise—in this case his ability to follow a case and to adjust treatment according to the meanders and caprices of the disease as it evolved—rendered anything like a clinical trial virtually impossible. Experts in radiotherapy such as Gilbert believed that the many clinical variants of Hodgkin's disease precluded not only clinical trials but also any attempt to standardize treatment. Read in the context of the time, therefore, the following quotation from Gilbert is less an expression of frustration than a statement of faith:

> By its polymorphism and by the great variation in its evolutionary features in different cases . . . the disease presents great diagnostic and therapeutic difficulties. Its treatment cannot be standardized; it must always be adapted to each particular case and, on the part of the treating physician, requires much clinical sense. (1939: 198)

Given that both pathologists and, now, radiotherapists concluded that Hodgkin's disease in all its manifestations defied simple categorization, one might expect that any attempt to bring order to the diagnosis and treatment of Hodgkin's would be destined to fail. Nonetheless, clinical researchers succeeded in standardizing the diagnosis and treatment of Hodgkin's very shortly after Gilbert's pessimistic conclusion. In a series of papers that began to appear in 1950 and that are now regarded as classics, Vera Peters (1911–1993), a radiologist and radiotherapist at the Toronto General Hospital, changed the way therapists and clinicians viewed Hodgkin's in a fundamental manner (Rosenberg, 1999: 47 n. 20).

The work Peters reported consisted essentially in a statistical evaluation of the patients she and others had treated at the Ontario Institute of Radiotherapy (McCulloch, 2003). The cases that formed the basis of her statistics had accumulated slowly in the period 1924–42 and, *contra* Gilbert, constituted an effort to draw some kind of generality out of the capricious nature of Hodgkin's (Peters, 1950; Peters & Middlemiss, 1958). In her analysis of the Toronto data, Peters discovered that the most significant prognostic factors in Hodgkin's disease turned out to be the extent of anatomic involvement upon presentation of disease. While this may seem rather unsurprising given that similar conclusions obtained in breast and rectal cancer, the unexpected nature of Peters's results can be more fully appreciated if we recall that at the time it was not clear whether or not Hodgkin's was a cancer. It was furthermore unclear whether or not it arose from a single site and then spread throughout the body or, instead, originated as a systemic disorder, that is to say, emerged in an already disseminated form.

Generalizing from her results, Peters concluded that rather than treating a seemingly infinite number of possible developments that required focus on the individual patient, one could simply divide all cases into three distinct stages based on the anatomic development of the disease. Thus Peters defined Stage 1 as disease that involved the lymph node region and showed that the five-year survival rate following radiotherapy for that stage was approximately 88%. Regardless, then, of the pathological fluidity or clinical meandering, consistent forms of radiotherapeutic treatment generated equally consistent results. This was new. Peters then took her analysis a step further in a form of bootstrapping or feedback. She used the stages derived from the data to reanalyze the data on treatment by comparing two forms of radiation therapy used at the Toronto hospital; one with and one without prophylactic radiation. By comparing patients at the same stage of disease, she was able to compare relatively uniform strategies of radiotherapy—impossible, we recall, according to Gilbert.

Thus by dividing the disease into stages and putting patients into the stages rather than drawing increasing complicated forms she made possible something like a standard treatment. Furthermore, by reporting treatment results in terms of her clinical stage classification, Peters described the evolution of the disease under specific therapeutic regimes: as a treatment history as opposed to a natural history. In sum, Peters's approach not only constituted a decisive step in the standardization of treatment but also overcame the initial obstacle to that standardization, the cloudy picture presented by the somewhat fluid pathology. The revolutionary nature of Peters's work will become even more apparent later on in this book when we learn that well into the late 1950s and early 1960s many radiotherapists resisted clinical trials in radiotherapy and continued to endorse Gilbert's view of radiotherapy as an intensely personal undertaking requiring unusual clinical skills, a view also widely shared by surgeons.

Clinical trials are themselves a postwar invention. While attempts to undertake a "fair assessment" of medical therapies date back to the eighteenth century, most historians of the practice have settled on the celebrated trial of streptomycin in tuberculosis—designed by the British researcher Austin Bradford Hill (1897–1991) in the late 1940s—as the first controlled clinical trial (Armitage, 2003; Marks, 2003). One of the central features of the trial was the assignment of patients to the current therapy (bed rest) or the new therapy (streptomycin plus bed rest) on the basis of random numbers (see First Interlude). Previous clinical assessments of therapy had, for instance, allocated patients alternately to study groups. This meant that treating physicians knew the treatment schedule and could, consciously or unconsciously, adjust their practice accordingly. Randomization concealed the allocation schedule and the treatment that the patient would receive from the treating physician. This did not blind the physician to the treatment, since in the present case no placebo injection was used. The patients remained unaware of the differing treatments, because Hill felt it unnecessary to ask them to sign consent forms. A second novelty of the trial lay in the assessment of the results of the treatment: a lung x-ray examined by nontreating radiologists served as the determinant of the success or failure of the treatment, removing both the patient's and the clinician's subjective judgment from the equation.

Although the success of the trial served as a proof of principle, controlled or randomized clinical trials did not gain immediate acceptance in the medical community. We will return to the continued resistance to this practice in cancer therapy in the 1950s later on in this book. For the moment, we need merely mention the special circumstances that made

Hill's trial possible. In a retrospective account of the trial, Hill (1990) noted that it was largely due to the fact that the supply of streptomycin was severely limited—there was enough of the drug for only fifty patients—that allowed him to overcome objections to the use of randomization to assign patients to the two arms of the trial.

Conclusion

On the eve of the institution of the NCI's chemotherapy program, radiologists, radiotherapists, gynecologists, dermatologists, hematologists, surgeons, and general physicians routinely treated cancer just as pathologists and surgeons graded and classified the tissues and tumors removed in the course of biopsy and treatment. In addition, clinicians staged a number of cancers based on the anatomic spread of the disease. In some rare cases like Hodgkin's and breast cancer, something like a standard treatment had begun to emerge. All of the above treatments—particularly chemotherapy, which remained experimental throughout the period— were largely confined to the rare cancer centers that specialized in specific kinds of cancer surgery or radiotherapy. Those centers accumulated sufficient numbers of cases to make something like standardization thinkable by the late 1930s, a process that occurred, as we have seen, in the case of Hodgkin's disease. Clinical trials would reverse this process. Rather than waiting for standardization to emerge, so to speak, from historically accumulating cases, the new mode of practice would seek standardization as way of creating a new history of cancer.

PART I

The Emergence of Clinical Cancer Research (1955–66)

CHAPTER TWO

A Landmark Clinical Trial

Curing Leukemia: The VAMP Trial

In the summer of 1961, two members of the NIH Clinical Center in Bethesda (Maryland), Emil Frei, the chief of the Medicine Branch at the NCI, and Emil J Freireich, a senior investigator in the same branch, decided to conduct a clinical trial to test a combination of anticancer drugs on children afflicted with acute lymphocytic leukemia (ALL). They launched the study, whose protocol they summarized under the evocative rubric VAMP, in early 1962. The results were so promising that they stopped the trial after treating a mere sixteen children. As a consequence and despite the subsequent notoriety of the protocol—considered the first combination of drugs to actually "cure" childhood leukemia—the only published trace of the original plan remains a brief abstract in the *Proceedings of the American Association for Cancer Research* (Freireich, Karon & Frei, 1964).

As described in the *Proceedings*, the protocol set out a rigorous regime of chemotherapy that combined four different substances. Frei and Freireich had chosen the drugs so that each had a different mechanism of action. Each agent thus destroyed cancer cells in a different way, creating, the researchers hoped, a cumulative, or even synergistic, effect. The protocol also stipulated different routes of administration and schedules for each drug: vincristine (V) and amethopterin (A, also called methotrexate) were to be given every four days intravenously; mercaptopurine (M) and prednisone (P) were to be administered orally on a daily basis. The protocol further specified that the treatment would continue until a patient achieved "remission," defined not only as the abatement of the leukemic symptoms but also as the reduction of the proportion of leukemic cells in the bone marrow to below 5% of the total number of cells contained therein. Once in remission, the protocol then submitted the patients to five more ten-day courses. A "rest period" that lasted either ten days or until toxicity subsided followed each course. In case of relapse following remission, treatment began anew.

Frei and Freireich used the amount of time the young patients spent in complete remission as the end point of the trial. When the trial finished, the median remission time stood at 150 days. To present-day readers, the decision to stop the trial may appear somewhat surprising given the small numbers. The sixteen "consecutive" patients—the first sixteen children with acute leukemia to come through the doors of the NIH Clinical Center—were the only patients studied. So why stop so soon after beginning? The simple answer is: 150 days was the longest remission yet achieved in the treatment of early childhood leukemia by either the study trialists or by others in the field.

The others in the field included members of several collective undertakings known as Cooperative Groups. In addition to being researchers at the NIH Center, Frei and Freireich also belonged to a research collective that conducted clinical cancer trials in leukemia known as the Acute Leukemia Group B (ALGB). Along with other members of that collective, Frei and Freireich had undertaken previous trials with each of the substances in VAMP. The group had thus developed what are often termed "historical controls," and Frei and Freireich were consequently able to compare the results of VAMP with the results obtained in the earlier trials. In the case of prednisone, for example, a trial completed the year before and known as "Protocol No. 3" had produced only sixty-day remissions (Frei & Freireich, 1965: 284). The remissions produced by VAMP, however, were not simply better than prior results: both the VAMP results and the results of the prior studies were all quite unprecedented.

Each of the drugs used in VAMP had been discovered in the postwar era, and three of the four had been developed in the previous decade. We have already discussed the oldest member of the combination, methotrexate. One of the more recent compounds, vincristine, was the first naturally occurring substance to be used as an anticancer agent. Derived from an extract of the Madagascar periwinkle, a plant used as a folk remedy for centuries, it had long been reputed to be a useful remedy in diabetes (Noble, 1990). Following up on these folk reports in a search for an oral compound to treat diabetes, the American pharmaceutical giant Eli Lily had begun examining extracts of the leaves of the periwinkle in the 1950s and discovered that they contained over seventy different alkaloids (Pearce & Miller, 2005). Canadian researchers had embarked on a similar quest but for logistic reasons were unable to compete with Eli Lily; to extract one ounce of the drug, researchers had to process fifteen tons of leaves. So even though the Canadians had enrolled Jamaican Boy Scouts to collect the leaves, they were in the end forced to rely on extracts produced by Eli Lily. Early tests of the extracts had shown, unexpectedly, anticancer properties. Thus despite the initial interest in an insulin substitute, the Canadians went on to perform the first successful human cancer tests in the late 1950s (Duffin, 2002a; 2002b) and subsequently forwarded the substance to the NCI for further examination (Carbone et al., 1963).

While the discovery of the anticancer properties of vincristine was partly serendipitous, prednisone emerged in 1955 following intensive research in the field of steroid chemistry that had begun with the isolation of the sex hormones in the 1920s. By the time of prednisone's synthesis, greatly stimulated by the discovery of an extremely lucrative and structurally similar compound, cortisone, in 1948, all the major pharmaceutical companies had established research programs in the field of steroids. Although Schering ultimately won the patent for prednisone in 1964, five other pharmaceutical companies (Merck, Squibb, Pfizer, Upjohn, and Syntex) also claimed priority for the drug in what was clearly a case of multiple discovery (Herzog & Oliveto, 1992; Heusler & Kalvoda, 1996; Hillier, 2007).

When prednisone entered the cancer research market as an experimental substance in the mid-1950s, it was not the only steroid in use. A similar hormone, ACTH, isolated in 1949, had already shown some activity in ALL. Prednisone's career as an anticancer agent thus began as a substitute for a known agent and initially did little to distinguish itself. As Sidney Farber commented in 1956 when reviewing cortisone, ACTH, prednisone, and fluorohydrocortisone: "In their effect, all agents are similar and may be discussed as one" (Farber et al., 1956: 13). Those effects were distressingly minimal. None produced remissions that lasted longer than seventy days and all had a tendency to produce "Cushing's syndrome," whose principal symptoms—weight gain, thinning of the skin, depression, and anxiety—merely added to the growing list of unpleasant side effects produced by anticancer compounds.

The fourth and final component of the VAMP combination, 6-mercaptopurine (6-MP), was also a product of systematic industrial research and earned its discoverers, Gertrude Elion and George Hitchings, the 1988 Nobel Prize. The two Wellcome Company researchers had begun the work leading to the discovery of 6-MP in the early 1950s in a search for compounds to block bacterial reproduction. They had found that bacteria could not reproduce without a class of compounds known as purines. They knew purines were among the building blocks of DNA (well before the discovery of the structure of DNA in 1953 by Watson and Crick) and consequently began hunting for compounds to interrupt purine production. 6-MP emerged in the course of this research. Early animal tests of the compound showed that it inhibited the growth of a variety of rat and mouse tumors, and by 1953, 6-MP had induced short-lived remissions in humans (Quirke, 2009).

Within a relatively short period then, industrial researchers had developed three different compounds that produced regressions in childhood leukemia. By themselves, none of the substances produced long-lasting remissions or anything that could be even remotely associated with a cure. Moreover, even when used sequentially or two at a time, although remissions were accelerated, all patients died eventually as resistant forms of leukemia emerged to defeat the drugs. Low expectations thus replaced the initial excitement that had surrounded the drugs. As one of the pioneers of leukemia therapy in the 1960s has noted, the failure of the anticancer compounds in the 1950s "led to a fixed notion among most pediatricians and hematologists that temporary remissions and prolongation of survival in comfort were the most one could expect from leukemia chemotherapy" (Pinkel, 2006: 12).

Two ancillary components of the VAMP protocol that remain crucial to the cure of acute leukemia in children were filtered air and the frequent transfusion of blood platelets. The transfusions combated the severe bleeding in acute leukemia patients that results from a loss of cells responsible for blood clotting. Provoked by the chemotherapy, the bleeding was sometimes so severe that it often became the immediate cause of death. One of the conditions of possibility for treating leukemia patients with chemotherapeutic substances was thus the control of bleeding so that the patients might live long enough for the drugs to have an effect.

Freireich and his colleagues had developed the blood platelet transfusion techniques just before the VAMP trial and had done so in part at the

behest of the head of the chemotherapy program at the NCI, Dr. Gordon
Zubrod. As Freireich recalled:

> I was in charge of the leukemia program and Frei ran the solid tumor and
> Dr. Zubrod was the godfather and he'd come on rounds every once in a while.
> They'd come on the leukemia rounds, and in the leukemia ward with these
> children there was blood all over. There was blood on the sheets, blood on the
> pillows, blood on the floor; the nurses were covered with blood.... Zubrod came
> out in the hall one day and said, "This is a bloody ugly mess, Freireich. Why
> don't you do something about hemorrhage?" So being that I was a young guy,
> respectful, I said, "Yes sir." I went to the lab and I did simple experiments....
> If you just take the children's plasma and take fresh platelets in the labora-
> tory, it is 100% corrected. So I said, gee-whiz, all we got to do is give them
> platelets.[1]

The process was slightly more complicated than Freireich's recollections
suggest. When the NCI Clinical Center researchers began their work on
the control of bleeding, whole blood transfusions had generated some
controversy since "classical hematologists had supposedly shown, using
labeled (and therefore damaged) platelets, that the infused platelets did
not last more than a few hours" (Zubrod, 1984: 11). So Freireich and Frei
first had to undo this doctrine. Moreover, while fresh blood transfusions
raised platelet levels more significantly than blood drawn from blood
banks, neither method seemed adequate because of the sheer amount of
blood that had to be transfused into patients (Freireich et al., 1959). To
overcome the blood shortage problem and to limit the size of the therapeutic transfusions, Frei and Freireich sought ways to concentrate platelets and to calculate the minimum number of platelets necessary to stop
bleeding (Gaydos, Freireich, & Mantel, 1962). This initiative also proved
controversial:

> These studies were vigorously attacked by classical hematologists because
> they required larger amounts of blood than blood banks could easily obtain.
> I recall a dramatic showdown at a clinical staff meeting in 1956 where a mo-
> tion was made to deny platelets to the NCI. In those days the entire clini-
> cal staff of the Clinical Center met monthly to discuss patient care problems.

1. Interview with E. J. Freireich, July 2006, Houston, Texas.

The other clinical directors came to our support and the motion was narrowly defeated. (Zubrod, 1984: 11)

In collaboration with members of the Division of Biological Standards at the NCI, Freireich developed a device that would allow them to take blood from donors, separate the platelets, and return the red blood cells to the donor, thus creating the possibility of multiple donations from a single donor (Kliman et al., 1961). While the platelet concentration system worked well enough for the VAMP patients, it was labor intensive and the search for a large-scale source of platelets became an ongoing concern both for Freireich, who continued working on the problem for the next twenty years, and other members of ALGB who pursued cooperative research with IBM to produce an automatic cell separator (Hester et al., 1985).[2]

The chemotherapy itself created further problems. Leukemia patients submitted to the rigors of methotrexate, 6-MP, and prednisone lost other components of their blood, most notably the white blood cells, and as a result suffered severe immunodeficiency and thus frequent infection. Like other researchers, Frei and Freireich treated the problem of infection through megadoses of intravenous antibiotics. They also resorted to other devices such as oxygen tents when, in spite of precautions, the children caught pneumonia. But even that was not enough, and so they went one step further: in order to combat antibiotic-resistant fungal infections caused by common inhabitants of hospital air ducts, the NCI researchers developed "life islands" equipped with special air filters (figure 2.1) (Nathan, 2007). In recognition of the importance of these "ancillary measures," *Life* magazine chose to illustrate its 1966 report on the "all-out assault on leukemia" with photographs showing not only bottles of anticancer drugs but also the accompanying infrastructure (figure 2.2).

When all was said and done, compared with the disappointments of the 1950s, the results of the VAMP protocol were quite spectacular. Of the sixteen patients treated, thirteen achieved "complete remission." More importantly, the time spent in remission lasted two to three times longer than any remission obtained using a single compound or any combination of two compounds. Finally, when Frei and Freireich reported their results in a review article (1965), two of the patients were still alive

2. NCI, Acute Leukemia Task Force, Platelet-Supportive Care Group, 24 April 1967, National Archives, AR-6704–001804. See also Freireich et al. (1965).

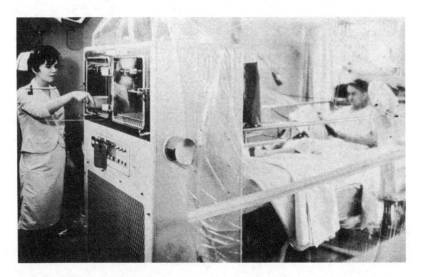

FIGURE 2.1. A "life island." Reprinted with kind permission from G. P. Bodey, P. Watson, C. Cooper & E. J. Freireich, "Protected Environment Units for the Cancer Patient," *CA: A Cancer Journal for Clinicians* 21 (1971): 216. © 1971, John Wiley and Sons.

and in complete remission after over six hundred days, a result heretofore unseen.

The VAMP trial illustrates three important aspects of the early development of chemotherapy. The first is that contrary to the unrealistic expectations maintained in some quarters, successful chemotherapy was not produced by "magic bullets." For while some chemotherapists of the 1940s and early 1950s entertained the possibility of finding a drug or a category of drugs that, as antibiotics did for bacterial infections, would wipe out cancer cells with no important side effects, cooperative group trialists quickly showed such fancies to be counterproductive. They noted, in particular, that "prospecting for the 'penicillin' of cancer [would be] wasteful of the opportunity to detect lesser degrees of activity and to validate or improve [the trialists'] abilities to select more useful compounds."[3] And indeed, therein lay yet another important lesson of the VAMP trial: not only were the prospective magic bullets less than supernatural, they also produced significant collateral damage. As a result, in order to transform chemotherapy into a viable option, an entire

3. Eastern Cooperative Group in Solid Tumor Chemotherapy, Over-all group research plan and progress report. May 1958–May 1963, 7.

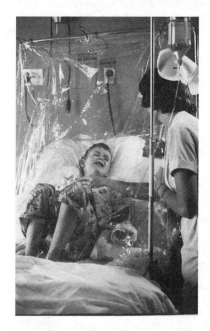

FIGURE 2.2. In an abundantly illustrated report titled "The All-Out Assault on Leukemia" and focusing on a young patient named Mike Parker, *Life* magazine highlighted both
anticancer drugs and the hospital infrastructure necessary to turn those drugs into a viable option. (a) The headline above the picture reads: "Of 170,000 drugs tested, eight might
help Mike Parker"; the eight drugs are in the foreground, and 6MP is the fifth from left.
(b) Mike Parker's immune system having been weakened by anticancer drugs, an oxygen tent is used to fight pneumonia. Reproduced with kind permission from W. Bradbury,
"The All-Out Assault on Leukemia: Two Views—the Lab, the Victim," *Life*, 16 November
1966; (a) pp. 92–93, © Fritz Goro/Time & Life Pictures/Getty Images; (b) p. 104, © Leon-
ard McCombe/Time & Life Pictures/Getty Images.

infrastructure—ranging from mechanical devices to other pharmacological substances—had to be put in place. The deployment and assessment of (candidate) chemotherapeutic compounds thus became dependent upon this infrastructure.

A second related aspect of the trial reminds us that, as noted in chapter 1, early chemotherapy was a "heroic" endeavor that some physicians regarded as not quite legitimate: a therapeutic delusion that far from curing or at least significantly prolonging the life of patients resulted in unnecessary suffering. VAMP, because it induced lasting remissions *and* incorporated strategies to tame the side effects of chemotherapy, represented a de facto rebuttal of this criticism. Last but not least, the "polychemotherapy" approach made it clear that what mattered were not single drugs per se (although, obviously, they were a condition of possibility for the whole endeavor) but their use in combination according to particular doses and administration schedules—that is, what trialists called a *regimen*.

A Collective Undertaking

Although initially inspired by a conversation between two individual researchers, the VAMP protocol participated in a number of concurrent collective and institutional endeavors. The aforementioned NIH Clinical Center, for example, had opened in 1953 as part of a concerted effort by the NIH to transfer knowledge generated in its labs, such as those maintained by the NCI, to the bedside. We have also noted that Frei and Freireich belonged to the Acute Leukemia Cooperative Group B (ALGB) that in 1976 became the Cancer and Leukemia Group B (CALGB) to signify the extension of its interests beyond leukemia. ALGB was one of several clinical trial groups that had been formed as part of the NCI's Cancer Chemotherapy Service inaugurated in 1955. In principle, the service had two arms, a screening service that tested tens of thousands of compounds annually in mice in the search for active anticancer compounds and clinical trial groups like ALGB that tested the compounds in humans. By the time of the VAMP protocol, the most important groups did less testing and more research in what had become an all-out research program. Although Freireich and Frei had been members of ALGB from its inception, given the complexity of VAMP they had decided to withdraw temporarily from the group and to conduct the trial as members of the NIH Clinical Center, their primary institutional attachment.

As members of ALGB, Frei and Freireich would go on to join yet another collective research organization, the Acute Leukemia Task Force set up in 1962, shortly after the completion of the VAMP trial, by the "godfather" of the Cancer Chemotherapy Service, Gordon Zubrod. Zubrod (1980: 113) has recalled the importance of VAMP for the task force in the following terms:

> Freireich, Karon and Frei designed the VAMP regimen and tried it at the NCI; it was apparent within six months that it was giving remissions with a duration not previously seen. The Task Force and Leukemia B (and later Leukemia A) then built from this, with strong input from Skipper and Schabel and their animal model, the regimens of induction, consolidation, and maintenance that provided the cornerstones for today's curative regimens in ALL and the principles that led to the curative combination chemotherapy of Hodgkin's disease, non-Hodgkin's lymphoma, and childhood cancer. From this work came the reorganization of the NCI chemotherapy program in 1965 and most of the principles that are now used in today's cancer chemotherapy.

In other words, a protocol such as VAMP, far from being an isolated event, was part of a complex endeavor whose raison d'être went far beyond the simple testing of a given drug or drug combination. As previously noted, this effort combined not only anticancer compounds and the infrastructure that came with them but also, as we will see later on, theoretical principles, animal models, and distinct organizations. It was not a case of a preexisting organization dictating goals and defining procedures. Results such as those produced by VAMP would lead to institutional reorganization.

The idea of a establishing a task force at the NCI had emerged at the beginning of 1962 and had prompted a visit to IBM—"a corporation that had used the Task Force approach with considerable success"—that same year.[4] Zubrod had then gone on to meet with "the key political players in the cancer field," Dr. Isadore Ravdin at the University of

---

4. "Background Information on NCI Task Force, Prepared for NACC, November 1966," NCI Archives, 1.

Pennsylvania and Dr. Sidney Farber at Harvard in a successful attempt to gain their approval for the task force (Laszlo, 1995: 102). Further stimulated by the success of the VAMP protocol, Zubrod mandated the task force to inquire into the "chemical control" of leukemia, the "total care" of patients, and the viral etiology of the disease. Highly interdisciplinary, the task force brought together members of the clinical trials groups like ALGB, NCI laboratory staff, and industrial researchers. Divided into subcommittees, the group met at monthly intervals to discuss progress and to plan further research. Zubrod later extended the task force approach to tackle chronic leukemia and myeloma (1964), lymphoma (1965), clinical trials in general (1965), solid tumors (1966), and breast cancer (1966).[5]

As a clinical trial organized by members of the NCI working at the NIH Clinical Center, the VAMP trial was conducted in a fluid and evolving institutional context. The distributed nature of the undertaking can be seen in the smattering of prizes later awarded to this growing collective. In 1983, twenty-one years after the VAMP trial, the trial's instigators, Frei and Freireich, won the prestigious General Motors Award for cancer research largely for work carried out as members of the clinical center, ALGB, and the Acute Leukemia Task Force. The GM news release observed that "these researchers, working together for 17 years, developed the first treatment for leukemia that truly cured patients. In the process, they established nearly all the principles of scientific chemotherapy trials for cancers of all kinds" (Frei, Freireich Share . . . , 1983). The same award had been given the previous year to Howard Skipper, a member of the Acute Leukemia Task Force and the Cancer Chemotherapy Service. Skipper had already shared the Lasker Prize for Basic Research in 1974 with Ludwik Gross, Sol Spiegelman, and Howard Temin.

Two years prior to Skipper's award, the Lasker Foundation had given out a joint award to a number of other clinical researchers involved in the cancer chemotherapy collective. The award citations show that the foundation had some difficulty in dividing up the specific contributions of the members of the different research groups (figure 2.3). The foundation cited Frei for his "contribution in the application of the concept of combination chemotherapy," Freireich for "contributions in combination

5. Ibid., 1–2.

FIGURE 2.3. The chemotherapy collective as rewarded by the Lasker Foundation in 1972 (name tags added). Reproduced and adapted from M. B. Shimkin, *Contrary to Nature: Being an Illustrated Commentary on Some Persons and Events of Historical Importance in the Development of Knowledge Concerning Cancer* (Washington, DC: U.S. Department of Health, Education, and Welfare, Public Health Service, National Institutes of Health, 1977), 415 (public domain source).

chemotherapy," and James Holland, one of the founders of ALGB, "for his outstanding contribution to the concept and application of combination therapy." Zubrod, on the other hand, received a Special Award "for his leadership in expanding the frontiers of cancer chemotherapy."[6] Neither GM nor the Lasker Foundation attempted to assign a single source to the cure of leukemia.

Participant histories also tend to distribute credit for the cure of leukemia quite widely. Nonetheless, despite the scattered nature of the attributions, all retrospectives converge on two specific protocols as milestones or turning points in the discovery process: the VAMP protocol and one developed shortly thereafter known as Protocol 6313. Largely the work of James Holland, 6313 (ALGB's thirteenth study of 1963) consisted of a study of a combination of anticancer agents similar to those used in VAMP. It differed significantly, however, in the sequence in which the compounds were administered (Holland, 1968). Describing its innovative sequence, Holland has recently suggested that not only was 6313 an improvement on VAMP, but that VAMP garnered more than its fair share of the spotlight:

6. At www.laskerfoundation.org.

> The outcome was that the sequence of vincristine, prednisone, cyclophos-
> phamide, 6-MP, methotrexate and BCNU gave a longer unmaintained remis-
> sion than did the NCI VAMP program, and that's published. But we didn't
> have a strong press reaction that characterized announcement of the VAMP
> data, and it never had the same impact on the field. (Laszlo, 1995: 223)

Regardless of the relative importance of each protocol, most observers
admit that both VAMP and 6313 contained a number of innovations that
had become typical of a new style of work in the field of clinical cancer
research, and that both protocols served as models for the work that lay
ahead. Joseph Burchenal of Sloan-Kettering and one of the twenty-one
original members of the Acute Leukemia Task Force concluded, for example, that, in addition to the novel combination and number of anticancer substances used, the VAMP protocol innovated in its increased intensity of application of those substances:

> Here for the first time was intensive intermittent combination chemother-
> apy with four drugs with different mechanisms of action. The short intensive
> courses were designed to cause massive leukemic cell kill, while the intervals
> between courses allowed the normal cells times to recover. The success of
> the VAMP program influenced the design of the MOPP protocol for dissemi-
> nated Hodgkin's disease and set the stage for most of the successful intensive
> intermittent therapeutic programs today. (Burchenal, 1975: 1125)

As these competing perspectives suggest—Freireich himself felt that
platelet transfusions made VAMP possible—participants easily drew
different conclusions concerning the importance of specific protocols
and specific elements of those protocols. Although other explanations
are possible, we understand this dissension as an indication that the development of an adequate treatment for childhood leukemia was less a
single event than a process. While both VAMP and 6313 were significant
in their own way, they can also be seen as part of a larger movement that
resulted in the establishment of clinical cancer research. In other words,
by describing these protocols as the products of an emerging practice of
clinical therapeutic research, we can avoid falling into the faux empiricism of the "clinical discovery of a cure" and the tendency to cast clinical
research as the mechanical application of the more "rational" results obtained in laboratory animals.

The Elements of Innovation

While, as we just saw, observers point to VAMP and 6313 as significant turning points or discontinuities, it is possible to represent them as
parts of an ongoing process that began, for instance, with the publication in 1958 of the first collaborative study conducted in three centers
by members of ALGB. The 1958 study compared the continuous versus the intermittent administration of a two-drug combination (notice
that the comparison centered on administration schedules, not different drugs) (Frei et al., 1958). Many passages in the article testify to the
novelty of the undertaking. For instance, the authors openly state, somewhat surprisingly for the present-day reader, that "it is proper to question whether patient welfare was compromised by a clinical trial such
as this," concluding, rather reassuringly, that there was "evidence that
a comparative therapeutic trial in acute leukemia can be accomplished
without recognizable compromise of patient welfare" (1958: 1141, 1145).
Even more revealingly, the authors devoted a paragraph of the discussion (not the methods) section to somewhat mundane (and today taken-
for-granted) organizational issues:

> The mechanics of a cooperative study conducted in different places by sev-
> eral individuals assumed critical importance. A printed protocol was distrib-
> uted which specified procedural steps in detail. The conclusions of this study
> are valid only insofar as the protocol was interpreted uniformly by all investi-
> gators throughout the study. Meetings of the principal investigators and stat-
> isticians were held about every two months and sometimes more often. In
> addition, telephone conferences were frequently used for consultation about
> procedure and interpretation. These factors should not be overlooked in the
> budgetary planning of cooperative clinical trials. (1958: 1142)

Additional innovations not listed in the preceding quote included
"clearly stated eligibility and exclusion criteria, required pre-study tests,
a description of the treatment plan, and a plan for randomization," the
latter using "a sealed envelope technique" (see First Interlude) (Schilsky
et al., 2006: 3553s).

The authors of a subsequent trial, carried out in eleven centers by another cooperative group and published in 1960, felt a similar need to describe organizational arrangements and to justify their endeavor by stating that the cooperative group that carried out the trial was established

"in order that [new chemotherapeutic agents] could be tested on a large series of patients with childhood leukemia within a short period of time" (Heyn et al., 1960: 350). One of the main conclusions of both studies was to demonstrate the "value of a cooperative study group in accumulating a large number of patients in a short period of time" (356). Since the 1958 and the 1960 trials did not produce substantive results, insofar as the comparison of two drug regimens showed no significant difference and, most importantly, no noteworthy improvement in the condition and survival of patients, their most interesting finding was the demonstration (the "proof of principle") that such trials could be carried out.

The VAMP and 6313 protocols, while owing much to previous practices, also incorporated a new type of statistics known as sequential methods, a new way of conceptualizing the process of leukemia, a novel combination of chemotherapy agents, their intensive and intermittent use, and an unprecedented deployment of cell kinetics principles. Cell kinetics drew mainly on the work of Howard Skipper and his colleagues at the Southern Research Institute (affiliated with the Sloan-Kettering Institute), one of the screening laboratories of the Chemotherapy Service. They had conducted the first cancer cell kinetics studies in mice in the early 1960s. In one series of experiments, after inoculating mice with suspensions of diluted leukemia cells, the researchers left half the mice untreated; the other half received a fixed dose of methotrexate for nineteen days. Throughout the experiment, Skipper and his associates (1964) selected mice at random and, following sacrifice, counted the white cells in the blood and blood-forming organs. They were thus able to estimate the speed with which the leukemic cells multiplied, the number of cells present at death, and the number of cells killed by the anticancer agent. They were, in other words, able to describe the kinetics of the cancer process and the effect of therapeutic substances on this process.

The Skipper team managed to calculate that on average the number of cells doubled every twelve hours. They then calculated the amount of time it took for the mice to die from leukemia and the approximate number of leukemic cells present at death. Analysis of the data showed something that, while perhaps obvious today, was not at all obvious at the time—namely, that "the life span of leukemic animals following a single drug treatment [was] proportional to the time required for the relatively small number of surviving cells, doubling in number each 0.55 days, to reach a lethal number" (1964: 27). Skipper's group had thus learned that it was possible for a mouse to develop leukemia starting with a few cells or even a single cell. Consequently, a cure (in mice) would necessarily entail "sterilization" or death of all leukemic cells. Did this apply to humans?

Skipper and his colleagues hoped so and proposed that future human research begin by taking into account the fact that a single leukemic cell is sufficient to cause death, and that successful therapy, therefore, must seek the complete elimination of leukemic cells. In this respect, one quantitative variable stood out as the most significant: how many leukemic cells survived a given dosage and schedule of drugs? As Skipper pointed out: "*For reasons already belabored, this is a critical value for guidance of therapeutic research*" (1964: 46, emphasis in original). In addition, researchers would have to calculate the doubling time of human leukemic cells. Finally, since the fraction of a population of leukemic cells killed by a single dose was proportional to the amount of drug, then "it appears that in general a high-level, relatively short-term maximum-dosage schedule has greater potential for cure than does a low-level, long-term maximum dosage schedule" (1964: 47).

To what extent did the VAMP protocol constitute an application of cell kinetics? It was certainly an application in the sense that the protocol explicitly set out to eliminate all leukemic cells and that this objective was clearly consistent with the cell kinetic work. In order to determine how far down the road to complete elimination of leukemia cells the VAMP program had led them, Frei and Freireich had to generate the same kinds of numbers as those generated in the mice (1965: 284). They had to know, for example, the number of cells that survived treatment, the time between treatment and death, the generation time of human leukemic cells, and the number of leukemic cells upon death. To acquire this knowledge they were forced to make several simplifications. To obtain a total leukemic cell count upon death, for example, they used the frozen organs of expired patients, thus limiting their counting of leukemic cells to the bone marrow, spleen, and liver on the assumption that the vast majority of the cells would be contained therein.

The generation of human leukemic cells caused further difficulty as different techniques had given different estimates. Studies using radioactive isotopes had suggested a doubling time of twenty-four hours whereas bone marrow studies of patients entering remission suggested instead a doubling time of four days (Frei & Freireich, 1965). Frei and Freireich confirmed the latter figure using bone marrow counts of patients who had relapsed and who had had a routine bone marrow biopsy two to six weeks previously. Given the two numbers, they were able to plot the increase on a graph and deduce that a four-day doubling time was the norm. Using that doubling time, they then estimated, by extrapolating backward from the number of leukemic cells that appeared upon remission, the number of cells that presumably had been left after treatment. Since the numbers of cells existing after treatment had been determined by working backward, the model was less predictive than retrospective. As Frei and Freireich admitted: "The method . . . of detecting small numbers of leukemic cells persisting after treatment is a retrospective one and involves a number of assumptions. More direct methods are desirable" (1965: 294).

Once again we are faced with the fact that a cancer clinical trial such as VAMP cannot be reduced to the "mere" testing of drugs, insofar as it also furthered the development of experimental models. Yet while the VAMP protocol was consistent with and replicated the variables in the mice model used for kinetic studies (as well as drug screening), it did not entirely depend upon that model. It was not an application of the model in the usual sense of the word. The model provided only an interpretation of the data generated by the protocol and in several instances dictated what data should be collected. The numbers generated offered, in turn, a possible explanation of what had happened with a particular dose and why, in certain cases, the time to relapse was the length that it was. Nonetheless, both Frei and Freireich recognized that other interpretations of the data were possible. One could have argued, for example, that the generation time for leukemic cells varied in the course of the disease and that the cells doubled at a slower rate following therapy. Frei and Freireich's use of the model did not allow them to eliminate this possibility.

In addition to the use of combinations of anticancer agents and cell kinetics, by the time of the VAMP trial clinical cancer researchers no longer tested substances on leukemia or even acute childhood leukemia per se but on different clinical stages of leukemia defined in terms of the treatment history of the disease—namely, induction of remission, remission maintenance, and relapse. In other words, clinical trials began coproducing the disease stages against which a particular chemotherapy regimen was tested, in the sense that they led to the clinical redefinition or reinforcement of nosological categories (e.g., "leukemia" was shown to consist of two distinct entities, childhood and adult leukemia, that responded differently to treatment and could be correlated to the involvement of different cell types) and to the production of clinically relevant subdivisions of those categories based on the (treated) disease trajectory. Maintenance of remission, for example, used therapies different from those used in remission induction. While the first such maintenance therapies had emerged in the early 1950s (Osgood & Seaman, 1952; Farber et al., 1956), members of ALGB went on to create a specific kind of trial for the evaluation of remission therapies. Put together in 1959 by Frei, Freireich, and an NCI statistical consultant, Edmund Gehan, who was responsible for its innovative statistical design, the trial allowed researchers to compare a placebo with chemotherapy after the induction of remission and thus draw conclusions about the natural history of the disease and the efficacy of the substance used (Freireich et al., 1963).

The trial, the Protocol No. 3 mentioned above, presented an interesting contrast to previous protocols. First of all, it used a novel statistical technique known as a sequential design. Ivan Bross of the Biostatistics Department at Johns Hopkins and statistician for Acute Leukemia Group A had introduced the method into clinical medicine in the early 1950s when, in an article titled "Sequential Medical Plans," he outlined the three advantages of sequential methods over the procedures then in use:

> First of all it allows for analysis of the data *as it comes in* (instead of waiting until the end of the experiment for the analysis). Second it may allow an appreciable reduction in the amount of data that has to be collected to reach statistically valid conclusions. Finally, it *may* eliminate ALL computation on the part of the research worker. (1952: 188)

To understand why sequential clinical trials have these advantages, we have to know how they work. Here, then, is a simplified version of the set-up. Suppose a physician decides to conduct a trial to compare a new treatment with an old treatment on the basis of a mutually exclusive dichotomy, "cured" and "not cured." Now suppose that pairs of patients are randomly allocated to the two treatments. It is then possible to treat each successive pair of patients as miniature experiments. Each experiment has three possible outcomes: (i) the old treatment is better, (ii) the new treatment is better, or (iii) the treatments are the same. The question then becomes how many "experiments," how many pairs of patients, must be treated in order to decide which of the three outcomes is valid. Since the experiments are performed in sequence, the results of

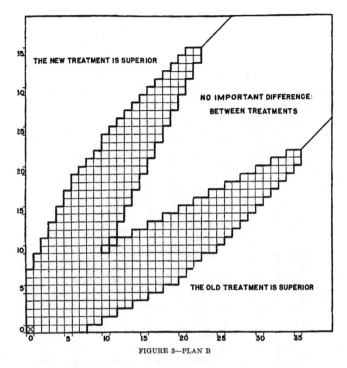

FIGURE 3—PLAN B

FIGURE 2.4. Plotting rules for a sequential experiment. Reprinted with kind permission from I. Bross, "Sequential Medical Plans," *Biometrics* 8 (1952): 198. © 1952 Wiley (UK).

each mini-experiment can be plotted on a chart with preassigned statistical "limits" that are calculated according to how stringent the physician wishes the test to be; crossing a line or limit determines which treatment is superior or if the two treatments are the same. The plotting rules for figure 2.4 are: (i) if the treatment results are the same, enter nothing; (ii) if the old treatment is better than the new treatment, place an *X* to the right of the starting point; (iii) if the new treatment is better, place an *X* above the starting point. Successive experiments follow the same rule and thus chart a path of *X*s toward better, worse, or the same. As can be seen in figure 2.5, in Bross's hypothetical experiment it took only five pairs of patients to "cross the line" and decide that the old treatment was indeed the better treatment. Most importantly, as the figure shows, no calculation was required after setting the initial statistical limits: Bross's approach is a *graphical* statistical method: application requires only pencil and paper. Figure 2.6 shows a nonhypothetical example; namely, the

Order of "little" Experiment	Outcome "old"	Outcome "new"
1	Cured	Not Cured
2	Cured	Not Cured
3	Not Cured	Cured
4	Cured	Not Cured
5	Cured	Not Cured

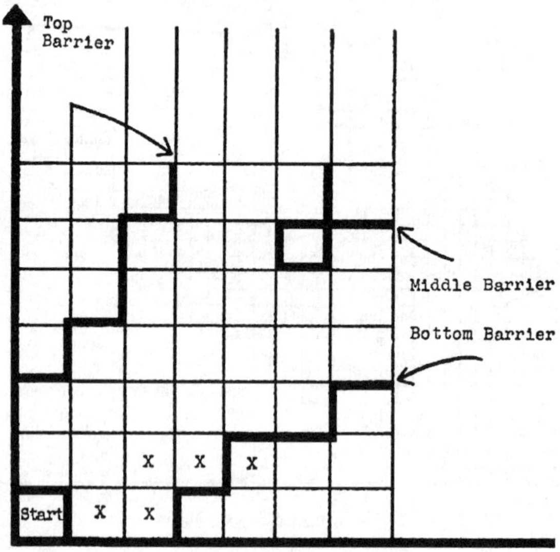

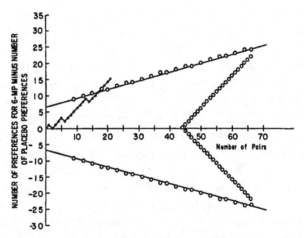

FIGURE 2.5. Hypothetical sequential experiment. Reprinted with kind permission from I. Bross, "Sequential medical plans," *Biometrics* 8 (1952): 192. © 1952 Wiley (UK).

FIGURE 2.6. Results of the sequential procedure for Protocol No. 3. Reproduced with permission of the American Society of Hematology from E. J. Freireich et al., "The Effect of 6-Mercaptopurine on the Duration of Steroid Induced Remissions in Acute Leukemia: A Model for Evaluation of Other Potentially Useful Therapies," *Blood* 21 (1963): 713. © 1963 by American Society of Hematology.

data generated by Protocol No. 3. Notice that the trial did not stop immediately after the "line" was crossed but continued for several more cases.

Unlike a number of previous protocols, Protocol No. 3 had not been inspired by prior laboratory work. It had been designed to overcome problems that had emerged in the course of the clinical evaluation of screened substances and that involved a mixture of emerging ethical concerns (due to the availability of anticancer substances of proven, albeit limited, value), logistical problems (the insufficient number of patients available as "study material"), and experimental design issues (linked to the difficulty of isolating relevant variables and parameters in human subjects). As the ALGB team explained:

> The existence of effective palliative therapy for acute leukemia has hampered the evaluation of new and potentially more effective therapeutic agents. Therapeutic trials with new agents are usually reserved for patients who have been treated with and have become refractory to the agents of proven value. Such patients have active leukemia at the onset of study. Because agents are studied for their ability to induce remission and are not always effective, many patients expire during treatment and thus the number of patients that can receive a new agent is greatly diminished. Moreover, the study of the therapeutic and toxic effects in such patients is frequently confused by the manifestations of the active leukemic process. (Freireich et al., 1963: 699)

Firstly, Protocol No. 3 circumvented this penury of patients by shifting the focus of drug testing from the patient per se to the different phases of the disease process, so that a single patient could be the site of more than one intervention within a single trial. Secondly, a division into "remission" and "relapse" created an altered natural history that allowed researchers to compare drugs with placebos during remission without denying patients access to the palliative care provided by "remission induction." A brief description of the protocol (outlined in figure 2.7) shows how.

The study began in April of 1959 by treating all patients with prednisone. This drug produced the highest levels of remission defined, as they would be later in VAMP, in terms of the percent of leukemic cells appearing in a bone marrow biopsy. The patients were then divided into three groups. Those who failed to achieve remission on prednisone were treated with 6-MP. Those who achieved remission were randomly

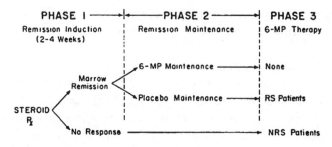

FIGURE 2.7. Schematic representation of the design of Protocol No. 3 (RS = Patients responding to steroids; NRS = Nonresponders to steroids). Reproduced with permission of the American Society of Hematology from E. J. Freireich et al., "The Effect of 6-Mercaptopurine on the Duration of Steroid Induced Remissions in Acute Leukemia: A Model for Evaluation of Other Potentially Useful Therapies," *Blood* 21 (1963): 700. © 1963 by American Society of Hematology.

subdivided into two groups. The first received a placebo and the second received 6-MP. In the case of a bone marrow relapse, the researcher opened a sealed letter to determine whether or not the patient had been on placebo or 6-MP treatment. If the patient had been receiving a placebo, then he or she was immediately placed on 6-MP treatment. If, on the other hand, the relapsed patient had been on the 6-MP treatment, then he or she went off the study. In April of 1960 the entire operation stopped, the limits set by the sequential method statistics having been surpassed. As part of the protocol a patient could fail to respond to either remedy or the placebo and yet remain in the study for further therapy. As remission and relapse were treated as independent events occurring in the same person with the same disease, an individual could be treated twice.

The understanding that intensive doses were required long after clinical symptoms had disappeared complemented this division of the therapeutic process and the disease into remission-induction phases and remission-maintenance phases. Clinical researchers currently credit this insight to M. C. Li, a researcher at the NCI, whose treatment of several choriocarcinoma patients with intensive, intermittent doses of methotrexate in the 1950s became known as the first cancer cure through chemotherapy. Even so, according to Freireich, Li's use of intensive doses went largely unnoticed at the time. As a result the use of intensive dosing in the VAMP trial was seen as yet another innovation: "So we broke the prevailing concept of giving continuous low-dose treatment by treating the children aggressively even after all evidence of the leukemia was gone—I mean we treated with *full* doses" (Laszlo, 1995: 145).

Clearly, then, the innovations incorporated into the VAMP protocol were heavily indebted to previous clinical knowledge and know-how. From Protocol No. 3 onward, the very idea of a clinical trial had become something more than a test of the efficacy of a drug. The therapy—drug, dose size, dose schedule—and the disease had become intertwined; the test had evolved into an inquiry. Just as the phenomena of induction and remission pointed research in the direction of resistance and mutation and showed how therapy intervened in the biology of the disease, the application of cell kinetics framed cancer as a quantitative, biological process of growth and resistance. The innovations, moreover, continued long after the VAMP protocol.

Figure 2.8, created by Emil Frei, attempts to summarize two decades

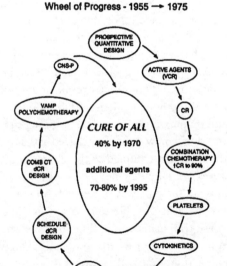

THE CURE OF ALL
Wheel of Progress - 1955 →→ 1975

FIGURE 2.8. The "cure of ALL." (ALL: Acute Lymphocytic Leukemia; CR: complete remission; dCR design: treatment during CR; CNS-P: central nervous system prophylaxis). It is difficult to decide whether the pun "The cure of ALL" is voluntary. Adapted and reprinted by permission from the American Association for Cancer Research: E. Frei, "Confrontation, Passion and Personalization," *Clinical Cancer Research* 3 (1997): 2555.

of innovation in the treatment of ALL: as it can be seen, it depicts VAMP as an important milestone and in a sense a break with past, but it also places it within a continuous process of innovation that had begun several years before. According to Frei, 1955 was a "watershed" year: the pre-1955 era amounted to a "Tower of Babel" marred by qualitative observations that did not rise above the anecdotal. In contrast, a new style of research emerged in the late 1950s characterized by quantitative "experimental design" deploying biostatistics, careful patient selection, the comprehensive definition of treatment procedures, and correlative response criteria (1997: 2555). Frei's exclusive focus on method ignored the material, institutional, and organizational innovations without which the methodological innovations would have had neither existence nor meaning. In addition to these methodological and institutional innovations, we must not lose sight of the fact that by 1963 clinical cancer research at the NCI and in the cooperative groups like ALGB had become a *biomedical activity* in the interactive sense of the term. Researchers like Frei and Freireich did more than simply test compounds. Statistical methods and the novel construal of old diseases into stages of remission and maintenance made it possible to study diseases while doing therapy. In addition, they used laboratory models for the construction of their protocols. They were not, moreover, constrained to simply apply the models to their research. Protocols tested variables such as complete remission that were only indirectly replicated in mice. Clinical research was thus not directly dependent on laboratory research. Rather, new protocols built on previous protocols thus giving rise to a distinctive line of experimentation, a new style of practice that, as we will see in the next chapter, did not remain confined to the United States but reverberated throughout Western biomedicine.

Frei's diagram is not the only way of representing the results obtained by cooperative groups. It is possible to display the work leading to and following the VAMP trial using co-citation analysis, which refers to the convention whereby two references are "co-cited" if a given article cites both of them. The assumption is that co-cited references contain contributions to the scientific literature that are either thematically and/or methodologically related. Co-citation maps of the most frequently cited references thus provide a map of the core contributions in a given domain and can therefore reveal its cognitive structure and evolution. Based on a comprehensive bibliography of cooperative group papers published between 1955 and 1975 (Hoogstraten, 1980), figure 2.9 shows

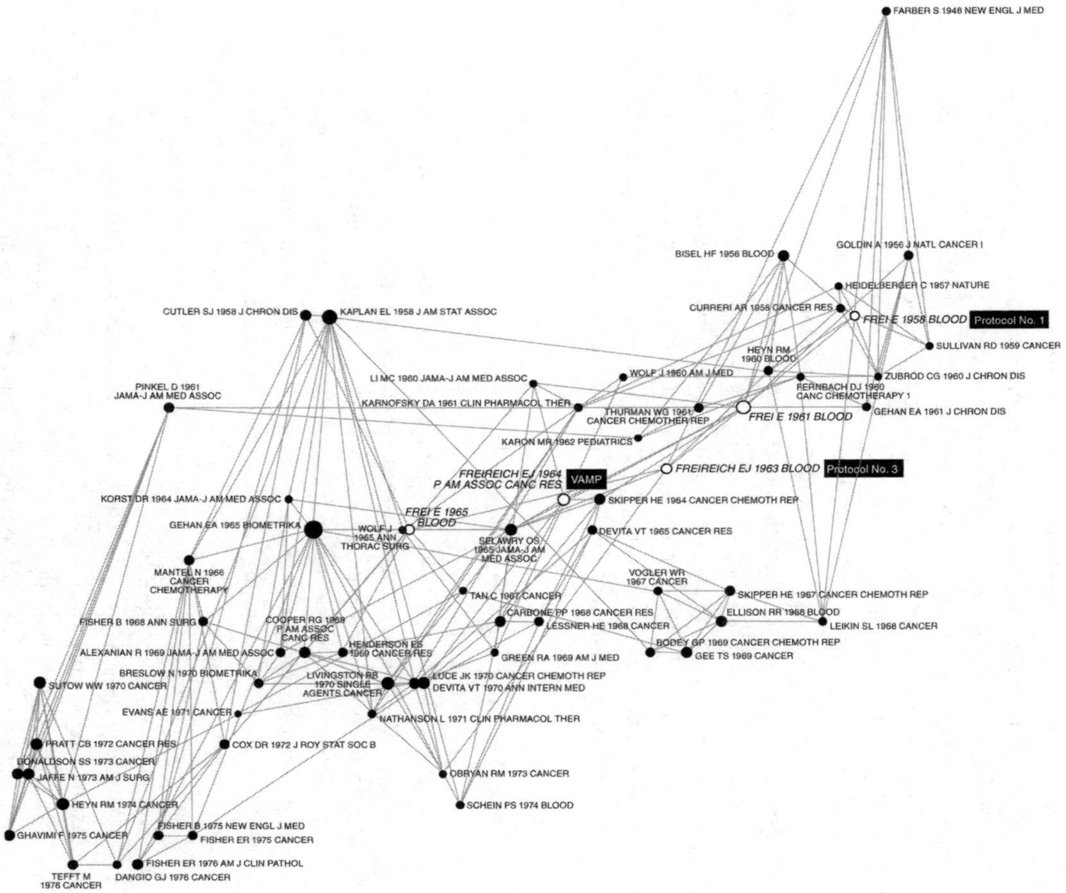

FIGURE 2.9. Co-citation map of cooperative group publications (1955–75). Data source: Web of Science.

the top sixty-five co-cited contributions. They are arranged vertically so that the oldest co-cited references lie on top and the most recent lie at the bottom.[7] The lines connect co-cited references, and the map can thus be read in two complementary ways: as a timeline from top (oldest) to bottom (most recent), and by looking at clusters of linked references. On the right side of the map we see a transversal cluster displaying the sequence from Protocol No. 1 to Protocol No. 3 and the VAMP trial, culminating in a 1965 review article. Notice that the oldest co-cited reference on the map is the classic 1948 article by Farber that recounted the first use of the antifolic compound methotrexate in children with acute leukemia. Just below lies a 1956 Bisel paper that discusses criteria for measuring drug response to treatment, a condition of possibility for the development of multicenter trials and a key component of protocols, as we will see in the next chapter. Finally, the map also highlights the important role played by statistics—namely, by Gehan's work mentioned in this chapter, and David Cox's work that will be further discussed in chapters 6 and 7, as well as, on the bottom left side of the figure, a cluster of papers on breast cancer that we will examine in chapter 6.

What about the Patients?

What about patients in all this? So far we have mentioned them only in passing, no doubt a glaring omission in the eyes of present-day readers attuned to patient activism. This "omission" is in part a reflection of the situation in which cancer patients found themselves during the immediate postwar period. First of all, patients that underwent chemotherapy were rarely in a position to take an assertive stance. In the early days of chemotherapy and radiotherapy, physicians recruited patients in the end stage of disease or, more dramatically, those "on their last gasp" (Bernard, 1967, cited in Rigal, 2003: 245).

Although those testing the compounds reasoned that patients with the least to lose had the most to gain, this moral calculus did not garner universal acceptance. With death rates in children suffering from leukemia running at 100%, early and consistently failed attempts at chemotherapy

---

7. Citation data were obtained from the ISI Web of Science database. We were able to retrieve 78% of the original 1,056 references, which were subsequently analyzed with the ReseauLu X2 program software (www.aguidel.com).

seemed to some clinicians highly suspect. As we noted in chapter 1, reactions to repeated failure sometimes bordered on hostility. One early
researcher observed in retrospect: "The general philosophy of care was,
'Let them die in peace.' To obtain repeated blood samples and to examine
the bone marrow seemed to some to be cruel" (Mercer, 1999: 409). A similar suspicion reigned at the NIH Clinical Center when cooperative group
members undertook their first trials. As Freireich recalled: "We had to
convince doctors and patients that we weren't experimenting on people.
So at the NCI our first two years were spent establishing our credibility."[8]

While surgery and radiotherapy had improved their survival rates
throughout the century, with the isolated exception of choriocarcinoma,
chemotherapy remained without any major success until the VAMP
trial. To shed positive light on such drastic measures as chemotherapy
and the ancillary investigations that accompanied it, early chemotherapists framed the undertaking as a subsidiary component of a larger enterprise known as *total care*. Coined by the Harvard physician and researcher Sidney Farber (1951) in the early 1950s, the term *total care*
referred to the treatment given to children suffering from leukemia, although the parents were quickly drawn into the circle of concern. Farber
had, in fact, originally developed the notion to accompany his free clinic
for childhood leukemia patients. Total care signified that the cancer clinician treated the family as a whole, factoring in its psychosocial and
economic needs. The term gained widespread acceptance if we judge
from the aforementioned 1966 report in *Life* magazine that described
the hospital trajectory of Mike, a leukemic child: after initially reacting
with hostility to the test and treatments that he had to endure (see figure 2.2b), he learned to function "as a normal child in the busy world
of homework and friendly chatter that [the cancer] Hospital was weav-
ing around him. 'It isn't a question of treating the disease,' one of Mike's
doctors said. 'It's total care'" (Bradbury, 1966: 97).

Total care subordinated clinical investigation to patient welfare to the
extent that "the clinical investigator must be prepared to terminate or al-
ter any course of treatment if the continued survival or comfort of the
patient requires the addition of other forms of therapy, because clinical
investigation in the field of cancer may be carried out only as part of the
total care of the patient" (Farber et al., 1956: 4). Total care ostensibly subtracted chemotherapy and other therapeutic strategies from the realm of

8. Interview with Emil J. Freireich, Houston, Texas, July 2006, 3.

human experimentation by combining therapy with enhanced supportive or palliative care such as antibiotics to combat infection. Nonetheless, summarizing almost a decade of chemotherapy in 1955, Sidney Farber and his associates at Harvard Medical School admitted:

> Experience with the study of the action of chemical compounds as part of
> the total care of 1700 patients with disseminated or advanced cancer during
> the past 10 years has demonstrated the importance of adequate preparation
> both in personnel and in facilities for the handling of emotional problems of
> the patient and the family. For them the greatest mental peace is obtained by
> the realization that a devoted doctor or a group of doctors, nurses and hos-
> pital workers are doing *everything that can be done in the light of available*
> *knowledge* for the comfort, treatment and happiness of the patient. (1956: 6,
> our emphasis)

The "available knowledge," however, was more laboratory knowledge
than clinical know-how. And as the reader learns further on in the same
paragraph, the "peace of mind" depended upon a rather sunny view of
the ease with which laboratory findings could be translated into clinical
reality:

> If the quarters in which the patients are cared for are bright and cheerful,
> and if the entire atmosphere is one of guarded optimism based upon actual
> achievement in laboratory research, fear is more easily dispelled and replaced
> by courageous handling of problems. (1956: 6)

Even though physicians pushed the research component of total care
into the background, the program remained controversial since, as Farber and his collaborators admitted, "no clear-cut line can be drawn between specific chemotherapeutic effects and supportive therapy" (1956:
17). Moreover, although the "natural" life span of children with untreated leukemia averaged four months, and although Farber's total care
claimed a survival time of eight months, other therapists claimed equal if
not better survival times with supportive care alone (e.g., Bierman et al.,
1950). Even without these numbers, however, the situation would have
remained difficult to judge: in an era prior to clinical trials, much of the
evaluation of therapy relied on clinical judgments exercised on a case-
by-case basis. Practitioners often saw attention to this variation more as
a virtue than a vice.

In the mid-1960s, following the constitution of successful protocols for the treatment of childhood leukemia such as VAMP, the notion of total care morphed into the concept of *total therapy*. Retrospective accounts sometimes blend the two, but there was no direct conceptual connection (Christie & Tansey, 2003: 14 n. 41). Donald Pinkel, a student of Sidney Farber, initially put together the practice of total therapy at St. Jude's Hospital in Memphis. The "total" part of the treatment (not care) referred to the fact that a variety of therapeutic modalities such as radiotherapy, chemotherapy, and blood transfusion were included in the treatment package.

Clinical researchers in this period operated largely outside the confines of the modern structures of informed consent. This does not mean that patients lacked voice or were constrained to submit to any and all interventions. In 1957 a California court ruled that patients did indeed have a right to consent prior to medical intervention, and the subsequent legal chill generated discussion in the medical literature and stimulated local practices of consent. A recent study of hospital records at a major teaching hospital has shown that despite the fact that the cancer diagnosis itself was often withheld from patients, consent forms had become standard procedure in radiotherapy by the late 1950s and that patients on occasion refused treatment (Lerner, 2004). Nevertheless, most observers agree that an overall tone of paternalism pervaded an era when patients were routinely referred to as "clinical material," and the description of patient populations appeared under the rubric "Material and Methods" (e.g., Lasagna 1960: 45). The situation would change radically in the 1960s following the disturbing revelation of unorthodox human experiments in 1963 that included the injection of live cancer cells into poor elderly patients at the Brooklyn Jewish Chronic Disease Hospital, and the NIH-funded findings published that same year showing that few institutions had guidelines covering clinical research.[9] As we will see later on in this book, the introduction of "informed consent" regulations and the rise of the patient activist is the product of a later time and significantly different circumstances.

9. "A Study of the Legal, Ethical, and Administrative Aspects of Clinical Research Involving Human Subjects: Final Report of Administrative Practices in Clinical Research," [NIH] Research Grant No. 7039 Law-Medicine Research Institute, Boston University, 1963 (ACHRE No. BU-053194-A).

CHAPTER THREE

The Collective Turn

Cooperative Groups as Epistemic Organizations

In chapter 2 we described the VAMP protocol and many of its constituent innovations. Some of these innovations bear further analysis. Chief among them are the cooperative groups set up by the trialists and the statistical tools developed to guide the trials. This chapter will focus on the groups and the collaborative nature of clinical cancer trials. Following a brief technical interlude, chapter 4 will discuss the development of statistics and the role of statisticians in clinical cancer trials.

Despite the considerable interest in cancer therapy displayed by historians and sociologists, surprisingly few mention the cooperative oncology groups. The reasons for this may be quite simple. First, given that historians privilege archival sources, they tend to investigate institutions such as hospitals, clinics, professional societies, or commercial enterprises that produce and conserve such documents. The distributed, flexible, and in some cases fleeting nature of cooperative oncology groups and the consequent absence of archives may have therefore led social scientists to overlook this novel form of research. Second, although multi-institutional and multi-disciplinary by definition, the groups consist of networks of individuals, not institutions (Fisher, 1979: 6). Thus since social scientists commonly view medicine as an activity organized solely through the workplace or through professional organizations, cooperative groups have quite likely slipped through this observational net. Their relative neglect, in other words, simply emphasizes the extent to which cooperative groups represent institutional, organizational, and scientific innovation and, as such, stand as an early illustration of the collective turn in biomedicine.

Clinicians and researchers had collaborated prior to the establishment of the first cooperative groups in the mid-1950s, although certainly not on a present-day scale (Marks, 2006). In fact the very first cooperative group—the ALGB mentioned in chapter 2—grew out of a traditional, "bottom-up" form of collaboration between clinical researchers at Roswell Park and the NCI who decided to join forces in order to carry out trials. This transformation—from traditional collaboration to a cooperative group—amounted to more than a change of name. The group institutionalized investigator-initiated collaborations and entrenched them in an expanding national organization. A more subtle difference lies in the fact that ALGB and subsequent cooperative groups inverted, so to speak, the standard research sequence: instead of defining a project and then, if necessary, creating collaborations in order to carry it out, cooperative group projects took the collaborative structure as a given. More than a condition of possibility for collective research, cooperative groups constituted a platform for the design of new projects; as pointed out several times so far, trialists built on previous trials when planning new trials. In this sense it is possible to refer to cooperative groups as "epistemic organizations."

It would be wrong to characterize cooperative groups as top-down institutions. In particular they should not be confused with the centralized projects established by the command economies of the Second World War to build the atomic bomb or to fight malaria. Nonetheless, numerous observers have described the postwar US cancer program, and in particular chemotherapy, as an outgrowth of wartime, centralized research activities (Bud, 1978). There is some basis to this assertion insofar as several leading cancer research institutes have claimed on occasion that they pursued cancer chemotherapy research as a form of industrial research.[1] Social scientists, eager to establish connections between "industry" and "science," have accordingly jumped to the conclusion that chemotherapy is the result of industrial research. As we will see below, the cooperative group approach was conceived *in opposition* to a centralized, industrial model of organization.

We should speak of cooperative approach*es*, for the cooperative group system did not flow from a single model, nor did it emerge fully formed from the postwar establishment of the US Cancer Chemotherapy Na-

---

1. See, e.g., "Sloan, Kettering to combat cancer," *New York Times*, 8 August 1945, 1, 40.

tional Service Center (CCNSC; see below). Trialists tinkered with the groups' structure and organization and created a variety of formats. European clinicians and researchers also experimented with a cooperative group approach. Although Europe lacked central institutions such as the NIH and the NCI, early members of GECA nonetheless dreamed of establishing a European equivalent of the NCI, and they naturally turned to institutions such as the European Common Market (created in 1957) and Euratom to support their initial efforts. Beyond this important difference, European trialists confronted choices and decisions remarkably similar to those of their US colleagues. Given that US and European trialists participated in the same international cancer research network, this is perhaps unsurprising.

Assembling the US Cooperative System

Cancer clinical trials became common practice at the National Cancer Institute in the middle of the 1950s following their sudden institutionalization as the testing arm of the Cancer Chemotherapy National Service Center (CCNSC). Combining both drug screening and drug evaluation through clinical trials, the NCI program would go on to become a widely imitated model, and key elements of the program remain mainstays of current clinical cancer research.

A series of events both within and without the NIH precipitated this transformation of clinical cancer research. Within the NIH an expanding budget had led to the creation of an extensive intramural program of research at the NIH campus in Bethesda (Maryland). As part of its program expansion, the NIH had opened its research hospital (later named the Warren Grant Magnuson Clinical Center) in the early 1950s. The NCI quickly became the largest user of the center, concentrating primarily on chemotherapy research (Nathan & Nenz, 2001). Putting together the screening and clinical research programs fell clearly within the overall mandate of the Clinical Center to translate findings from NIH laboratories into clinical reality. In 1954 the Center took the initial steps in this direction by carrying out the "first randomized clinical trial at the National Cancer Institute" (Gehan & Schneiderman, 1990: 871). Organized by James Holland, the trial sought to replicate the results obtained the previous year at the NCI laboratories that had shown (in mice) that a combination of two anticancer drugs was more successful than either drug used separately (Holland, 1995; Law, 1952). Holland had joined the NCI in 1953, and when he left for Roswell Park shortly thereafter, in 1954, he established collaborative ties with his former NCI colleagues that quickly became the first cooperative group, ALGB.

As the NIH sought to hasten the movement of facts and findings from the laboratory to the clinic, NCI grants to chemotherapy clinical trials rose spectacularly so that by 1954, of the $7 million disbursed in NCI grants, $2 million went to cancer chemotherapy. Outside the NIH the concurrent and well-documented lobbying by Mary Lasker for increased funding for cancer research led the 1953 Senate Appropriations Committee to believe that given the wartime success of the antimalarial and antibiotics programs, a similar undertaking could be "engineered" to defeat cancer (Patterson, 1987; Löwy, 1995a). With a rising number of clinical researchers interested in testing experimental chemotherapy, the early organizers of the NCI program confronted a situation in which:

> the cancer chemotherapy screening capacity was extremely limited, and fewer "new" drugs existed than clinicians who were ready to evaluate these drugs. This led to continuing pressure from industrial, academic, laboratory and clinical sources for an expanded program in drug screening that would meet the needs of a growing number of interested investigators. (Zubrod et al., 1966: 352)

As shown in figure 1.4 (chapter 1), the "expanded program in drug screening" included not only the initial screening of compounds in the laboratory and in animal models but also the evaluation of the resulting candidate drugs in human clinical trials. Figure 3.1 displays, in a distinctive "engineering" mode, the structure of the chemotherapy program. Following the 1955 inauguration of the CCNSC, between mid-June 1955 and mid-July 1956, members of the NCI's newly formed center set up a series of interlocking committees to direct the complex program. The screening panel ostensibly ran the show but did so in concert with a number of other panels—namely, the chemistry, pharmacology, endocrinology panels and the clinical studies panel that funded human clinical trials. This combination of clinical trials and screening set the CCNSC apart from previous drug screening programs. More than a simple juxtaposition of autonomous activities, the clinical trials functioned as an extension of the screen by incorporating a sequence approach inaugurated

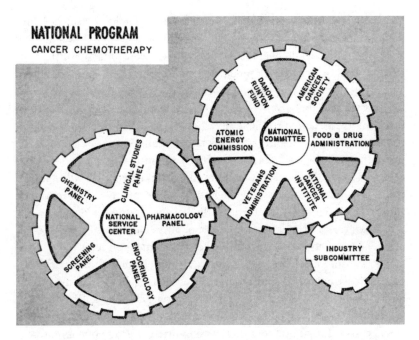

FIGURE 3.1. An "engineering" representation of the US Cancer Chemotherapy Program. Reprinted with kind permission from K. M. Endicott, "The Chemotherapy Program," *Journal of the National Cancer Institute* 19 (1957): 278. © 1957 Oxford University Press.

in the animal screen. In turn, the animal screening contained many features that made it more like a clinical trial than a simple drug screen.

The director of NCI, Kenneth Endicott, agreed that the new endeavor required unprecedented levels of collaboration between a heterogeneous group of actors:

> There is probably no field of medical research which requires more varied skills than drug development. One must have available biologists, chemists, pharmacologists, biochemists, endocrinologists, engineers, microbiologists, statisticians, nurses, pharmacists, and clinicians. Furthermore, their efforts must often be coordinated. Such extensive teamwork is rare in academic research where the emphasis is on independent, free-roving, individual investigation. (1958: 172)

Yet while politicians and researchers, such as the director of the Sloan-Kettering Institute Cornelius P. Rhoads—who, after his wartime experi-

ence with chemical warfare agents had established a major drug development program at the institute—favored the aforementioned "engineered
solution" to this challenge, researchers at NCI were "fearful of the wartime models of drug development for malaria and penicillin" (Zubrod,
1979: 493). As Endicott explained, these models stipulated that "an expert committee [be] given control of funds and [be] assigned the job of
directing the research of those investigators and laboratories willing to
participate"; such an approach "was rejected almost immediately as being both psychologically and scientifically unsound." Psychologically because "neither the academic nor the industrial researcher willingly ac-
cepts bureaucratic direction by the Federal Government except in time of
war or similar crisis," and scientifically because of the "evident need for
a high degree of originality, which is a quality seldom displayed by committees" (Endicott 1958: 172–73). The NCI clinical director then added
a more substantive, biomedical argument; namely, that unlike malaria or
tuberculosis, cancer was not a single disease. This fact alone invalidated
any analogy with previous chemotherapy programs. Explicitly criticizing
screening programs such as those initiated by the Sloan-Kettering Institute, he argued that no "centralized committee or group of committees"
could possibly "cope with the analysis of data on so many diseases."[2]
NCI researchers thus promoted a more diffuse model—namely, a network of hundreds of independent investigators who were not required to
work together and did so only on a voluntary basis.

In 1954 the Senate succumbed to the pressure and adopted the NCI
solution, instructing NCI to set up a "cooperative system" to screen and
test chemotherapeutic compounds for cancer (Zubrod et al., 1966: 355).
"Voluntary cooperative basis" did not mean, however, a total lack of
central initiative, as made clear by an early description of the cooperative approach set up by the CCNSC's clinical panel:

> Small groups of 6 to 12 investigators are called together to explore the or-
> ganization of a cooperative group. Usually they are selected on the basis of
> medical specialty, geographic location, and demonstrated interest in certain
> forms of cancer. After explaining the problem and the reason for calling the
> group together, the Service Center leaves the rest to the group. (Endicott,
> 1958: 174)

2. Office Memorandum from Clinical Director, NCI, through Associate Director in
Charge of Research, NCI, 22 December 1959. NCI Archives, 2–3.

A year later, the NCI director provided a more flexible description: "Under the program a cooperative study group arises out of encouragement from the Service Center or from the spontaneous interest of individuals" (Endicott, 1959: 105). Endicott's view differs from the view of cooperative group members, according to whom "the original impetus for the Cooperative Groups was truly investigator-initiated with the National Cancer Institute a partner with the scientific community" (Davis, Durant & Holland, 1980: 386). These accounts need not be reconciled: they reveal the essential tension between centrifugal and centripetal forces that characterized the development of cooperative oncology groups on both sides of the Atlantic.

As previously mentioned, shortly after conducting the first randomized clinical cancer trial at the NIH Clinical Center, James Holland moved to Roswell Park (Buffalo, New York), the United States' first cancer research and treatment center. In order to pursue clinical trials in leukemia, he and his former NCI colleague, Gordon Zubrod, agreed to continue their collaboration, which they extended to Holland's successor at the NCI, Emil Frei. Shortly thereafter the alliance "quickly grew into Leukemia Group B" (Zubrod, 1979: 494; Laszlo, 1995: 220). Under Frei's leadership the group conducted nineteen clinical trials between 1955 and 1965, including Protocol No. 3 discussed in the previous chapter. A second cooperative group, Leukemia Group A, emerged in 1955 from discussions between two other clinicians (Burchenal and Farber) on how to establish common criteria to measure drug responsiveness in leukemia trials, an issue increasingly perceived as a prerequisite for conducting multicenter cooperative trials. At this stage the clinical panel adopted a more proactive attitude and building on these investigator-initiated collaborations established the Eastern Solid Tumor Group in 1955 to carry out a comparative study of two drugs against three kinds of cancers (Zubrod, 1979; Hoogstraten, 1980: iii). Then the floodgates opened. By 1956 the NCI had created eight groups of clinical cancer researchers. Six years later a further ten groups had been formed.

Figure 3.2 provides an overview of the groups established by the cooperative program between 1955 and the mid-1970s; it visualizes the rise and decline of numerous groups bearing witness to the NCI's continued experimentation with different forms of cooperative research. We can identify three main categories of cooperative group research. The first draws together groups such as ALGB and the Eastern Solid Tumor Group that continue to operate under new names—that is, Cancer

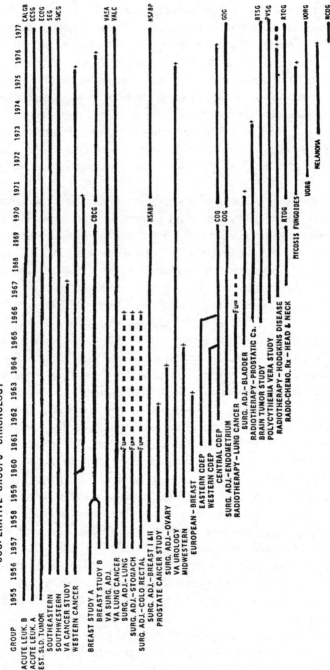

FIGURE 3.2. Timeline of cooperative oncology groups in the United States. Reprinted with kind permission from B. Hoogstraten ed., *Cancer Research: Impact of the Cooperative Groups* (New York: Masson, 1980). © 1980 Elsevier.

and Leukemia Group B (CALGB) and Eastern Cooperative Oncology Group (ECOG). Retrospectively designated "multiprotocol" groups— insofar as they "perform[ed] many studies simultaneously, continuously generate[d] new studies, conduct[ed] studies in all phases of clinical chemotherapeutic research, and frequently undert[ook] related preclinical research"—they became permanent institutions and diversified their activities (Foye, 1970: 14). In contrast, single-protocol groups (such as the Surgery-Adjuvant-Lung Group) "confined their cooperative efforts to a single study of a single disease" (Foye, 1970: 16). As ad hoc groups, the single-protocol groups had a finite existence that spanned an initial "active entry phase" lasting two to five years and during which patients were recruited, and another two- to five-year follow-up phase that gathered recurrence and survival data. Groups of both types rose and fell throughout the 1960s so that by 1968 over thirty-two cooperative groups had been formed, twenty-two of which were still in existence that year. At the time, administrators classified twelve of the twenty-two groups as multiple-protocol groups.[3]

In addition to the single- and multiple-protocol groups, the CCNSC initiated a short-lived Clinical Drug Evaluation Program (CDEP) in 1961. Devised to accelerate the clinical evaluation of the materials flowing from the drug screening program, the CDEP divided the country into three regions and created three clinical oncology groups, the Eastern, Midwestern, and Western CDEP groups. In 1966 the NCI folded all three groups into one, the Central Oncology Group, which, itself, folded tout court a decade later. In the interim, the NCI had endowed the groups with a central biostatistical office and recruited investigators to enforce a uniform Phase II protocol. The Eastern CDEP group, for example, consisted of twenty-seven investigators who maintained a central office at Roswell Park Memorial Institute in Buffalo. The trials usually included hundreds of patients afflicted with up to thirty different cancers. Figure 3.3 shows how the system worked: if an investigator believed that a patient was eligible for a trial (i.e., had inoperable cancer but would live at least four months), he or she submitted a request to the regional central office. The office then either assigned a drug to the patient or offered the investigator "his choice of 2 or more drugs which

3. Laurance V. Foye, "The National Cooperative Program for Clinical Cancer Research," undated, ca. 1970, document prepared for the 10th International Cancer Congress, AR-1000–010646, NCI Archives.

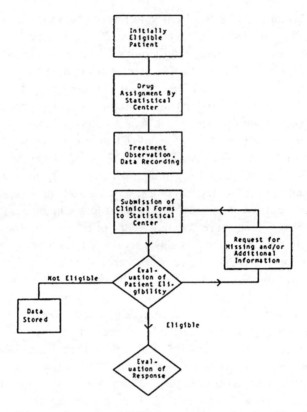

FIGURE 3.3. "Diagram of structure of CDEP for processing each patient." Reprinted from G. E. Moore et al., "Effects of 5-Fluorouracil (NSC-19893) in 398 patients with cancer," *Cancer Chemotherapy Reports* 52 (1968): 643 (public domain source).

[were] being investigated in the CDEP at that time" (Moore et al., 1968: 643). Subsequently, regardless of the disease or the drug, the patient in question was given "at least the minimal drug dosage within a specified maximum length of time" and was kept under observation for at least sixty days.

The CDEP one-size-fits-all approach was a continuation of the screening approach that, as we will see in the next section, was used to test potential anticancer drugs in animal models: the imposition of a single national protocol sought to do with humans what seemed so successful with mice. This approach was destined to fail as, by the mid-1960s, the cooperative groups had redefined their work not as screening but as clinical research. Initiated in part by the turn to combination therapy, of which VAMP was a paradigmatic example, clinical research rendered screening intellectually empty.

Screening for Drugs: From Mice to Humans

We have already suggested that in some respects the designers of the CCNSC system conceived of clinical trials as an extension of the drug screening process. In addition to the explicit attempt to articulate screening and clinical trials, NCI planners used an architecture modeled on the animal screen and that divided the entire screening process into structured phases. This fact is often overlooked in the history of clinical trials where the animal and human portions are commonly viewed as separate and often incomparable undertakings. Since it combined screening with clinical trials, the program sought more than an active substance; it sought a *cure*. And this remained the program's horizon in spite of the difficulties that soon became apparent:

> We have asked, "What do we seek?" and we have answered, "Obviously a cancer cure." Being realists, we have admitted that there is no satisfactory cure or cures at hand, so that we then seek improvements in our patients while we examine additional leads for development of cures, and, if cures are found, we seek means of applying these. We believe this program is the means by which it is best done. (Lee, 1960: 83)

In the ideal-typical representation in Figure 1.4 (chapter 1), NCI planners expected the cure to emerge from a sequence of operations that tied together mouse and human screening. It is thus necessary to have a closer look at the mouse component of this continuum: once again, we will see that the CCNSC activities cannot be reduced to the transposition or adoption of an industrial model, although industry quickly became one of the principal sources of candidate anticancer drugs.

While based on an almost random input of substances, the material screened by the NCI did not turn up by chance. Within five years of its initiation, the CCNSC had developed sources and systematic procurement policies for three kinds of compounds—synthetic compounds, natural compounds, and fermentation products—thus providing three distinct entry points into the screen. Initially, university and industry furnished most of the synthetic compounds (95%). As one participant recalled: "We had staff who were going out to universities and . . . at some university chemical departments, they would simply have permission to go down to the store-room and take samples from everything that was there."[4] The remainder (5%) entered the system through contracts drawn up by the NCI to synthesize compounds that had a "high probability of positive activity."[5] Similar contracts developed in 1956 launched the program to screen industrially produced fermentation products for antibiotics with antineoplastic activity. Finally, a fourth source and a third type of compound emerged in 1961 when, in the wake of the discovery of vincristine, the Center began the systematic sourcing of plant products (Sessoms, Coghill, & Waalkes, 1960; Zubrod et al., 1966).

Although the CCNSC first sourced synthetic chemical compounds in its own back yard—NIH labs alone accounted for almost 20% of the submissions for 1959—industry participation grew substantially following the institution of the cooperative groups who, by 1958, conducted over one hundred studies a year (CCNSC 1959–60a). The consequent demand for drugs in bulk and the need for a steady supply of new compounds led the NCI to quickly develop an intellectual property policy that would protect companies and allow the surgeon general "march-in" rights to ensure reasonable costs in the unlikely event of the discovery of a penicillin-like substance. With a new law approved and funded by Congress in 1957, industry quickly joined the quest for anticancer agents by supplying compounds for screening, developing screening programs funded by the NCI, establishing bulk production programs, and, finally, instituting a national advisory committee composed of industrial research executives (NPCCR, 1959–60). As the members of the original screening panel recalled: "Once it started, industry was very happy to send those compounds under the screening agreement whereby they retained patent rights. We were doing screening at no expense to them."[6]

By 1959 over 150 industrial suppliers produced more than half of the synthetic chemical compounds submitted to the program and all of the antibiotic culture filtrates (NPCCR, 1959–60: 18). The Pfizer Company

4. Group interview with John M. Venditti, Saul Schepartz, Melinda Hollingshead, Joseph G. Mayo, and Michael C. Alley, Frederick, Maryland, 29 October 2003.

5. Memorandum, Dr. Kenneth Endicott (Director, NCI) to Dr. James A. Shannon (Director, NIH), "Preliminary Comments on the Report of the Richardson Committee," 5 October 1966, NCI Archives, 2.

6. Group interview with John M. Venditti, Saul Schepartz, Melinda Hollingshead, Joseph G. Mayo and Michael C. Alley, Frederick, Maryland, 29 October 2003.

alone, for example, contributed almost 5% of compounds reported in early 1962.[7] Once sourced, the compounds had to be screened. The number of compounds screened was truly industrial strength: by the close of the first decade of screening in 1965, researchers had screened 90,000 synthetic compounds, 120,000 fermentation broths, and 20,000 plant extracts (Schepartz, 1964). Carried out in a dozen university and industrial laboratories, the screen consisted primarily of mice inoculated with standard cancer tumors according to standard protocols (Geran et al., 1962).

The choice of tumors had resulted from an unusual process. Prior to the selection of common tumors for the screen, researchers had had to screen the different types of screening materials that had been used in previous programs. This "screen screening" had been carried out before the establishment of the NCI program when, in 1952, the American Cancer Society funded a study led by Alfred Gellhorn to determine which of the drug screening systems then in use were most sensitive to anticancer compounds. The group had originally hoped to find a non-tumor system, since such a system would be cheaper than maintaining mice and their transplanted tumors. Having tested twenty-seven different compounds of known therapeutic value in seventy-four different screening systems, the extensive study concluded that mouse tumor systems constituted the best screen for anticancer compounds. Rather than choosing a single mouse tumor, the CCNSC chose three, reasoning that in order to maximize its chances of isolating active compounds the most effective screen would be a combination of mouse tumors derived from the two principal tissue categories (a carcinoma and a sarcoma) and a fast-growing and more systemic cancer called leukemia L1012 (Gellhorn & Hirschberg, 1955).

While the CCNSC conceived of mouse and human studies as a continuum of activities modeled on each other, the issue of whether data derived from animal models had much bearing for humans remained controversial and continued to haunt the chemotherapy program. More than a mere topic of debate, it became the focus of a series of NCI initiatives to investigate human-mouse connections. Gordon Zubrod elegantly summed up the issue during a 1959 meeting:

The former Director of the National Cancer Institute . . . when presented with a brilliant piece of work by one of the staff would often ask, "What does this

7. Figures based on entries 27621–29738 in Leiter et al. (1962a; 1962b).

have to do with cancer?" This question has been asked by a number of speakers, in this form: "What do the transplanted rodent tumors have to do with
human cancer?" I shall try to answer this question. (Zubrod, 1960b: 136)

The answer, according to Zubrod, lay in the difference between the relatively uncomplicated nature of infectious diseases and the complex con-
stellation of diseases that fell under the heading of cancer. As we saw,
this distinction had been one of the reasons for rejecting an engineered
approach to the search for a cancer cure. Accordingly, Zubrod also distinguished the more effective (semi-) rational screens that had been used
for infectious diseases and that reproduced essential elements of those
diseases from the more empiric screens such as the transplanted rodent
tumors that unfortunately led "to useful drugs slowly and hesitatingly."
While admitting that empiric screens "frequently fail to predict accu-
rately for human cancer," the reason for such a failure was to be sought
less in the fact that the screen was "intrinsically poor" than in the lack
of quantitative parameters allowing researchers to define and control the
experimental variables of the animal systems. A research program focusing on those parameters and variables, Zubrod believed, would establish robust correlations between rodent and human cancer. Other participants at the 1959 meeting remained unconvinced. Their position—a
"minority report" as they themselves admitted—was:

> The data obtained in rather esoteric and complex systems, rendered familiar
> but not less bizarre by common usage, cannot be used to support interpreta-
> tions of clinical results. . . . Human data must stand on its own feet, separate
> and distinct from mouse results that may have stimulated the clinical study.
> (Karnofsky, 1960: 142)

Gellhorn's inaugural study, despite its thoroughness, was essentially a
bootstrapping operation. As researchers at the time clearly recognized,
in order to test the systems, Gellhorn's group had been forced to rely on
anticancer substances of unproven quality. Since none of the substances
then in use had been submitted to a controlled clinical trial, Gellhorn
admitted that some of the substances used to test the systems "may have
been of doubtful value" (Armitage & Schneiderman, 1958: 897). In recognition of the recursive nature of the operation, the CCNSC immediately began research into alternative systems and placed the mouse tumor system under continuous review. In the late 1950s, for example,

approximately 10% to 20% of all materials screened in the three-tumor mouse screen were also screened in tissue culture in order to determine the relation between the two kinds of screen (NPCCR, 1959–60: 12). Less a robust technology than a temporary protocol, the screen produced contingent data that were continuously subject to further scrutiny and revision. In other words, the two-way connection between human and mouse variables was an ongoing concern.

As planned, the CCNSC contracted out the actual screening to academic and industrial laboratories (Leiter et al., 1962a; 1962b). With Pfizer, Upjohn, and Bristol leading the way, by 1962 eight of the twelve screening laboratories were industrial (Zubrod et al., 1966). The laboratories followed protocols that covered all aspects of the screening process down to the kinds of bedding to be used for the mice (MINIMUM Standards, 1959; CCNSC, 1959). In the laboratory, technicians injected the sourced substances into the mice bearing the tumors. If the substance had an effect on the tumor, then it was deemed a potential anticancer agent. If not, as was the case for 99% of the substances tested, it was deemed inactive. Under this simplified description, the process might be described as random shots in the dark, and it was indeed derided in some quarters as wholly empirical and quite unscientific. But closer examination of the program shows this charge to be exaggerated. To begin with, the screen consisted not simply of testing a compound or substance in cancer-bearing mice but rather of a series of tests that ran sometimes in tandem, sometimes in parallel, and that generated information (negative and positive) that could be fed back into the system. The screen was thus more than a simple empirical test and used techniques similar to and provided a model for subsequent human therapeutic protocols.

By the late 1950s the pharmaceutical industry furnished the CNSCC approximately 8,000 synthetic compounds and 17,000 antibiotic culture filtrates a year.[8] The growing screening staff of the Service Center reflected the intensity of the activity: by 1957, the Service Center had over sixty members. The travel time from submission to the beginning of a clinical trial for a typical compound was four years (Zubrod et al., 1966). The number of mice used each year to test these substances rose quickly

8. K. M. Endicott, "Interim Report of the Cancer Chemotherapy National Service Center to the Cancer Chemotherapy National Committee," 25 January 1957, NCI Archives, 2.

into the millions (NPCCR, 1959–60: 11). Each stage had a rejection level based on the test substances performing twice as effectively as the control substance. The rejection level was not arbitrary and had been chosen by a British statistician, Peter Armitage, who had been on loan to the NCI from the UK Medical Research Council in 1956–57 and who had established the initial three-stage sequential scheme. He and his colleagues used two different terminologies to describe the screening process: an "acceptance-rejection" system and a series of "screening experiments" (Schneiderman, 1961: 237; Armitage & Schneiderman, 1958: 897). The different terminologies reflect the fact that the system could be conceptualized in two different and slightly antagonistic fashions— namely, experimental and industrial.

In the simplest scenario, new compounds moved through the system in four steps (CCNSC, 1959–60b). If material showed more than a prespecified degree of activity in one of three different mouse test tumors, then it passed stage I. Stage II consisted of repeating the test, and if the material came within 20% of its original value, then it passed stage II. In stage III, screeners retested the material yet again, this time with the proviso that the average values of the three tests fall within a specified range (Leiter et al., 1962a). At each level, screeners tested the compounds at four or five different dose levels. Stage IV consisted of preclinical toxicology studies for entirely new compounds and comparative trials for analogues of compounds that were already in use. A single test did not determine whether or not a compound was active. Rather, the determination relied upon the *evaluation* of the performance of the substance within a system according to a predefined protocol that generated *evidence* for the evaluation and not simply a *result*. Thus given the different responses from animal to animal and from test to test, each compound was submitted to a miniature clinical (mouse) trial. In addition the screen was embedded in a larger information system implicating inputs from a variety of disciplines.

As far as synthetic chemical compounds were concerned, knowledge of chemical structure enabled researchers to eliminate much of the guesswork and a sort of intuitive chemical information system emerged, described in more prosaic terms by the original screeners as follows:

After a while the selection of materials going into the screen was no longer random. There was an intellectual input into this . . . if we looked at 150 compounds of a particular type we would probably reject the next one in favor of a novel structure.[9]

Promising compounds and groups of compounds could also be produced by the CCNSC's own laboratories located in the "Pharmacology-Biochemistry Section" that had been set up toward the end of the 1950s to study structure-activity relations of the anticancer compounds (NPCCR, 1959–60: 15). Finally, the screen itself evolved through continuous experimentation. In the ten years since its inception the primary screen was augmented and then reduced from more than a dozen different animal tumors to the L1210 leukemia tumor system and the Walker 256 intramuscular solid tumor of the rat.

Although represented as a somewhat mechanical translation from the laboratory to the clinic, the movement from random sources through the mouse screen and on into the human clinic was quite novel—not entirely empirical and not completely rational. Endicott described the process in 1958 as:

a mixture of empiricism, exploration of existing leads, and active support of that basic research which appears to offer the possibility of yielding useful information. In the present state of knowledge the search for chemical cures must proceed to a large extent on an empirical basis. The random screening of large numbers of chemicals and natural products turns up "leads," and their exploration results in the synthesis and testing of related materials. From such empirical research there gradually emerges a body of data that permits generalizations and rational approaches. At the same time, biochemical studies of cells and of growth and reproduction, though not directed toward chemotherapy, yield information which permits the chemotherapist to experiment on the basis of testing rational hypotheses.[10]

Nonetheless, Endicott also recognized that:

selection of drugs for preliminary clinical workup rests on a tenuous scientific basis since none of the screening techniques now in use is known to predict

9. Group interview with John M. Venditti, Saul Schepartz, Melinda Hollingshead, Joseph G. Mayo, and Michael C. Alley, Frederick, Maryland, 29 October 2003.

10. K. M. Endicott, "Interim Report of the Cancer Chemotherapy National Service Center to the Cancer Chemotherapy National Committee," 25 January 1957, NCI Archives, 2–3.

clinical experience. Factors influencing the selection are almost as numerous as those who make the final decision.[11]

Even when a promising compound turned up, results either in the mouse or human were unspectacular. Unlike the decisive and long-lasting effects of, say, penicillin on pneumonia, in cancer chemotherapy the effects sought were less dramatic: "We are often comparing one drug against another to demonstrate that one causes remissions in 20 percent of the cases as opposed to 15 percent for the other." Consequently, according to Endicott, chemotherapy's relatively low clinical expectations further reinforced the need for double-blind, randomized clinical trials:

> The highly organized, controlled clinical trial of today has been accepted as the best known means of selecting the most effective agent from a large group of compounds which possess only marginal activity—compounds that relieve discomfort, reduce the size of large neoplasms, extend life, or in some way modify the manifestations or the course of neoplastic disease.[12]

The muted effects of chemotherapeutic agents also called upon cooperative clinical trials in order to generate the large number of patients needed to detect the less than obvious differences. Again, according to Endicott: "A single hospital can rarely make enough observations in . . . highly selected patients to give adequate data in a reasonable time; hence, collaborative research becomes essential."[13] Zubrod, who had earlier run a screening program for antibacterial substances, was quite aware of the novelty of the undertaking, especially the idea of systematically conducting clinical trials:

> Most clinical trials, not all clinical trials, in cancer and in other diseases were hit or miss, haphazard, you know, you'd treat twenty patients and see what happened without any kinds of controls. . . . We fortunately attracted mem-

11. "Annual Report of Program Activities," 1958, *I*, National Institutes of Health, National Cancer Institute, 52. See also Endicott (1958).

12. K. M. Endicott, "Interim Report of the Cancer Chemotherapy National Service Center to the Cancer Chemotherapy National Committee," 25 January 1957, NCI Archives, 10.

13. "Annual Report of Program Activities," 1958, *I*, National Institutes of Health, National Cancer Institute, 50.

bers to the Clinical Panel who were also convinced that [the cooperative approach] was the way to go.[14]

Tinkering with the Cooperative Group Structure

From today's point of view, CDEP and single-protocol groups were a short-lived organizational solution. Their temporary nature called for centralized, permanent administrative structures to provide the necessary expertise. Consequently, in 1968 the NCI bureau in charge of the cooperative groups recommended that the NCI establish a "Coordinating Center" for the existing and future groups and staff the center with "statisticians, programmers and clerical personnel."[15] In contrast, multiple-protocol groups developed their own central operations offices. For instance, while the administrative headquarters of both ALGB and the Eastern Solid Tumor Group had been operated jointly, by 1963 the Eastern Group had established its own operations office to provide "medical supervision, control and coordination" of the studies, a task that involved checking adherence to protocols, maintenance of all patient records, statistical evaluation of data, distribution of the new agents to the cooperating clinics, and liaison with the CCNSC's clinical panel and between individual investigators.[16]

Using the Eastern Solid Tumor Group as an example, we will explore the early structure of a multi-protocol group in greater detail in order to provide a window into both the entities and technologies generated by this novel form of collective work. While it is tempting to describe cooperative groups as a distributed institutional network—a group like ALGB that began with several investigators in two institutions had by 1970 developed into an organization housing over one hundred investigators in forty-one institutions—such a description overlooks the key

14. "Interview with C. Gordon Zubrod," NCI Oral History Project, History Associates Incorporated, 27 May and 27 June 1997, 69.

15. "Report of the Cancer Clinical Investigation Review Committee on the 'Cooperative Studies Workshop,'" Williamsburg, Virginia, 20–23 October 1968, NCI Archives, AR-6810–006645, 5.

16. Eastern Cooperative Group in Solid Tumor Chemotherapy, "Over-all group research plan and progress report. May 1958 to May 1963," Frontier Science and Technology Research Foundation Archives, Boston, 1.

role played by the central operations office. The clinical director of the NCI had promoted the development of this institution toward the end of the 1950s by arguing that responsibilities for cooperative groups such as ALGB that had initially been lodged at the CCNSC be "given back to the operating scientists." This did not mean turning over responsibility to individual scientists. Rather, Zubrod had in mind the "centralization (of authority) in smaller [than the CCNSC] units," called "large operating units" or, in other words, cooperative groups.[17] The supervisory office of the "large operating unit" was to serve as "a central point where all current and historical data for the Group [could be] maintained, thus creating a convenient and useful place for the Drug Study Chairmen to meet with the Statistical Office staff or Chairman of the Group, etc., to review general problems relating to studies or other Group activity."[18]

Neither the precise activities nor the organizational structure of a cooperative group were clearly defined when the first groups were established. Organization and content were, moreover, mutually constitutive. The first publication by the Eastern Solid Tumor Group clearly set out the relations between these two aspects of the groups:

> The members of the group undertook a critical examination of the methodology of clinical cancer chemotherapy trials and, based on this review, began a study of cancer chemotherapy in man. *It was hoped that this study would permit an understanding of what was required* for quantitative comparisons between the effectiveness of two anticancer drugs. (Zubrod et al., 1969: 6, our emphasis)

The authors went on to explain that they had set out to study "the feasibility and usefulness of collaborative clinical research" and to "solve the problem of common criteria and other semantic traps." They tested a mixture of methodological choices (e.g., "use of objective measurement of tumor size"), decision-making technologies (e.g., "final determination of effect by all cooperative investigators by vote"), and organizational solutions (e.g., "frequent meetings of the investigators to ensure common application of the conditions agreed upon and common solution of unanticipated difficulties") (Zubrod et al., 1969: 6–7). The authors pre-

17. Office Memorandum from Clinical Director, NCI, through Associate Director in Charge of Research, NCI, 22 December 1959, NCI Archives, 4.
18. Ibid.

sented collaborative research as a solution to the problem of evaluating an increasing number of drugs both rapidly *and* adequately and argued that their results showed that this mode of research was feasible despite recognized objections and shortcomings. Some objections, moreover, were rejected out of hand as moot, in particular, the claim that "protocol research [was] too rigid and inflexible" (19–20).

The Eastern group's organizational structure thus grew out of these investigations and included standing subcommittees for membership, new agents, and protocols. The membership committee not only reviewed applications for membership but also collected detailed information on member institutions and made site visits to determine the degree to which members implemented group criteria. In addition, the group periodically reviewed the quality of participation and the productivity of member institutions. The protocol subcommittee reviewed study protocols "for uniformity, clarity, and purpose," making "firm and virtually final recommendations."[19] During its first eight years of existence, the group also examined other methods of collaborative research, developing its own modus operandi that included the development and application of "a standard drug study protocol and uniform quantitative methods of characterizing drug effect . . . commonly accepted statistical techniques at all stages of drug evaluation . . . uniform criteria of drug effect to the data derived from individual drug trials by a group of investigators who remain unaware of the nature of the treatment employed . . . [and] 'double unknown' techniques."[20]

In order to understand the challenge of conducting cooperative research, let us take a closer look at the issue of common criteria. The importance of common end points for the conduct of clinical research had become apparent through a number of retrospective studies. In 1959 at a "Conference on Experimental Cancer Chemotherapy," Emil Frei analyzed a review of previous studies of chemotherapy in acute childhood leukemia published in the early 1950s (Bross, 1954). A series of thirteen different chemotherapy studies had produced response rates ranging from 10% to 70%. Focusing on those studies involving 6-MP, Frei concluded that absent any difference in the mode of administration of the drug or in the experience of the investigators the only factor that could account for such a disparity in results was the criteria for remis-

19. Ibid., 2.
20. Ibid., 6.

sion. Given the current efforts to standardize remission criteria, Frei surmised that "it is probable that the high varying remission rate for a given
drug in 1954 would not occur today" (Frei 1960: 282).

The establishment of common criteria for diagnosis, treatment, staging, and evaluation confronted research collectives like ALGB in the
mid-1950s as a problem of considerable urgency. Different members of
the groups worked in different institutions, and criteria varied from hospital to hospital and in some cases from practitioner to practitioner. It
was not, however, a question of imposing preset or arbitrary standards
simply to create uniformity. It was impossible to know in advance which
standards were the most accurate and, moreover, which standards would
function across different clinical settings. Common criteria for remission, cure, relapse, and so on thus became subjects of research. Unsurprisingly, then, the first meeting of ALGB held in 1955 in Boston and
hosted by Sydney Farber focused on this problem. Similarly, the first
trial carried out by the group, although it had initially set out to compare
two drugs, showed not which drug worked better but, rather, that the
best predictor of survival of the several predictors that had been used
was the bone marrow biopsy. Prior to this point, the clinical end point of
remission had usually been determined by counting the number of leukemic cells in the circulatory blood (Laszlo, 1995: 100). In the years that
followed, all groups adopted a common definition of complete remission
in leukemia: less than 5% leukemic cells in the bone marrow, a standard
that remains valid to this day (Bisel, 1956).

The adoption of common definitions did not occur immediately nor
did it appear as urgent to some group members as it did to others. In
1959, for example, all of the cooperative groups funded by the NCI
met in Washington to compare notes. While it had become apparent
that individual groups maintained standards, it was also obvious that
there were few intergroup standards. Meeting participants thus asked:
were intergroup standards worth the effort? The answer to this question depended on how different practitioners viewed clinical trials. Were
therapeutic protocols merely tools for the discovery of active compounds—an extension of the screening activity carried out with mice—or
were they research devices? According to John Louis of the University
of Illinois:

Developing objective criteria is one of the most important problems we have
had in Leukemia A and also in the Midwest groups. At present we are at-

tempting to get uniform criteria in all cooperative groups and I believe in a few months we will succeed. (Burchenal et al., 1960: 1967)

Responding to Louis, Emil J. Freireich of ALGB remarked: "My own point of view is that criteria are unimportant. What we want to know is what drugs are active in leukemia" (Burchenal et al., 1960: 1967). Freireich's ultrapragmatic stance belied the fact that at the time there was simply no other way of knowing which drugs were active without first establishing common criteria for the meaning of the term *active*. In other words, the question of activity presupposed the question: active compared to what? As the director of the program, Gordon Zubrod concluded: notwithstanding Freireich's opposition the dominant theme of the meeting had become "that of criteria: criteria for diagnosis . . . criteria for remission, criteria on every score" (Zubrod 1960a: 280). Indeed, without common standards there could be no cooperation. Despite appearances this was not a conflict between pathologists and clinicians or a fight between empiricists and rationalists. In retrospect the tension can be seen as part of the process whereby the research potential of the clinical trial protocol slowly moved to the forefront of cooperative group preoccupations.

Cooperative groups faced other challenges in addition to standardization, coordination, and control. Maintaining interest among group members constituted a significant problem. Trialists occasionally admitted that their line of work was "not as exciting as research of an individual scientist," in particular because it called for "obedience to rules, democratically achieved and agreed to" (Ravdin, 1960: 6). Moreover, investigators participating in cooperative studies quickly realized that despite initial optimism about finding a cure for cancer, the cooperative program had not immediately led to major breakthroughs. Once confronted with the long-term nature of some of the studies, organizers feared that some investigators might develop "a considerable degree of boredom, fatigue, or lackadaisical attitudes" (Greenberg, 1961: 189). As a result, groups sought to develop strategies to "promote high morale and to avoid loss of interest" (189). Early suggestions included regular meetings held in rotation among participating institutions, feedback mechanisms, the selection of only promising drugs whose investigation was likely to promote new knowledge ("clinical investigators cannot be expected to find a continued challenge in routine screening programs"), and plans for publication with at least one coauthor from each participating institution (190).

These suggestions implied that as long as clinical trials remained simple screens for promising agents, the root problem of maintaining scientific interest in a quasi-scientific enterprise could not be resolved. As we have seen with the VAMP protocol and Protocol No. 3, the idea that clinical trials are experiments implicating both drugs and disease processes had largely been implemented by 1963.

Meanwhile in Europe . . .

In November 1961 a group of about twenty researchers and clinicians from six European countries all belonging, with one exception, to the European Common Market (Belgium, France, Germany, the Netherlands, Italy, and Switzerland) met in Paris to discuss the development of some sort of organization in the field of cancer chemotherapy. Their discussions issued in the establishment of the aforementioned Groupe Européen de Chimiothérapie Anticancéreuse. Key participants, in particular those who subsequently became presidents of GECA/EORTC, had a number of characteristics in common. They had lived at some point in their career in the United States, where they personally experienced the burgeoning new field of cancer chemotherapy. They shared a desire to reform medical research that would eventually lead to important changes in their country's biomedical research system. They also had a European outlook: "There was a very strong European spirit at that time. I think that was one of the most important motives at that time. We were impressed, we were inspired by the European idea."[21]

Consider, for instance, Henri J. Tagnon (1911–2000), a Belgian clinician who, fleeing the Nazi invasion of his country, began a career in oncology at the Memorial Sloan-Kettering Cancer Center in the 1940s. Returning to Brussels in 1953, he sought to develop clinical oncology as a discipline based on treatment *and* research, and by 1964 he had become the first professor of clinical oncology in Belgium (Kenis, 1977; Tagnon, 1977; Meunier & van Oosterom, 2001). His attempts to reform Belgian medical research led him to conclude that it would be necessary to expand the field of play to include the European continent as a whole: overcoming national boundaries, he believed, would create a critical mass of clinicians and researchers that would allow Europeans not so much

21. Interview with Dirk van Bekkum, Rotterdam, 15 March 2005.

to compete as to collaborate with Americans on an equal footing. Another participant, the French clinician Georges Mathé (1922–2010), had initially trained under key figures of the modernist wing of French medicine before becoming one of the leaders of that movement and a protagonist of the establishment in 1964 of the Institut National de la Santé et de la Recherche Medicale (INSERM), the French medical research council (Picard, 1994). In 1955 Mathé spent a year at the Memorial Hospital in New York working with Joseph Burchenal and David Karnofsky, two pioneers of chemotherapy,[22] and upon his return to Paris he created a journal programmatically titled *Revue française d'études cliniques et biologiques*. Finally, Silvio Garattini (b. 1928), an MD originally trained as a chemist, found himself at the relatively young age of thirty-five at the head of the Mario Negri Institute, a private, not-for-profit biomedical research organization that stood out within the Italian medical and university system, run by long-standing dynasties of so-called university barons (Clark, 1977).[23] On the occasion of his first visit to the United States in 1957, Garattini met Dr. Charles Heidelberger (1920–1983), who had just synthesized the anticancer compound 5-fluorouracil (Heidelberger et al., 1957). In the years that followed, Garattini visited the United States regularly. He freely admitted the American influence when asked about the creation of GECA: "We've been very much influenced by the Americans. There is no doubt. In fact, the idea to establish a sort of network, a European network came from what I learned in the United States."[24]

The Paris meeting resulted in a preliminary exchange of ideas for these reform-minded clinicians and scientists and a follow-up meeting was held in Milan in 1962. As in the United States, a tentative, tinkering spirit infused these initial steps insofar as the very nature of the future organization had yet to be defined. The first question to be resolved was: Should the new organization be modeled on a "scholarly society" or a "working group"? As a scholarly society, the group would have followed the model of the International Society of Chemotherapy for Infection and Cancer, established that same year (1961) as a nonprofit organization "to advance the education and science of chemotherapy," and organized international congresses and conferences.[25] The group rejected

22. http://picardp1.ivry.cnrs.fr/Math%258e.html.
23. Interview with Silvio Garattini, Milan, 11 May 2005.
24. Ibid.
25. H. Lettré (Heidelberg) to Garattini (1 January 1962). EORTC Archives (Brussels).

this option, however, and opted instead to create a collective of scientists and clinicians engaged in cooperative research projects. To allow the working group to become operational as soon as possible, the new organization limited its membership.[26]

But in order to do what? The answer seemed easy enough: as Garattini put it, to stimulate and coordinate experimental and clinical research on cancer chemotherapy in Europe. That answer immediately raised further questions: What disciplines, specialties, and domains should be included under "chemotherapy"? To what extent should GECA members actually collaborate and according to what division of labor? And how would these activities be financed? In their answer to the first question, GECA, like the CCNSC, adopted a broad definition of chemotherapy that included "all chemical, clinical and biological disciplines concerned with the discovery and testing ('experimentation') of anti-cancer chemical substances." The group, in other words, decided to tackle all stages of chemotherapy running from the initial screening of potential anticancer drugs (and related pharmacological studies) to human therapeutic testing in clinical trials. This choice is illustrated in figure 1.5 (chapter 1), which appeared in a promotional brochure produced by GECA soon after its establishment. The brochure represented GECA's activities as a six-stage "chain of tests" running from the synthesis of potential anticancer agents, to their screening through a series of in vitro systems and a number of animal models, and finally to clinical trials in human patients. Individual GECA members, depending on their clinical or scientific specialization and the nature of the products being tested, took responsibility for different kinds of tests, whose results underwent statistical treatment before being discussed by the group.

The broad definition of chemotherapy as scientific *and* clinical research that included chemical, biological, pharmacological, and clinical disciplines clearly implied a division of labor at the European level, although it was not then evident whether responsibility for a given project should be assigned to a given country, whether the same work should be repeated in two or more countries, or whether, as in the case of therapeutic trials, work should be carried out in several countries. In spite of these ambiguities, the notion of chemotherapy used by the group entailed a Europe-wide participation as individual European countries lacked the funds to purchase the necessary research equipment and ani-

26. Mathé to Garattini (undated), EORTC Archives (Brussels).

mals and the personnel needed to establish collaborations among phar-
macists, biochemists, cytologists, geneticists, immunologists, radiobiolo-
gists, virologists, clinicians, and statisticians. Moreover, clinical cancer
trials frequently called upon a large number of patients, larger, at any
rate, than those available in a single hospital and, sometimes, in a single
country.[27]

GECA attempted not only to overcome the limitations of national
boundaries but also to create an explicit alternative to existing organizations that practiced something more akin to "scientific diplomacy"
than actual research, such as the Union for International Cancer Control (UICC; Maisin, 1966). According to Henri Tagnon (1976: 239–40,
our emphasis):

> [in Western Europe] one important factor for success in many research areas
> is lacking: this factor is the *continental*, as opposed to the *national*, dimension
> of cancer research as exemplified for instance by the United States. National
> responsibility for research may be quite enough for certain programs but by
> and large, training in clinical medicine and research is more successful when
> carried out in a broader environment. . . . At the other extreme, *world wide
> organizations* have their special usefulness which is not primarily the train-
> ing of research personnel or the production of original data and new con-
> cepts. In contrast, the rapid circulation of people and exchange of ideas with
> the resulting cross fertilization of minds and elimination of "parochial" types
> of planning and working is characteristic of the *continental* dimension as is
> well illustrated by medical science in the United States.

Tagnon's comment begs the question as we can still ask why the group
chose such a broad definition of chemotherapy in the first place. Part
of the answer lies in the fact that, as previously noted, the founders of
GECA used the US system as a model. As we have seen, the CCNSC established a continuum between (animal) screening and (human) clinical trials. But GECA founders also sought to erase the boundaries between clinical medicine and biological research. Tagnon, for instance,
echoing the comments of the chairman of the CCNSC clinical studies
panel rejected any and all distinctions between fundamental and applied

27. Le Groupe Européen de Chimiothérapie Anticancéreuse (GECA), typed manuscript (ca. 1962–63), EORTC Archives (Brussels), (GECA-EORTC Proceedings of the
Council, 1962–69).

research (Ravdin, 1960: 6). He furthermore denounced attempts to separate research on diagnosis, pathogenesis, and therapy, arguing that "clinicians have been proclaiming, louder and louder, their once contested
and now recognized right to design and to perform their own research
programs." Accordingly, "the time has finally come when we can happily
watch clinicians operate electronic microscopes or electrophoresis equipment, use spectrophotometry, or engage in organ culture" thereby demonstrating that "clinical research blossoms and best renews itself by an
assiduous acquaintance with the preclinical disciplines" (Tagnon, 1964:
284, our translation). Clinical cancer research, in other words, was—or
at least ought to be—a "seamless medical act" involving a daily confrontation between clinical medicine and laboratory disciplines such as biochemistry, biology, chemistry, and histology (284, our translation).

GECA did not follow the NCI model slavishly. For obvious financial, organizational, and staffing reasons, GECA established something
smaller and therefore different. The group thus sought new approaches
to screening based on a rational search for substances such as those that
inhibited metastases rather than destroyed tumors, and different screening criteria, such as a system based on a primary screening with tissue
cultures, biochemical tests, virus-induced tumors, and immunological
tests. The source of the substances to be screened by GECA posed an
immediate problem: would they be substances obtained from US and
British laboratories, original products derived from GECA's own chemical research and screening program, or molecules prepared by university
laboratories and pharmaceutical companies for GECA?[28]

Last but not least, in the absence of a European NCI, there were no
obvious sources of financial support for GECA. Given the relatively
small scale of the group's early activities, its members initially relied
on local resources. Since the point of the operation was to expand research to the European level, they were forced to seek additional support. European institutions provided a target of choice and Euratom offered to fund selected activities.[29] The UICC also provided funds during
the early years, but it was not until 1979 that the European Community
finally offered "partial support" for EORTC's research program (News

28. Letters by H. A. van Gilse, 30 December 1961, D. Schwartz (undated) and S. Garattini (undated), EORTC Archives (Brussels).

29. GECA, "Compte-rendu de la réunion du 29 janvier 1966," EORTC Archives
(Brussels).

from E.O.R.T.C., 1979). In the meantime, the NCI had stepped in and since 1972 had become one of the major sources of funding for the organization. Given its nongovernmental nature, funding remained a major problem for the EORTC for many years, and its leadership continuously struggled to ensure funding through such creative initiatives as the establishment of a foundation.

GECA's structure changed drastically in the late 1960s and early 1970s following the nomination of Maurice Staquet (an internist who had subsequently received a degree in biostatistics at Columbia University in New York) as a central clinical trial coordinator and the establishment of a data center in Brussels, both made possible by NCI support. Now endowed with a more formal structure and with norms and procedures that governed the cooperative groups, the institution of the data center once again followed the American model. GECA's development, however, departed from the American paradigm in important ways. Whereas the US cooperative groups, initially centralized at the NCI, decentralized, GECA's loosely connected network of national members adopted a more centralized structure.

GECA's novelty cannot be overstated. With the exception of the United Kingdom, there were no multicenter cancer clinical trials in Europe in the early 1960s. As recalled by one of the early presidents of the EORTC:

> They [GECA's precursors] just did trials. And sometimes . . . mostly they followed a leader: if the professor said we give so much methotrexate because he had heard it at a conference then you gave it. That was the idea. If somebody wrote a nice article about it with good results . . . that was that.[30]

Even though the French chemotherapy pioneer, Jean Bernard, participated in trials organized by ALGB in the 1960s, his own trials remained somewhat parochial and did not include randomization (Rigal, 2008). GECA chose a route that differed not only in its European dimension but also in its deployment of the multicenter approach. The route chosen was scattered with material and financial obstacles, but it also allowed relatively young clinical researchers to work around medical mandarins: members from the different countries "were chosen not because of their position in the university or hospital hierarchy, but because of their qual-

30. Interview with Dirk van Bekkum, Rotterdam, 15 March 2005.

ifications in their discipline and the usefulness of their discipline for the Group."[31] It was, in particular, learning by doing:

> We considered it as an ongoing process of learning, of understanding what's going on . . . against various forces, some statisticians saying that this is not good, some clinicians who did not want to join certain trials because they didn't believe this was the right way to proceed. There were continuous discussions but the main thing was that we met every month or every six weeks. . . . It was a very practical dedicated, intelligent, interesting, multidisciplinary group and everybody liked to be there and take part in discussions or listen to discussions. That was the big binding for us.[32]

The creation of the *European Journal of Cancer* in 1965 created further room for discussion for European oncologists. While not the official organ of the group, the journal's editor (Tagnon) and most of its founders were members of GECA.

As in the American cooperative groups, the protocol lay at the core of the work championed by GECA. In 1966 the *European Journal of Cancer* published two "official" GECA trial protocols, one for leukemia and one for breast cancer (GECA, 1966a; 1966b). More than protocols for specific trials, these two documents constituted a master plan for the clinical screening of anticancer drugs at different stages of development. Devised by Tagnon's group in Brussels, the breast protocol began by stating that it had been inspired by protocols developed at the NIH and went on to outline two types of trials. The first (Type I) tested substances successful in mice on a maximum of fifteen patients, with no controls. Successful substances then moved on to the second type of study (Type II) that established the percentage of remissions produced (as defined by "objective," quantitative criteria). These trials used a control group and a double-blind, randomized approach. The leukemia protocol also screened prospective substances but adopted a somewhat more convoluted scheme: the "stage 1" test was similar to the first type of breast trial. The "stage 2" trial depended on whether the substance belonged to a known category of anticancer drugs; if so, it was tested on patients resistant to other substances of the same category using again a

---

31. Le Groupe Européen de Chimiothérapie Anticancéreuse, undated, circa 1963–64, brochure, EORTC Archives (Brussels), 7.

32. Interview with Dirk van Bekkum, Rotterdam, 15 March 2005.

stage I approach; if not, it underwent a test to measure its activity in a group of patients (with no control group). Clearly, these initial protocols involved considerable "tinkering." Moreover, while the breast protocol might look familiar to modern oncologists, the leukemia protocol would probably strike them as rather suspect. Yet these protocols were a clear sign of the progressive entrenchment of the new style of practice.

Conclusion

The cooperative chemotherapy programs in the United States and in Europe were not greeted with unanimous consent and remained controversial for many years. We will examine some of the early criticisms in chapter 5, after analyzing the statistical component of trial design in chapter 4. For the present we conclude this chapter with several remarks concerning the original organizational structure established to conduct the clinical evaluation of anticancer drugs. As mentioned above, cancer chemotherapy grew out of an interlocking sequence of activities that ranged from the procurement of substances to their testing in animal models and the evaluation of selected compounds in human trials. This complex program has often been branded as an industrial approach to medical research, but as we have also seen, to reduce it to such a model is to miss several of its novelties and the fact that one of its key components, cooperative groups, was explicitly designed as an alternative to industrial or other "engineered" forms of research. Industrial firms were of course involved in the overall endeavor, but they did not run it: they followed the evolving organization and content of the chemotherapy program.

We have characterized cooperative groups as "epistemic organizations" to emphasize the intimate connection between research and the organizations devised to carry them out. European clinical researchers clearly understood this, and taking their cue from the United States, explicitly rejected traditional models such as scholarly or professional societies, or international organizations more involved in scientific diplomacy than in actual research. More traditional activities—such as the creation of new journals (*Cancer Chemotherapy Reports* in the United States and the *European Journal of Cancer* in Europe) to report the results of the novel approach, and the establishment of professional organizations (the American Society of Clinical Oncology and the European Society of Medical Oncology) drawing together the bearers of the new

style of practice—would follow, but they were the downstream consequences of the emergence of the new style, rather than its cause.

Any attempt to define the nature of cooperative groups using existing models of organization—organizational sociologists have suggested
a number of denominations to us, including: Schumpeterian entrepre-
neurs, quasi-firms, coalitions of alternative or reform clinicians, etc.—is
bound to fail, not because they exclude elements of those models, but
because they are simultaneously all and none of the above: borrow-
ings from different domains were connected and rearranged producing a novel, emergent institutional form. Even the recent notion of a distributed network only partly captures the topology of the cooperative
groups since they are simultaneously centralized and distributed organizations. Moreover, the groups evolved along with the transformation
of bioclinical practices and the understanding of cancer pathology and
treatment for which they were partly responsible. They also responded
to the evolution of cancer itself insofar as treated patients are no longer
the same biological and social entity as patients of the prechemotherapy
era. In other words, with the new style of practice more or less firmly in
place, its configuration evolved. The rest of this book will analyze these
transformations, but first, we need to take a closer look at the biostatistical practices generated by the new style.

Clinical Trial Statistics

This interlude is no substitute for a textbook description of clinical trial statistics. A mixture of historical snippets and simplified technical notions, it is designed to provide readers with some understanding (although obviously not a working knowledge) of common statistical notions and practices mentioned in this book. In all there are truly only two ideas that require a more extended explanation than would be appropriate within the confines of any given chapter: the notion of a random sample and the concept of statistical significance.

What Is a Random Sample?

Cancer researchers conducting a clinical trial usually begin their investigation by randomly allocating a sample of cancer patients to two competing treatments. The resulting trial seeks to determine which treatment is better for a given disease. When completed and if successful (failed trials are not reported in the general press as often as successful trials, if at all), a typical newspaper summary of the clinical trial might read: "Researchers tested two treatments (A and B) on a random sample of 100 patients suffering from X. They found treatment B to be significantly better than treatment A." Newspaper reporters (and their editors) probably assume that notions like random, sample, and significant are self-explanatory. Most readers may not be aware of the complexities surrounding not only their definition and use in actual practice but also their evolution as scientific concepts subject to continuous debate and revision. So some explaining is in order. We begin with an example of a clinical cancer trial that we discussed in chapter 2. It is, appropriately, the very first cooperative group clinical cancer trial carried out by NCI researchers.

The trial compared two different treatments of acute leukemia. Both regimens used the same two drugs, methotrexate and 6-mercaptopurine (6-MP), but administered the 6-MP in two different ways: a daily dose (Treatment 1 or T1) versus a triple dose every third day (T2). The researchers hypothesized that there would be a significant difference between administering the drug on a continual rather than on an intermittent basis. Planned and carried out in 1954–55 at the NIH, the trial tested the two regimens on a group of sixty-five patients treated in four different institutions (Frei et al., 1958). The researchers did not randomly assign the sixty-five patients to the two treatments, nor did they gather together a random assortment of sixty-five patients suffering from acute leukemia. Rather, they chose acute leukemia patients according to three different criteria: age (older or younger than fifteen years of age), morphological type of acute leukemia (lymphocytic or myelogenic), and extent of prior treatment (untreated versus previously treated). They then randomly assigned members of these subgroups, each containing approximately twenty-one patients, to one of the two treatments. Let us be clear about what is random in this particular random sample: the selection of individuals for the sample was carefully planned; the assignment of members of the sample to one of the two treatments was random.

In order to ensure that the treating clinicians assigned patients to the two regimens in a random fashion, the researchers in our example used a "sealed envelope technique": individual clinicians received a set of sealed envelopes numbered according to the number of patients that they would be treating. Upon receiving the first patient in the series, the clinician opened envelope number 1 in order to discover which of the two treatments the patient would receive. For the second patient, the clinician opened envelope number 2 and so on. Notice that the clinician opened the letters in order, not randomly. The treatment assignments in the envelopes, however, followed a random order. How was this random order produced?

To understand this stage of the process, let us suppose that a given clinician treated a group of patients defined as: patients older than fifteen, suffering from acute lymphocytic leukemia, and with no prior chemotherapeutic treatment. Let us further suppose that he treated six such patients. In order to be able to make a fair comparison, the clinician would have to treat three of the patients with T1, and three of the patients with T2. In other words, before randomly assigning patients to one of the two treatments, the researchers overseeing the treating clinician had to balance the treatments. Only then could they set up a system to randomly assign patients to treatment. In this case the random assignment consisted in ensuring that the *sequence* of the six treatments (three to T1 and three to T2) occurred randomly.

To create such a random sequence, the researchers first organized the treatments into a systematic order. For example: T1, T1, T1, T2, T2, T2, is a systematic ordering of the six treatments. Other systematic orderings are possible (e.g., T1, T2, T1, T2, T1, T2). The researchers then assigned each treatment a random number chosen from an official table of random numbers (see, below), like this (the numbers in brackets are random numbers): T1 (0.608), T1 (0.739), T1 (0.831), T2 (0.016), T2 (0.759), T2 (0.877). Finally, they arranged the random numbers in order of increasing size, creating, as a consequence, a random order to the sequence of treatments. Like this: T2 (0.016), T1 (0.608), T1 (0.739), T2 (0.759), T1 (0.831), T2 (0.877). Thus, patient 1 received T2, patient 2 received T1, and so on. Notice the symmetry: ordering a series of random numbers transforms the systematic order of the treatments into a random order. Notice also the recourse to convention: the random order was underwritten by the fact that the random numbers used in the randomizing process were certified random numbers. So where did the random numbers come from and who certified them? A short answer to the question is, as indicated above, from an official table of random numbers. But this only pushes the question back one step. To fully understand the origin of the random numbers we must briefly consider the notion of randomness.

Despite common perceptions, it is not easy to do something randomly. Much effort has been and continues to be expended to produce the kind of randomness contained in the above series of random numbers. Even mechanical means of randomization such as coin tossing and roulette wheels do not produce purely random outcomes. Mechanical devices and their products such as coins have all the limitations of existing on an earth with an oxidizing atmosphere and a gravitational pull that varies according to the distance from the earth's poles. As a result the coins and the wheels not only begin as imperfect objects, but, worse, through use they wear down. These imperfections are thus the sources of variations that are anything but random; even mechanical set-ups come to favor one result over another. The founder of modern statistics, the British mathematician Karl Pearson (1857–1936), demonstrated convincingly at the turn of the last century that standard methods for generating random outcomes such as dice or roulette wheels produced series of numbers that on close inspection deviated from random (although they are certainly random enough to make gambling a losing proposition) (Pearson, 1894).

Given the inherent and often unknown biases involved in generating homemade random events and related numbers through such activities as coin tossing, statisticians and clinical trialists have long preferred ready-made, official random numbers. Since the late 1920s and the appearance of the first book of random numbers, L. H. C. Tippet's *Random Sampling Numbers* (1927), researchers have had a variety of printed sources at their disposal. It has not always been smooth sailing. Tippett, for example, generated his random numbers by taking digits "at random" from British census reports (Kendall & Babington Smith, 1938: 157). Statistical examination of the numbers in the 1930s by another prominent statistician found the series to be "patchy" (Yule, 1938: 167), and subsequent authors concluded that the numbers did not truly have the status of random (Horton, 1948).

Statisticians sought solutions in a number of directions. The British statisticians Maurice Kendall and Bernard Babington Smith attempted to improve on Tippett by selecting numbers "at random" from the *London Telephone Directory*. Subsequent examination of the resulting 10,000 digits showed them to be, for reasons unknown, "significantly deficient in fives" (1938: 157). So the statistical duo invented an electric randomizing machine whose functioning defies easy description and need not detain us here. Suffice it to say that the operation of the machine required a skilled operator "moving the stylus with one hand and writing down the digits with the other" (1939a: 52). Despite its Rube Goldberg overtones, the randomizing human/machine hybrid worked well enough to become the source of a new book of random numbers (Kendall & Babington Smith, 1939b) that quickly superseded Tippett's publication.

The rise of methods that analyzed random physical processes in simulation experiments (so-called Monte Carlo methods) in physics and engineering in the postwar period created a great demand for random numbers (Galison, 1997: 689–780). Whereas early random number books contained tens of thousands of digits—Tippet's original publication had 40,000 random numbers—the new forms of experimental simulation re-

TABLE OF RANDOM DIGITS 1

00000	10097	32533	76520	13586	34673	54876	80959	09117	39292	74945
00001	37542	04805	64894	74296	24805	24037	20636	10402	00822	91665
00002	08422	68953	19645	09303	23209	02560	15953	34764	35080	33606
00003	99019	02529	09376	70715	38311	31165	88676	74397	04436	27659
00004	12807	99970	80157	36147	64032	36653	98951	16877	12171	76833
00005	66065	74717	34072	76850	36697	36170	65813	39885	11199	29170
00006	31060	10805	45571	82406	35303	42614	86799	07439	23403	09732
00007	85269	77602	02051	65692	68665	74818	73053	85247	18623	88579
00008	63573	32135	05325	47048	90553	57548	28468	28709	83491	25624
00009	73796	45753	03529	64778	35808	34282	60935	20344	35273	88435
00010	98520	17767	14905	68607	22109	40558	60970	93433	50500	73998
00011	11805	05431	39808	27732	50725	68248	29405	24201	52775	67851
00012	83452	99634	06288	98083	13746	70078	18475	40610	68711	77817
00013	88685	40200	86507	58401	36766	67951	90364	76493	29609	11062
00014	99594	67348	87517	64969	91826	08928	93785	61368	23478	34113
00015	65481	17674	17468	50950	58047	76974	73039	57186	40218	16544
00016	80124	35635	17727	08015	45318	22374	21115	78253	14385	53763
00017	74350	99817	77402	77214	43236	00210	45521	64237	96286	02655
00018	69916	26803	66252	29148	36936	87203	76621	13990	94400	56418
00019	09893	20505	14225	68514	46427	56788	96297	78822	54382	14598
00020	91499	14523	68479	27686	46162	83554	94750	89923	37089	20048
00021	80336	94598	26940	36858	70297	34135	53140	33340	42050	82341
00022	44104	81949	85157	47954	32979	26575	57600	40881	22222	06413
00023	12550	73742	11100	02040	12860	74697	96644	89439	28707	25815
00024	63606	49329	16505	34484	40219	52563	43651	77082	07207	31790
00025	61196	90446	26457	47774	51924	33729	65394	59593	42582	60527
00026	15474	45266	95270	79953	59367	83848	82396	10118	33211	59466
00027	94557	28573	67897	54387	54622	44431	91190	42592	92927	45973
00028	42481	16213	97344	08721	16868	48767	03071	12059	25701	46670
00029	23523	78317	73208	89837	68935	91416	26252	29663	05522	82562
00030	04493	52494	75246	33824	45862	51025	61962	79335	65337	12472
00031	00549	97654	64051	88159	96119	63896	54692	82391	23287	29529
00032	35963	15307	26898	09354	33351	35462	77974	50024	90103	39333
00033	59808	08391	45427	26842	83609	49700	13021	24892	78565	20106
00034	46058	85236	01390	92286	77281	44077	93910	83647	70617	42941
00035	32179	00597	87379	25241	05567	07007	86743	17157	85394	11838
00036	69234	61406	20117	45204	15956	60000	18743	92423	97118	96338
00037	19565	41430	01758	75379	40419	21585	66674	36806	84962	85207
00038	45155	14938	19476	07246	43667	94543	59047	90033	20826	69541
00039	94864	31994	36168	10851	34888	81553	01540	35456	05014	51176
00040	98086	24826	45240	28404	44999	08896	39094	73407	35441	31880
00041	33185	16232	41941	50949	89435	48581	88695	41994	37548	73043
00042	80951	00406	96382	70774	20151	23387	25016	25298	94624	61171
00043	79752	49140	71961	28296	69861	02591	74852	20539	00387	59579
00044	18633	32537	98145	06571	31010	24674	05455	61427	77938	91936
00045	74029	43902	77557	32270	97790	17119	52527	58021	80814	51748
00046	54178	45611	80993	37143	05335	12969	56127	19255	36040	90324
00047	11664	49883	52079	84827	59381	71539	09973	33440	88461	23356
00048	48324	77928	31249	64710	02295	36870	32307	57546	15020	09994
00049	69074	94138	87637	91976	35584	04401	10518	21615	01848	76938

FIGURE INTERLUDE I A. The first page of the Rand Corporation's 600-page book, *A Million Random Digits* (1955). Reproduced with kind permission of the Rand Corporation.

quired many more. The Rand Corporation's celebrated *A Million Random Digits* (1955) thus replaced Tippet and his successors as the standard source of random numbers (figure Interlude 1, A). The Rand Corporation researchers had produced their numbers via an electronic roulette wheel that had originally produced biased numbers and that had required considerable correction. The postwar era also saw the advent of computer-generated random numbers known as *pseudo-random numbers* since the numbers depend upon specially constructed algorithms and cannot thus logically be considered random. All such random algorithms tend to produce regular as opposed to random series in the long run, although the resulting regularity is so faint and the bias so small that the string of numbers thus generated can be considered random for most practical purposes (Bennet, 1999).

To sum up: when the researchers in our example sought to randomize the order in which the patients received their treatment, they resorted to one of several standard works containing lists of random numbers. While humans cannot create purely random numbers, as the devices used to produce them inevitably bias the numbers, for the practical purposes of clinical trials a book of official random numbers or a computer program generating pseudo-random numbers fulfills the needs of clinical researchers. This holds only in principle in the case of clinical cancer trials. Producing randomness in practice raises a number of technical and mundane questions. Clinical trial organizers, for example, were on occasion confronted by "de-randomizing" physicians. For instance, in the sealed envelope technique, physicians could, if they chose, cheat the system: "What the clinicians used to do was they would hold [the envelopes] up to the window to see what the triage was, or they would wait until they had a handful of patients and opened up all the envelopes and chose the treatment that they felt fit the best."[1] Clinical trial organizers responded by centralizing randomization thus forcing physicians to call the central office for treatment assignments.

Now that we understand the random dimension of the notion of random sample, we can ask: what is the sample in this example? The answer to this question depends on how we depict our sample. Statisticians usually describe a sample as a "subset of a population," wherein a population is defined as "a collection of items of interest in research" (Simon, 2006: 138). On the basis of this very general definition, we might conclude that the sample consisted of the sixty-five patients drawn from the population of patients suffering from acute leukemia. We would not be entirely wrong and we would not be entirely right. Recall that the individual members of the group of sixty-five patients came from a variety of subpopulations defined according to age, prior treatment, and type of leukemia. Each of these subpopulations can also be considered "a collection of items of interest." There are thus two different kinds of sample if we subscribe to the definition above. The sixty-five patients taken as

1. Interview with Marvin Zelen, Boston, 5 November 2004.

a whole certainly constitute a subset of all patients suffering from acute leukemia, as do the subgroups within the sample of sixty-five. The second kind of sample (the age, etc. subgroups), however, functioned as the working sample, the sample on which the principle of randomness operated, insofar as members of these subpopulations underwent random assignment to the two treatments.

The decision of "what a sample is a sample of" is based on bioclinical considerations and has important implications for the interpretation of a trial's results. For instance, our example, chosen for its authenticity, is obviously quite dated. Fifty years after the original clinical trial, the characteristics used to define the sample population in the trial no longer make sense. Researchers today would be loath to submit adults with acute myelogenous leukemia and children with acute lymphocytic leukemia to the same treatment. Largely because early trials like the one in this example and more importantly the trials using combination chemotherapy in the early 1960s succeeded in producing lasting remissions in acute lymphocytic leukemia, researchers now treat the two diseases as entirely separate entities and not just as subpopulations within the larger population of acute leukemia patients. Defining a sample either in statistics in general or for the purposes of clinical cancer trials ultimately depends on comparing "like with like," and, as we have just seen, comparability judgments evolve in tandem with the transformation of the biomedical understanding of disease entities and processes.

What Is Statistical Significance?

We now need to examine a somewhat more thorny issue—namely, statistical significance. Let us begin by using again the acute leukemia trial as an example: of the sixty-five patients who entered the trial, researchers assigned thirty-two to the intermittent treatment and thirty-three to the continuous treatment. Five patients in each treatment group underwent complete remission of their leukemia. The researchers concluded that there was no *significant* difference between the two treatments, a conclusion that, as we will see, should not be taken to imply that the two treatments were of equal efficacy, since a failure to detect a difference does not mean that there is no difference. In the present case the aforementioned conclusion required no rocket science, since both arms of the trial had the same number of patients and the number of remissions were the same. But let us now suppose that ten patients in T1 rather than five had undergone remission, while T2 produced the same five remissions. Would ten remissions as opposed to five remissions mean that T1 was *significantly* better than T2? While we might be tempted to answer yes based on intuition, a proper answer to the question—that is, one that will convince trialists—can only be given by the application of a statistical *significance test* to the data.

At this point things start to get murky. To begin with, there are many kinds of significance tests. Clinical researchers choose tests according to a number of criteria. The size of the sample may determine the kind of test used just as the kind of variable measured may enter into its determination. If, for example, researchers compare patient reactions (e.g., did the patient go into remission or did the tumor shrink), they will use a different test than if they measure and compare the length of time that the patients survive following therapy. In the example above, the researchers used a test called Fisher's exact test to compare differences between samples because of the relatively small size of the samples.

A significance test is designed to make sure that the result of a given trial is not due to chance, or, to put it differently, that were clinicians to repeat the trial they would obtain the same positive or negative result. We have to avoid a possible confusion here. Multiple variables enter into the conduct a trial: for instance, researchers might perform two different trials comparing the same two regimens in different patient populations (different age, different subspecies of disease . . .) and in this case the outcomes might be different, and legitimately so. This is why some trialists and, especially, patient activists have campaigned in favor of staging trials on populations representative of the patients who will later receive the successful treatment. Deciding what kind of population to target for a trial is a question of *external* validity: Are the trial results relevant for the patients "out there." In what follows we discuss the issue of *internal* validity: Did the experimental treatment work in the population on which it was tested? As we saw in our hypothetical example, if T1 leads to ten remissions as compared to only five in T2, should we conclude that T1 is better than T2 (that there is a *significant* difference between T1 and T2), or could this difference be the result of chance variation?

Given that a clinical trial is a statistical operation, trust in the validity of the trial's results cannot be secured on anything like an unimpeachable foundation; trial results have only a greater or lesser probability of being true. There are two ways of describing the probability that the results of a trial are valid. In question form, they are: (1) What are the chances that a positive answer was reached when in fact it should have been negative (T1 works better than T2, when in fact there is no difference: a *false positive* result)? (2) What are the chances that a negative answer was reached when in fact it should have been positive (there is no difference between T1 and T2, whereas in fact T1 works better: a *false negative* result)? The parameters involved in these kinds of calculations, as we will see below, have esoteric names such as α (the Greek letter alpha), β (beta), and power. Before addressing the history and meaning of these parameters, we discuss two other related parameters—namely, sample size and effect size.

If the effect of T1 is overwhelmingly greater than the effect of T2 (e.g., if you are comparing the effect of antibiotic with that of a sugar pill in the case of a bacterial infection), you will need only a few patients in your trial to demonstrate that there is difference between the two treatments and convince your fellow trialists that the result is trustworthy. If, on the other hand, the difference between two treatments is minimal (as is often the case with two anticancer regimens), a large number of patients will be needed to show that there is a difference between the treatments and that the result is to be trusted. Statisticians refer to the expected difference between two treatments as the *effect size* and to the patient population subset entered in the study as the *sample size*.

Sample size can be modified at will in the abstract world of mathematics. In the real world, however, the size of a patient sample places major constraints on clinical research since "patient material" is not always readily available, especially in relatively rare forms of cancer. Before conducting a trial, clinicians must therefore consider the size of the difference between the two treatments that they expect the trial to produce. Should they foresee that the difference will be minimal, then they must plan to recruit a large number of patients. If the trial design does not plan a sample size that is commensurate with effect size, then the trial organizers will be saddled with a faulty design and a trial that lacks a sufficient number of patients to reach a valid conclusion. Such a trial is highly unlikely to be approved by one of today's many protocol review committees that judge clinical trial protocols. Following complaints by group members that a sizeable number of early trials were not published or were interrupted before completion due to faulty design, the cooperative groups themselves established review boards to address the problem at its root.

To facilitate our understanding of statistical significance, let us have a look at a present-day software program used to juggle with the parameters involved in these calculations (Faul et al., 2007). More specifically, figure Interlude 1, B shows a screen shot of the data-entry window of a computer program used to perform the calculations that determine the relations between significance, sample size, and size effect. On the left side, under "input parameters," we entered a value of 0.3 (i.e., 30% difference) in the effect size box (we will discuss the other parameters such as α and β below), and we obtained as a result a sample size of 111 patients on the right side under "output parameters." Now let us suppose that we wish to conduct a trial on pheochromocytoma, a rare adrenal gland cancer, and that we expect the two treatments we wish to compare to differ by 30%. Physicians in the United States diagnose less than 1,000 cases of pheochromocytoma a year, and given that for a variety of reasons only 5% of those patients are likely to participate in clinical trials, this creates an annual sample size of no more than fifty patients. Since by our calculations, we would need 111 patients in order to ensure that the results of the trial are valid, we would have to plan for a trial lasting several years . . . or perhaps arrange for an international

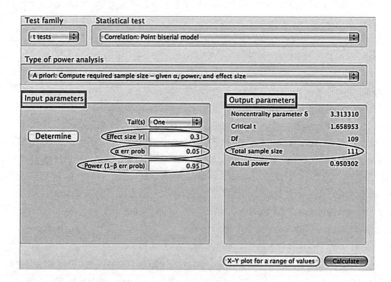

FIGURE INTERLUDE I B. Screenshot of the data-entry window of a computer program used to perform statistical significance calculations. Source: http://www.psycho.uni-duesseldorf .de/abteilungen/aap/gpower3/.

collaboration. Things would be even worse should the expected effect size be lower . . . or they could be better if there are grounds to believe that the regimen we plan to test will prove to be even more effective than its competitor. For instance, suppose we expect a difference between the two treatments to be 40%: by raising effect size to 0.4, we will need only fifty-nine patients. As can be seen, when conducting clinical trials, clinicians must take into account a number of biological (i.e., effect size) and institutional (i.e., patient availability) variables that above and beyond all statistical considerations may affect sample size. Clinical trial planning thus often requires trade-offs between statistical, biological, and institutional parameters.

What about the other two data entry boxes that are listed under "input parameters"? Suppose, for instance, that instead of fiddling with the effect size we modify one of the values of the other two "input parameters" and lowered "power $(1 - \beta)$ err prob" from 0.95 to 0.90? That simple change drags the required sample size down to eighty-eight. This would obviously reduce the time to complete our pheochromocytoma trial. Sample size thus appears to also depend on these other parameters. We can now examine what they are and how they relate. Understanding these relations requires a short historical detour through a controversy over the very meaning of clinical trials.

Let us begin with α. According to the British statistician R. A. Fisher, who developed the earliest significance tests, before conducting an experiment such as treating acute leukemia patients with two different regimens, the researcher must have a specific question in mind. In our example, the *scientific* question was quite straightforward: Is there a difference between treating patients on a continual or an intermittent basis? In order to apply a test of significance, the scientific question must be restated as a negative assertion (the opposite of what one is trying to prove) called the *null hypothesis*; that is, in the present case: "There is no difference between the two treatments." The opposite, positive statement—"There is a difference between the two treatments"—is called the *alternative hypothesis*. By reframing the scientific question as a negative assertion, the statistical test provides a measure of the evidence *against* the null hypothesis: if the clinical trial results pass the test of significance, then researchers can reject it. As already noted, that does not mean the alternative hypothesis is necessarily true; that is, that the statistical test or the clinical trial that furnished the data for the test proved there is a difference between the two treatments. The difference in the formulations is important. While there may be a difference between the two treatments, no single clinical trial can serve as a complete demonstration; a single trial can only show that two treatments are *probably* not the same. It is quite possible that a second or third trial might reach a different conclusion. The question that the test of the hypothesis answers, then, is: What are the chances that with this sample we will be led astray by our results?

To carry out that calculation, researchers decide in advance how much better than chance the result should be. Do you wish to calculate the possibility of producing those observations by chance as being one chance in twenty, one chance in one hundred, or one chance in a million? Fisher settled, somewhat arbitrarily as even he admitted, on the possibility of one chance in twenty. His original choice has remained a sort of implicit standard and is often chosen for no other reason than it carries the weight of a standard. This threshold or cut-off point is termed α. If we set the level as one in twenty, then that probability is 0.05. Once the significance level is set and the experiment has been completed, the observations are fed into the formula for the significance test. The number that results, a probability calculation, is termed a p-value. If, like Fisher, you set the significance level at 0.05, if the outcome of the tests produces a p-value of 0.05 or higher (i.e., between 0.05 and 1), you accept the null hypothesis (there is no difference). If the p-value turns out to be less than 0.05, you reject the null hypothesis.

Ever since Fisher, statisticians (and now their computers) have calculated p-values. Thousands of p-values are published every year in clinical studies and thousands more are calculated and not published. They are, in other words, routine calculations. Their interpretation, however, is not straightforward, for there is a second way of understanding the above calculation. In the first interpretation, which we just saw, the p-value had to remain below the threshold set by α in order for the data collected to stand as significant In contrast, and on the second, dominant interpretation, rather than simply accepting the data as statistically significant, the test may also be viewed as a means of avoiding a statistical error. By construing the statistical significance test as a means of avoiding error and not as a way of weighting evidence, a second possible error immediately arises. Statisticians refer to this kind of mistake as the Type II error, or β error as opposed to the Type I or α error. In our clinical research example, a Type II error would arise if researchers concluded that there was no difference between treatments when the treatments were, in reality, different, or, in other words, what we have referred to as a *false negative*.

Setting a threshold value for β means answering the following question: What are the chances that I am willing to take of making a Type II error? Clinical researchers often set β at 0.2, which means that they are willing to take a 20% or 1 in 5 chance of making a Type II error. In other words, the chances of a Type II error are considerably larger than a Type I error. By terminological convention, the *power* of a statistical test is said to be $1 - β$. So if researchers settle on a β of 20%, the power of the statistical test is 80%. Clinical researchers can therefore speak of a clinical trial as being high-powered or low-powered depending on the size of the β value (the lower the β, the higher the power) set for the significance test.

The articulation of the means to avoid the problem of the Type II error—the mathematics that lie behind the above verbal formulations— was largely the result of work carried out by Jerzy Neyman (1894–1981) and Egon Pearson (1895–1980), who worked together in direct opposition to Fisher's views of significance testing in the 1930s, producing in the process an alternative view (Pearson & Neyman, 1933; Pearson, 1966). Neyman and Pearson objected, in particular, to the fact that in Fisher's world, one simply rejected the idea that there was no difference (or, alternatively, that there was) between the two procedures and moved on to another experiment. Given the existence of Type II errors, Neyman and Pearson claimed that Fisher's significance test did only half the job. True, they conceded, Fisher's test allowed one to minimize the Type I error, but setting α did nothing to help researchers avoid a Type II error. How, Neyman and Pearson asked, could one reduce the possibility of both types of error?

Notice that we have just changed our vocabulary. Rather than weighing the evidence provided by the data (Fisher) with a view toward rejecting a null hypothesis, we are now looking for ways to avoid prespecified errors when testing a hypothesis (Neyman-Pearson). The difference between the two formulations may seem minimal. Even at the time, however, the protagonists clearly saw the difference between the two approaches as resting on radically different interpretations as to what the enterprise was supposed to accomplish and the status of the data collected in the course of the clinical trial or any other data gathering exercise: were the data *test* data or *experimental* data? For Pearson and Neyman, the answer was clear: statistical tests, whether used in clinical trials or in quality control studies carried out in factories, by seeking to minimize type I and type II errors allowed the statistician to come to some decision with regard to the results of the test. Unlike Fisher, Pearson and Neyman did not conceive of the data gathering exercises that preceded the statistical tests as experiments; they were simply tests leading to specific decisions (in our example, which treatment to use).

Once set up as an exercise in error reduction, a significance test must now juggle two variables (two error types). The variables are not, however, independent: by increasing the chances of avoiding a Type I error, for example, one automatically increases the chances of committing a Type II error and vice-versa. The situation is further complicated by the fact that unlike α, researchers cannot set a value for β through convention. Researchers must rely instead on bioclinical knowledge. To see where bioclinical knowledge intervenes, recall that the Type II error consists in concluding that there is no difference when there really is a difference. The chances of making such an error depend partly on the size of the difference one hopes to detect and the sample size used to detect the difference.

Imagine, then, a relatively small difference between two treatments such as the one in our initial leukemia trial example. Changing the rhythm of the treatment from continual to intermittent might have some effect on clinical outcomes, but it is surely no magic bullet. In this case the size of the difference researchers may hope to observe might be little more than 10% to 20%. To detect a 20% difference requires a larger sample size than that needed to detect a 40% difference. Not twice as large, as the relationship is not linear, but certainly larger. In any event, the magnitude of the difference varies according to the clinical situation; it cannot be set once and for all but must be estimated for each clinical trial. The size of the difference depends, in the end, on previous bioclinical knowledge and what results researchers expect their intervention to produce.

Once clinical trialists settle on the size of the difference they hope to uncover, they can set a value for β. In turn, they can settle on a value for α should they wish to depart from the conventional 0.05. But once these three values have been set, one by convention and one based on bioclinical knowledge, the size of the sample for the trial is also set. This is because, as mentioned above, both the value of β and sample size and the value of α and β are interdependent. As we saw for clinical trialists, the important number is often not α or β or the power of the test $(1 - \beta)$ or even the exact size of the difference expected between the two treatments (although there has to be some difference expected, otherwise there would be no point in conducting the clinical trial). Rather, the important number is often sample size,[2] since sample size determines the resources that will have to be devoted to the clinical trial, the length of time the trial will take, and so on. No matter how high-powered a trial a researcher might wish to undertake, facts on the ground create stringent limits.

While the hypothesis testing–error reduction approach advanced by Neyman and Pearson is today dominant, much of the debate went under the radar and textbooks commonly teach elements of each even today. In the postwar era, the two positions were reconciled more by fiat than by mathematics. As in other disciplines, textbook writers simply ignored the controversy and presented the two approaches as anonymous techniques. It is quite telling, for instance, that clinical trial publications, instead of using Fisher's terminology of α, have opted for the Neyman/Pearson terminology of *p-value*: a p-value $= 0.05$ (the equivalent of α $= 0.05$) has become a conventional threshold of statistical significance, meaning that there is less than one chance in 20 of the result being a false positive. Consequently, two somewhat surprising facts have been repeatedly pointed out in the historical and scientific literature in the last twenty years. First: despite the mathematical similarities, these differences are fundamental and they do make a difference. Second: nobody seems to care (Sterne & Smith, 2001; Salsberg, 1992). Since the 1950s textbooks for budding statisticians treat the two interpretations as if they were the same. As Steven Goodman, a professor of biostatistics at Johns Hopkins, pointed out in a series of articles on the issue published at the end of the 1990s in the *Annals of Internal Medicine*: "There is little appreciation in the medical community that the methodology is an amalgam of incompatible elements, whose utility for scientific inference has been the subject of intense debate among statisticians for almost 70 years" (1999: 995).

This interlude has touched on elements of the debate primarily to underscore the historical and practical role of statistics in clinical trials. We

2. In the case of time-to-event studies such as the survival studies discussed in chapter 6, the parameter on which the power and duration of the trial depend is the number of events that are observed.

have presented the notions of random sample and statistical significance in an evolutionary fashion in order to emphasize the fact that they are not timeless concepts. They are embedded in practices whose meaning and use vary as those practices evolve. Consider the problem of sample size. The mathematics for such calculations has been available since the 1930s. And yet just because the mathematics was available did not mean that statisticians and clinicians thought their application necessary. Clinical researchers and statisticians associated with the early cooperative groups in the 1950s did not see the need to settle on a fixed sample size as an absolute and counseled considerable flexibility given the absence of knowledge of effect size when testing anticancer substances. As we will see, this attitude underwent profound transformations in the subsequent decades.

CHAPTER FOUR

Statisticians, Statistics, and Early Cooperative Clinical Trials

Clinical statistics is today a lively specialty just as "clinical trial statistician" denotes a person and a career within the field of biomedicine and biostatistics. Neither of these statements held true at the time of the VAMP trial. Statisticians worked on clinical trials either as outside consultants or as hired hands expected to do a lot of the work that would today be consigned to secretarial staff and data managers. In this chapter we will see how statisticians were transformed into full-fledged collaborators in clinical cancer trials. As collaborators, part of their work consisted in the adaptation of statistical techniques to the exigencies of clinical research and also in the development of original solutions to emerging issues. Previous work carried out on the NCI's animal screen provided a model for the initial work on human trials insofar as clinicians and statisticians understood them as an extension of the animal trials and viewed both of them as a series of experimental screens designed to sort out the relatively good substances from the relatively bad. Only after the human trials became independent from the screening process did clinical researchers and statisticians develop their own problems, their own solutions, and their own style of practice.

Statisticians at the NIH

Statisticians played a key role in the establishment of the different components of the Cancer Chemotherapy National Service Center (CCNSC). Relatively new kids on the block, they had begun working at the NIH just after World War II when Harold Dorn became head of the newly formed Division of Statistical Methods of the US Public Health Service. Known both as an epidemiologist and as a skeptic with regard to the early case control studies that showed a positive relation between smoking and lung cancer, Dorn began recruiting the first NIH statisticians shortly after his appointment (Parascandola, 2004). By 1950 he had installed five statisticians in a single room at the NCI. At a time when university mathematics departments had at most one and often none, five statisticians made the NIH a major center for statistical research. A member of the original cohort, Marvin Schneiderman, recalled that when the group attended American Statistical Association meetings in the 1940s and 1950s it quickly became apparent that their work spanned the field. At one meeting in particular, he observed: "There isn't a subject that these people have been talking about that we haven't worked on" (Simon, 1997: 101).

The NIH statisticians initially worked as consultants for any and all NIH researchers who had statistical problems. As such, they avoided attachment to any particular disease or problem set. In keeping with their consultant status, each statistician was given a desk and a telephone, and since the office was equipped with a single telephone line whoever answered the telephone first became the consultant for the caller regardless of the nature of the problem. The absence of specialization meant that NIH researchers regarded the statisticians as experts in methods, not subjects. As a result, the NIH statisticians became involved in cancer research more by chance than by design. It all began when Samuel Greenhouse, one of the original five statisticians, picked up the phone one day and began collaborating with a well-known NCI laboratory researcher, Abraham Goldin, in a study of how to statistically determine the therapeutic efficiency of anticancer substances (Greenhouse, 1997). So it went in the early years when most of the statistical problems were those raised either in the laboratory or in epidemiology. As a consequence and despite statistical involvement in "mouse trials," interest in human clinical trials did not begin in earnest until the formation of the CCNSC, at which point some statisticians became what might be termed permanent consultants to the cooperative groups. In contrast to the 1960s, when the groups hired their own full-time statisticians, at the end the 1950s a consultant statistician could work for several groups; Schneiderman, for instance, provided statistical support to four groups.

The head of the CCNSC, Gordon Zubrod, had acquired a taste for statistical methods during the Second World War when he had worked on the development of antimalarial compounds using high-throughput screening methods. Together with Jerome Cornfield, yet another of the original five statisticians, Zubrod had participated in the design of the first randomized human trial at the NCI in 1954. Shortly thereafter, clinical trial design received a further boost following a visit from Austin Bradford Hill, the British medical researcher who had organized the first clinical trials in Britain. His presentations so enthused the NCI researchers that they invited him to stay for a year. Unable to accept the offer, Hill proffered one of his students, Peter Armitage, as a possible collaborator. Armitage and the NCI agreed. Shortly thereafter Armitage developed the statistical techniques for the mouse screen described above (Simon, 1997).

In the meantime, statisticians encountered a number of nonstatistical problems in the course of the clinical cancer trials conducted in the 1950s and early 1960s. Clinicians' unfamiliarity with procedures such as randomization and data collection often acted as obstacles to the very acceptance of the notion of a clinical trial. Historians and medical commentators, both then and now, have discussed these issues at length (Marks, 1997). Other more mundane problems associated with *multicenter* clinical trials also created obstacles to the spread of clinical trials. An early discussion of some of these issues appeared in 1960 when one of the original organizers of multicenter clinical trials, Donald Mainland of the Department of Medical Statistics of New York University, drew up a list of things that could and did go wrong in such endeavors. His inventory of pitfalls and quagmires was, in many respects, self-explanatory: "1) choice of investigators, 2) time for planning, 3) realism in planning, 4) carrying out the plan, 5) practice, 6) permanence of investigators, 7) sample size and case loads, and 8) the policeman" (1960: 486). The last and potentially most intriguing problem, the policeman, deserves further elaboration.

The function of the statistician as a policeman was a common trope in the early days of multicenter clinical trials. Unlike the NIH consultant statistician, statisticians working for a clinical research group like one of the cooperative oncology groups or a medical school were less consultants than employees. The idea that they policed clinical trials arose from the fact that they were also responsible for a variety of enforcement operations to ensure the trial's objectivity. Statisticians had acquired this responsibility not through methodological prowess but because they were cheap. As Mainland (1960: 494) pointed out: "A medical school can hire an experienced statistician for less than half the salary of a clinician who would be suitable for the job." Some of these duties would no doubt make present-day statisticians smile, but they also reflect the complex organization that presently underlies the conduct of a clinical trial. In the situations described by Mainland, statisticians had to monitor departures from protocol, missing observations, and deal with all events not foreseen by the protocol. To get some idea of what those events might be, consider Mainland's examples:

1. According to trial protocol, all patients must submit to a biopsy, and a pathologist must identify the cancer as a specific type. Here is the problem for the statistician policeman: a patient has finished treatment, but the biopsy report has still not arrived. How many times should the pathologist be called before removing the patient from the protocol series?
2. A medical secretary from one of the participating centers calls to say that all the drug bottles containing the placebo have actually been labeled "placebo" by the geniuses at the drug distribution center.
3. Patients are supposed to be assessed at the end of the fourth week of therapy. The reports from the mid-west centers participating in the trial have not yet arrived. It appears that the patients have gone out into the fields for the harvest. Should they be labeled as "improved," "worse" or "no change"? (494)

As the examples show, the statistician carried out many of the tasks nowadays distributed throughout the clinical trial organization. Combined with the statistician's potential encroachment on clinical expertise, this unruly mix of the secretarial, the disciplinary, and the analytic may account for the relatively low status sometimes accorded the statistician in clinical circles and for the fact that "we all know that it is not unusual to encounter active hostility to statisticians in clinical groups" (Greenberg & Grizzle, 1969: 1).

In 1961 the NCI held the "First Conference of Statisticians Associated with the Chemotherapy Program." Reviewing the problems confronting statisticians in cooperative group trials, conference participants sought to shift the chemotherapy statisticians out of the consultant and policing modes to become full collaborators in the clinical trial process.

Low morale loomed large as a concern and framed much of the discussion. Five years into the campaign to uncover anticancer agents, few such agents had come to light, and even those deemed promising did not cure. We have already discussed Bernard Greenberg's (1961) presentation during that conference (see chapter 3), portentously titled "Maintaining Interest among Members of Cooperative Clinical Study Groups." Suffice it here to mention that among the morale boosters promoted to remedy the boredom generated by long-lasting multicenter trials, Greenberg proposed regular visits to participating sites to meet fellow researchers. These visits would also afford statisticians an opportunity to "secure missing or deficient data." Such an activity had to be pursued with tact and diplomacy, for while "the statistician is assumed to have particular responsibility for the quality of the data collected," this "responsibility must be shared with other members of the group . . . [o]therwise the role of the statistician degenerates from that of a partner in a scientific investigation to that of a policeman" (Greenberg & Grizzle, 1969: 1).

Statisticians in Paris

We saw that shortly after its founding, GECA published a diagram (figure 1.5, chapter 1) outlining the six steps for testing candidate anticancer drugs. According to the diagram, the group discussions that followed each step were in most cases preceded by a statistical analysis conducted by Daniel Schwartz and his collaborators. Trained as a *polytechnicien* and engineer, Schwartz worked from 1942 to 1959 for the French state tobacco company before joining the National Hygiene Institute (INH [Institut national d'hygiène])—the forerunner of INSERM, the French medical research institute established in 1964—as director of the Statistical Research Unit, and the Gustave-Roussy Institute (IGR [Institut Gustave-Roussy]) as chief of statistical services. Armed with a degree in biology in addition to the mathematical and statistical training from the École Polytechnique, Schwartz became interested in the application of statistical methods to medicine and in particular cancer, conducting a 1954 study on cigarette smoking and lung cancer, followed by an epidemiological study of the causes of gynecological cancers (Berlivet, 2000; 2002; 2005).

With INH financing, Schwartz expanded his IGR unit and purchased a small electronic computing machine. By 1960 the unit contained fifteen members, including two physicians, three statisticians, and seven technicians executing tasks such as card punching. By 1971 the unit had grown to fourteen researchers and thirty-eight technicians, becoming one of the larger INSERM units, and had acquired a UNIVAC.[1] The unit's early work focused on cancer but soon incorporated other domains (cardiology, pediatrics) as well as requests for statistical consultations from a number of hospital services. The services ranged from ad hoc technical support on a specific question to the establishment of more permanent collaborations, including the design of experiments and the analysis of results. In the 1950s and early 1960s Schwartz's team launched multiple initiatives including the introduction of courses in medical statistics in Paris medical schools, all geared toward the development and promotion of statistical approaches in medicine. As an indication of Schwartz's considerable influence in clinical research circles in France, discussion of a late-1960s report on the promotion of clinical research (written by Georges Mathé) included a warning by one of the speakers that "clinical research should not be left in the sole hands of statisticians."[2]

Given that the IGR team had became the hub for French medical statistical work, it came as no surprise that when Schwartz joined GECA as one of the original members, he became the group's leading expert on all statistical matters. In a revealing expression of the exploratory nature of clinical trial work in this period, an undated (early 1960s) report on the team's activities mentions that "in spite of the difficulties of all sorts one encounters in order to conduct a correct clinical trial," the team had managed to design three trials. The team also noted in its 1962 report that given the onerous constraints imposed by clinical trials, hospital services could be convinced to participate only after careful initiation, via other types of studies, into the methods and approaches used by the statistical team.

At the conceptual level, Schwartz and coworkers tried to develop an understanding of the clinical trial process as a whole (Schwartz, Flamant & Lellouch, 1969; 1981). To this end in the 1960s they introduced a distinction between *pragmatic* and *explanatory* trials and described the sta-

---

1. Annual Reports of Unit 21, INSERM Archives (Le Vésinet), Box 42 (1960–1974) and Box 43 (1975–1986).

2. Procès verbaux du Conseil scientifique de l'INSERM, INSERM Archives (Le Vésinet), Box 9449, Folders 1–8, Meeting of 17 December 1969.

tistical apparatus that accompanied them (Schwartz & Lellouch, 1967). Here is how they drew this distinction:

> In the pragmatic approach, the type of patients is defined in advance so as to correspond, on the whole, to the patient population to which the results will be applied. In the explanatory approach, the type of patients can be chosen so as to select the population that is likely to better express a well-defined biological problem. To use a somewhat shocking simplification, one could say that patients are, in one case, treated as ends and, in the other, as means. In reality, the patients' best interests are always the ends—but either the immediate or the long-term ends. (Schwartz et al., 1981: 62–63, our translation)

Another way of explaining the difference is to say that a pragmatic approach tests the effectiveness (i.e., the risks, benefits, and costs) of a given therapy in routine clinical practice, whereas an explanatory approach answers pathobiological questions under experimental (or ideal) conditions. While Schwartz and his colleagues also admitted that "most real problems contain both explanatory and pragmatic elements" (Schwartz & Lellouch, 1967: 647), they also maintained that each approach required a different statistical design insofar as pragmatic trials, for example, aimed at comparing therapies in a heterogeneous population and a variety of practice settings, taking into account a broad range of health outcomes.

The distinction was not motivated by purely intellectual concerns. Schwartz and his colleagues believed that results of the clinical trials conducted for many years by their British and US colleagues had been largely underused in spite of the work invested in these trials. The primary cause of such a poor return on investment lay in the fact that clinical trials implicating humans used methods transposed from animal experiments such as screening. The principal interest of human trials—how to make a clinical decision and not, as in animal trials, how to test a hypothesis—had been overlooked in the process of transposition.

While the French statistical approach to clinical trials might have led to a divergence in the development of clinical trials on both sides of the Atlantic, in the end it did not. As we will see later in this book, the establishment of a data center in Brussels staffed by statisticians trained in the United States led to a fundamental convergence of European and American methods. The French approach, while recognized as distinctive, remained an outlier: discussing the impact of French clinical statistics from the 1960s, a former EORTC data center statistician admitted that although they were aware of the work, it had had little purchase on their research.[3]

Fabricating the Tools of the Trade

Researchers at the NCI had organized the drug screen according to relatively new statistical theories that belonged to a field of statistics known as sequential analysis (Wald, 1945; 1947). Developed during the Second World War, present-day statisticians generally attribute the invention of the technique to Abraham Wald (1902–1950), an Austrian mathematical economist forced to flee to the United States in 1938 (Darling, 1976; Wolfowitz, 1952). In the thirty years following the war, sequential analysis found wide application and by the mid-1970s had earned the epithet of "one of the most powerful and seminal statistical ideas of the past third of a century" (Wallis, 1980: 322). Although highly technical, Wald's work was in part the product of a solitary mathematician and in part the result of work carried out while he was a member of a wartime Statistical Research Group. The group had been organized by Warren Weaver (1894–1978) of the Rockefeller Foundation and Harold Hotelling (1895–1973), the head of the Columbia University Department of Statistics where Wald had found employment upon his arrival in America (Klein, 2000).

Weaver's Statistical Research Group consisted of eighteen statisticians—four of whom would go on to become president of the American Statistical Association—and statistically oriented economists such as Milton Friedman (1912–2006) and George Stigler (1911–1991) as well as a support staff of about sixty that included thirty Vassar and Hunter College graduates in mathematics hired to do the tedious calculations and known as computers (Galison 1997: 199, 375). Concerned with the perennial wartime problems of the quality control of supplies and ordnance and weapons testing, the group's statisticians sought ways to reduce the number of samples needed to conduct a statistical test. In one particular instance, work on acceptance tests for bombsights led the head of the Group, W. A. Wallis (1912–1998), to devise a method of ex-

3. Interview with Marc Buyse, Paris, 25 May 2005.

amining bombsights in a series rather than all at once. This sequential test appeared particularly powerful, in the mathematical sense of the word, and seeking a more abstract formulation Wallis turned his preliminary results over to Wald, a theoretician. Initially skeptical, Wald realized in the process of providing a general mathematical framework for Wallis's calculations that he had developed an approach of fundamental importance. Consequently, barely a month after starting work, Wald contacted Wallis and asked him to write a memo outlining how he had arrived at his original formulation so that when the time came to distribute credit for the discovery, the honors could be properly shared. Wallis complied, thus providing a written record of what, given the wartime circumstances, was supposed to be a secret process (Wallis, 1980).

In his historical memo to Wald, Wallis recounted how early ideas of sequential analysis took form in discussions with Milton Friedman in 1943, in the course of which Wallis became convinced that it would be possible to reduce the number of samples needed in order to apply classical statistical tests "when the observations are treated sequentially instead of *en masse* as in the classical procedure" (1980: 328). Further discussions with army personnel had prompted several ordnance officers familiar with quality control methods developed in industry in the 1920s to suggest to Wallis that he devise "a chart on which results would be plotted continuously, with two lines drawn on the chart in advance. Crossing one would mean rejection, crossing the other acceptance" (328) (figure 4.1). Wallis thus drew on the tradition of quality control

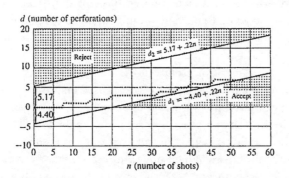

FIGURE 4.1. A 1945 sequential procedure chart for inspecting protective material. From Statistical Research Group, Columbia University, *Sequential Analysis of Statistical Data: Applications.* SRG Report 255, AMP Report 30.2R, New York, Columbia University Press, 1945, 2.24. © 1945 Columbia University Press. Reprinted with permission of the publisher.

statistics and the associated quality control charts that had initially been developed by statisticians like Walter Shewhart (1891–1967) at the Bell Laboratories in the 1920s and 1930s (Shewhart, 1926; Bayart, 2005) and that were widely used in the army (Klein, 2000). As a result, much of the practical work had already been accomplished before the problem and its initial solution was presented to Wald.

This is not to suggest that Wald merely put the formal icing on the practical cake. The fundamental nature of Wald's formulation, drawing as it did on recent work on the theory of statistical inference by Egon Pearson and Jerzy Neyman (see First Interlude), ultimately justified the wide-ranging applications of sequential analysis and served as a bridge for the manifold permutations of sequential analysis in the postwar era (Wald, 1945). In the meantime, the initial and most immediate uses of sequential analysis depended to a large extent on the work of Harold Freeman (1909–1997), professor at MIT and member of the group, who, basing himself on Wald's formulation, wrote an in-house manual outlining the applications of sequential analysis. The manual in turn became the foundation of a series of lectures given to members of the Quartermaster Corps of the US Army who immediately adopted Freeman's version of sequential analysis for their quality control programs (Enrick, 1946). Shortly thereafter the navy followed suit. The techniques of sequential analysis remained, however, classified material throughout the war.

Within several months of Wald's work and in comparable circumstances, British statisticians attached to the Ministry of Supply developed similar techniques (Barnard, 1946). Headed by George A. Barnard, the British group was composed of eight statisticians who, as in America, would go on to become prominent members of the field in the postwar period. We have already met one of them, Peter Armitage. The British work had been prompted by problems similar to those that had confronted the American group. In the British case the specific question was how to reduce the number of tests used on proximity fuses. The quality control tests then in use were considered too expensive in that they required too many tests in order to be applied (DeGroot, 1988). Once again a version of sequential analysis was invented to reduce costs. Although the groups remained unaware of each other's work until the end of the war, the uncanny convergence of the results obtained by the two teams of researchers remains part of statistical lore.

The surveys and extensive bibliographies published in the statistical journals of Great Britain and the United States in the early 1960s

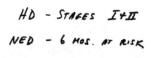

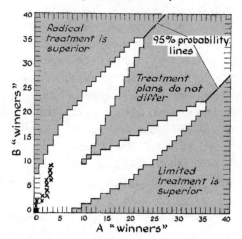

FIGURE 4.2. Sequential chart used by a study initiated in the early 1960s at Stanford University by Saul Rosenberg for the treatment of lymphoma. S. A. Rosenberg (personal files).

attest to the spectacular postwar growth and spread of sequential analysis (Jackson, 1960; Johnson, 1961). Initially confined to the detection of defective samples, in the early 1950s the techniques were adapted to clinical trials. Figure 4.2 is a typical sequential plot similar to the one described in our discussion of Protocol No. 3. This plot charts the results of a study initiated in the early 1960s at Stanford University by Saul Rosenberg for the treatment of lymphoma. As can be seen, the trial stopped before the stopping line was actually crossed.

Screening for pharmaceuticals emerged as a second zone of application of sequential methods in the 1950s. Prior to the establishment of the NCI screen, statisticians in pharmaceutical companies had explored the statistics of screening programs on both sides of the Atlantic. In Britain at Imperial Chemical Industries (ICI), O. L. Davies was among the first to adapt sequential methods to the problems of drug screening (Davies, 1958). The advantages of Wald's sequential methods in drug screening were not immediately obvious to practitioners. As J. T. Litchfield at American Cyanamid pointed out in a review of the issue: "It may well appear that there is no real difference between the sequential method and conventional methods for screening. In principle this is correct" (1960: 479). The only difference, according to Litchfield, lay in the ability of the sequential approach to lower costs. As we saw in the First Interlude, when screening for pharmaceutical compounds, researchers confronted two pitfalls. They could mistakenly reject an effective compound for a number of reasons that have nothing to do with the effectiveness of the compound. These *false negative* errors are unavoidable in any kind of screening or testing set-up. Equally as unavoidable are *false positive* errors: the screener could be led astray and mistakenly select a compound that does not have the sought-after effect. Sequential methods make these two types of errors visible and quantifiable. The probability of these errors varies according to the size of the sample used to conduct the experiment. The larger the sample, the less likely either error will occur. Since sequential methods seek to optimize sample size, the usual procedure is to set the size of the probability of making an error that one is willing to accept and then determine the appropriate sample size. The opposite procedure is also possible: determine the sample size that one is willing to pay for and then calculate the probability that errors will occur. In either event, sequential methods triangulate the number of samples and the two types of error.

When the CCNSC set up its drug screening program, they turned to the Lederle division of American Cyanamid "for their advice and help" (Schneiderman, 1961: 233). The statisticians in charge of the NCI screen, Marvin Schneiderman and Peter Armitage, worked out the mathematics of the multistage screening system, first calculating that a "two-stage scheme would give 10% 'false negatives,'" which seemed "too high a price to pay" (Schneiderman, 1961: 239). A three-stage scheme gave half as many false negatives and was in the end adopted by the CCNSC. Sequential analysis thus enabled the CCNSC to minimize the number of observations that had to be made based upon predefined criteria of acceptable levels of error (George, 1980). In this respect Wald had formulated statistics in a way that viewed the calculations as a means for making decisions under conditions of uncertainty with minimum cost. One can see how such techniques would recommend themselves for use in a massive, brute-force undertaking such as the screening of compounds for anticancer properties.

That such techniques could or should be used in human clinical trials, however, was not as immediately evident. Those who conceived of clinical trials as scientific experiments rather than industrial tests tended to think that while perhaps appropriate in industry, such techniques should not be used in the analysis of clinical trials.[4] Who should be in charge of deciding whether a given trial merited pursuit or had meanwhile turned into a futile exercise, and who should weight its results: a clinical researcher or the impersonal mechanism of a statistical device? Some saw sequential methods as an answer to one of the ethical problems posed by human experimentation—namely, how to conduct experiments as efficiently as possible when humans are at issue. As Gordon Zubrod noted at a 1959 conference: "I think the sequential design technique . . . is one of the possible solutions to the *moral* problem of discontinuing a study as soon as a significant difference appears" (1960a: 288; our emphasis). As we will see in the next section, not everybody shared Zubrod's enthusiasm.

Controversial Methods

Despite their ubiquity and the fact they had already entered standard statistical textbooks such as David Cox's *Planning of Experiments* (1958), the use of sequential methods in clinical research became somewhat of a cause célèbre when Peter Armitage published his *Sequential Medical Trials* in 1960. One review by the Princeton statistician F. J. Anscombe, a former member of the British wartime statistical group and thus a former colleague of Armitage, came to be cited as often as the book itself in the 1960s and 1970s. Such was the force of the critique that in the second edition of the book, Armitage felt obliged to devote an entire chapter of the book responding to Anscombe.

Anscombe (1963) had criticized Armitage's statistical sequential rules that determined how to calculate the boundaries of the charts such as

4. The opposition between Wald and Fisher can be viewed in even wider terms as part of the long-standing opposition between frequentist (Fisher) and Bayesian (Wald) views (Howie, 2002). For present purposes, the scientific experiment versus industrial test divide is sufficient to understand the terms and consequences of the debate.

the one displayed by figure 4.3. According to Anscombe, the boundaries were "simultaneously two things, stopping rules and decision rules; that is, they indicate how long the observation process should continue, and they also indicate what verdict should then be given" (371). The duality of the rules was quite appropriate when testing military ordnance or other products of industrial production, for here one sought a rule that offered both a reason to stop testing and a decision as to whether or not to accept the batch under scrutiny. In the field of scientific experimentation, Anscombe argued, researchers separated the reason to stop the test from the conclusion to be drawn from the results of the test. Specifically: a reason to stop the trial did not necessarily result in a decision to accept or reject the therapy under trial. Nonstatistical, clinical reasons could come into play. Perhaps a given therapy gave the same number of cures or remissions as a competing therapy, but was milder. Anscombe believed that the trial constituted evidence that had to be weighed and evaluated by experts as one piece of evidence among others. It did not by itself entail a decision. Anscombe then drew a contrast between science and industry, distinguishing between two kinds of repetition:

> There is indeed repetitiveness in science, but it is more subtle than that of industrial inspection. If an experiment shows some interesting result, that result will no doubt be confirmed by later experiments, but quite possibly not by any simple repetition of the first experiment. (369)

Viewed in this light, clinical trials in medicine produced results whose meaning depended in part on their relation to the results produced in past and future trials. More specifically, since each clinical trial was different, the data produced were entirely different in nature from data obtained by testing a homogeneous batch of commodities. Clinical trial data sampled the field of human variation in one of an almost infinite variety of ways in the sense that many different protocols were possible to answer a wide range of different questions about the same two treatments. To repeat: the data and their statistical analysis provided evidence for a decision about the effectiveness of a particular treatment; they did not automatically say what the decision should be.

As one of the founders of clinical cancer trials at the NIH, Jerome Cornfield, put the point: "The question is this: Do the conclusions to be drawn from any set of data depend only on the data or do they depend also on the stopping rule which led to the data?" (1966: 18). Both Anscombe and Cornfield believed, for statistical reasons, that the statistician qua experimentalist retained final judgment about the interpretation of the data; that in the end, only the data, not the stopping rule, mattered. Consequently, as far as sequential medical trials that used stopping rules—the preset limits described above—were concerned, Anscombe, like Cornfield, believed:

> Sequential analysis is a hoax. . . . The experimenter should feel entirely uninhibited about continuing or discontinuing his trial, changing his mind about the stopping rule in the middle, etc., because the interpretation of the observations will be based on what was observed, and not on what might have been observed but wasn't [e.g., a preset number of cures]. (1963: 381)

Armitage's sequential methods and their use in Protocol No. 3 fueled more general concerns regarding the extent to which statisticians should be able to interpret the data generated by clinical trials. Were the statisticians simply consultant/technicians who handed off the results of clinical trials in binary terms: yes or no? Or, did statisticians, working full time with the cooperative groups, acquire a form of expertise that meant that as members of a research collective—as collaborators, not consultants— they participated in trials resembling experiments that in turn generated evidence whose meaning required interpretation? Briefly, should they, as collaborating researchers, feel free to interpret the evidence? To broach these issues, a group of NIH statisticians organized a seminar in the summer of 1965. Here is a general statement of the type of problem that confronted statisticians associated with clinical cancer trials:

> The investigator starts with a question which is reformulated in the form of a null hypothesis that treatment A is no more effective than treatment B, where treatment B refers in general to another treatment or control. He then sets the error risks . . . of rejecting the hypothesis when it is true or accepting it when it is false, determines sample size, selects subjects who are then randomly assigned to two treatment groups, observes responses to the treatments, analyzes the data, and concludes whether treatment A is or is not better than treatment B. . . . The purpose of this symposium is to explore the extent to which an expensive and complicated experiment like a clinical trial should rigidly conform to this procedure. (Cutler et al., 1966: 859)

In practice, few investigators (statisticians included) conformed rigidly to this simplified view of the clinical trial process. Rather, the more pragmatic dilemma discussed at the meeting concerned how far could the clinical trial process drift from idealized mathematical procedures into experimentalism and still maintain credibility. Not only were a number of positions expressed at the meeting, but also the answers to this question—which obviously contained a host of subsidiary interrogations—evolved over the years. Consider the following example. At the time of the conference, the initial choice of sample size followed fixed rules and few serious clinicians would consider changing sample size in midtrial except in extraordinary circumstances. Yet, here is what Bernard Greenberg had to say about sample size in 1959:

> In most cases of cooperative investigations, the use of a fixed sample size will be the logical procedure to follow. Although the sample size may be fixed in advance, it is not necessarily good practice to rigidly adhere to this sample size as if the number were sacrosanct. There is even a strong likelihood that this fixed sample size was determined out of ignorance about the magnitude of the variation to be encountered. (17)

Such attitudes did not survive the 1960s. As the published discussions of the statisticians' meeting shows, however, they were not replaced with a slavish devotion to statistical rules. As we will see in part 2 of this book, the adoption of sequential methods and related techniques for testing the significance of statistical data in clinical cancer trials entailed the adaptation of those techniques to a research context; in other words, statisticians who functioned as research collaborators would find it necessary to develop techniques adapted to the practical context of their work.

One of the principal subjects of discussion at the 1965 symposium was Protocol No. 3. Marvin Schneiderman opened the discussion presenting the raw clinical trial data gathered during that trial. As he explained, and as the trial statistician (Edmund Gehan) confirmed in the course of the subsequent discussion, the trial had been set up according to a sequential design and the statistician had adhered to the stopping rules as derived from Peter Armitage's theoretical calculations. Gehan, however, and this is where the practical world of clinical cancer trials intervened, in addition to collecting data directly related to the hypothesis—Did the patients go into remission?—had collected a second type of information: How long did the remission last? The second datum was not, strictly speaking, entailed by the trial hypothesis and as such, according to the rules of sequential analysis, could not be used to decide whether or not to stop the trial. Before sufficient data on remission had accumulated for the stopping rules to kick in, the data on length of remission had already reached the point where Gehan had become convinced that the answer to the larger clinical question—Which treatment was better?—had already been answered. In spite of this, Gehan stuck to the sequential trial protocol and consciously ignoring the length of remission evidence allowed the trial go to the conclusion dictated by the stopping rules (and, as we have seen, slightly beyond) (Cutler et al., 1966).

Following the initial presentation of the data, Schneiderman pointed out that he and other cooperative group statisticians like Marvin Zelen had used the data as teaching devices in their clinical statistics classes. Specifically, they had each performed a statistical game with their students, clinical researchers at the NIH, by presenting them with the *two* sets of data—the number of remissions and the duration of each remission—in sequential form so as to reproduce Gehan's position with regard to the data. They then asked the students to say when they had seen enough data to stop the trial. In almost all cases the students stopped the trial well before the calculations laid out in the stopping rules would have stopped the trial. Schneiderman thus concluded his presentation to the symposium with the following rather devastating remark with which, in discussion, Gehan agreed:

> The strict adherence to the hypothesis-testing routine delayed a decision longer than an overwhelming majority of a sophisticated group of people interested in clinical trials would have taken to stop. It thus appears that the best available experiment design for early stopping was intuitively dissatisfying, because it led to too long a trial. . . . The results of this game imply to me that only statistical tunnel vision insists on the strictly limited (limited by the specific hypothesis previously announced) collection of data confined only to what the hypothesis "needs" for its testing. (Cutler et al., 1966: 871)

Marvin Zelen expressed similar sentiments contrasting statistics as taught with the real world of statistical practice:

> The student learns that statistical methods consist of a body of formulas and fixed sets of rules, which once memorized can be used throughout one's lifetime in drawing inferences from data. We've learned one has only to deter-

mine whether to reject at the 5 per cent or 1 per cent level [of significance]. Then the statistician can grandly draw obvious conclusions about data from any scientific field by proclaiming significance or non-significance. Such nonsense is usually taught by professors who have had minimal contact with the application of statistical methods to scientific problems. (Cutler et al., 1966: 873)

The theme of these discussions is quite clear. Statisticians working within the cooperative groups were not simply technicians hired to crunch numbers. Knowledge of the field in which they worked did make a difference in how they evaluated and analyzed the data. Knowledge drawn from clinical research itself, evolving knowledge of the disease and its treatment, for example, had to be explicitly integrated into the way the statisticians handled the data and drew conclusions from that data. Statisticians, in other words, were researchers who used the tools of statistics without being tools of statistical rules and methodologies. The matter did not end here. As we will see in part 2 of this book, group statisticians would ultimately open themselves up to accusations of caring too much about the data they treated; having moved from a generalized disinterest in the data per se as all-purpose statistical consultants to full-fledged members of clinical cancer researcher teams with both a knowledge of and proprietary interest in the data, statisticians came to be accused of entertaining a conflict of interest with regard to the data.

The Emergence of the Phase System

The sequential methods debate concerned the design and performance of clinical trials in the 1950s and 1960s. While sequential methods subsequently faded from clinical trial practice, ultimately to be replaced by techniques such as the *group* sequential methods, another statistical innovation introduced at this early stage remained one of the hallmarks of the cancer clinical trial apparatus: the categorization of trials into different *phases*. Today's canonical distinction between Phase I, II, III, and IV trials first emerged and evolved within the cooperative group system. Although eventually adopted on both sides of the Atlantic, it was not without competitors. In chapter 3 we described the two standard protocols published by GECA in 1966 that set out two types of studies that divided trials into phases. While these Type I and Type II studies framed much of the early work of GECA/EORTC trialists during the 1960s, such was the weight of the American phase system that toward the end of the decade EORTC officials began wondering how their own categories compared to the US phase system, and soon after aligned their terminology with US terminology.[5]

The origin of the American phase system can be traced back to the fact that the NCI saw the clinical cancer trials as an extension of the screening process. The transition from the animal screening system to human clinical trials formed a seamless whole. Following animal screening, the CCNSC organized the human experiments into a series of increasingly larger clinical trials so as to constitute a sort of human screen for the drugs uncovered by the animal arm of the program. While this observation should be self-evident thus far in our study, it has been obscured over the years by the misleading idea that the present-day organization of clinical trials into a sequence of Phases I–IV trials in the United States was a by-product of the 1962 Kefauver-Harris Drug Efficacy Amendment. The amendment made clinical trials of any new drug or new use of an old drug mandatory for approval by the US Food and Drug Administration (FDA) and mandated the four phases of clinical trials as the path that had to be followed to obtain that approval.

This much is true: The Kefauver-Harris Amendment followed hearings precipitated by the FDA's concern over patient safety in the wake of the thalidomide scandal of the early 1960s. The central doctrine incorporated into the amendment stipulated that no drug is truly safe unless it is effective and that clinical trials are the only way to demonstrate drug effectiveness. The historical image of the amendment, largely diffused through the ethics and legal literature, portray them and the resultant phase sequence as regulations whose main purpose was and remains patient protection (Beauchamp & Childress, 2001). The idea that the invention of a phase system for drug approval was precipitated by concerns for patient safety inverts the historical sequence of events.

Less the beginning of a new era, the 1962 amendment signaled the end of a period of significant reform and innovation in the FDA. Recent work has shown, in fact, that despite persistent portrayals of the 1950s as a sort of dark ages, the agency underwent fundamental changes dur-

5. M. Staquet, "Summary of the Minutes of the Meeting of the Chairmen and Secretaries of the EORTC Cooperative Groups (Bruxelles, October 1970)," *EORTC Newsletter*, no. 8, January 1971.

ing those years. Members of the agency, most notably recent appointees in the Bureau of Medicine, had adopted the new "clinical pharmacology" as practiced by the cancer chemotherapy pioneers Louis Goodman and Alfred Gilman and as codified in their *The Pharmacological Basis of Therapeutics* (1941). This text focused on animal experimentation, molecular chemistry, and human clinical trials as the basis for the evaluation of therapeutic agents. In addition, the 1950s saw the agency become increasingly concerned with linking toxicity testing to its original function, therapeutic efficacy. In other words, prior to the major reforms of the early 1960s, the FDA had already embarked upon a course that would make it receptive to the innovations developed in the field of clinical cancer research. More specifically, the agency had become particularly open to the notion of a sequence of clinical trials that would bring together both the evaluation of toxicity and therapeutic function (Carpenter, 2010).

The CCNSC introduced the practice of combining toxicity and efficacy testing within a single series of human clinical trials in the late 1950s. They did so in order to speed the flow of compounds from the animal screen through the human screen. The earliest definition of these different phases of human drug testing appears in the first five-year report (1958–63) of the Eastern Cooperative Group in which the group mentioned that during the period 1955–58, it had "evolved three types of sequential clinical drug trials."[6] The first type, Phase I studies, sought "information relative to the absorption, excretion, pharmacologic disposition, and toxicity in the human." Although qualified as a secondary objective, Phase I studies simultaneously tried to determine whether or not the therapeutic potential expressed in animal tumor systems in the NCI animal screen could be maintained in the humans. From the beginning, then, Phase I trials were endowed with some small measure of therapeutic promise if only because the compound studied had shown promise in animal trials. The second type, Phase II studies, characterized as "essentially screening operations," tested substances in a number of different human tumors. Like Phase I studies, Phase II studies were also two-step operations in which "compounds with a clearly demonstrated low grade of effectiveness are discarded. Those which demon-

6. "Research Plan," Overall Group Research Plan and Progress Report, May 1958 to May 1963, for the Eastern Cooperative Group in Solid Tumor Chemotherapy, 2, Frontier Science and Technology Research Foundation Archives (Boston).

strate a significant effect are studied further with particular emphasis placed on delineating their optimal clinical usefulness."[7] The third type of trial, Phase III trials—"definitive therapeutic trials"—compared treatments. Thus long before the Kefauver-Harris Amendment, the cooperative groups of the CCNSC had organized a sequential clinical trial system within the field of cancer. The system, moreover, sought efficiency in screening. Patient protection through the use of smaller samples of patients was the result, not the cause.

The different phases evolved over the years before assuming their final form; in other words, they were yet another result of tinkering within the new style of practice. Phase I and II trials, for example, had direct connections with animal laboratory studies. Phase I trials, in particular, initially bridged the animal screen and the human screen and their protocols and thus, unlike Phase II/III protocols, fell under the purview of the New Agents Subcommittee in Cooperative Groups rather than the Protocol Committee charged with the oversight of human clinical trial protocols.[8]

The phase terminology grew out of an earlier and for a time parallel vocabulary that may be aptly described as planning terminology. A short description of the planning terminology shows the extent to which the early days of the chemotherapy clinical trials constituted not only a search for useful drugs but also a search for useful methods to determine how and under what circumstances a potentially therapeutic agent could move from one phase or stage to another. Edmund Gehan, ALGB group statistician and head of the CCNSC's Biometrics Section, offered the clearest definition of planning terminology in a 1961 article. Symptomatic of the overlapping nature of the planning and phase terminologies, ECOG's five-year report (1958–62) cited above refers the reader to Gehan's article as a source for its Phase II methodology even though the paper does not contain any mention of Phase II or of any other phase for that matter. Instead, what we find in Gehan's article is the description of a larger and somewhat looser process than what would these days be described in terms of phases.

The ostensive purpose of Gehan's article was to introduce a sequen-

7. Ibid.

8. "Brief History and Structural Development of the Eastern Cooperative Group in Solid Tumor Therapy," Overall Group Research Plan and Progress Report, May 1958 to May 1963, for the Eastern Cooperative Group in Solid Tumor Chemotherapy, n.p., Frontier Science and Technology Research Foundation Archives (Boston).

tial method technique known as double sampling into the design and analysis of a kind of clinical trial that might retrospectively, and with some imagination, be termed a Phase II clinical trial. In order to adapt the technique to Phase II clinical trials, Gehan had to describe the clinical trial process as a whole and thus began by describing the planning system then in use. The system consisted of four types of clinical trials: early, preliminary, follow-up and comparative trials (figure 4.3). While there are similarities among these four different types of trial and the subsequent ECOG/FDA phases, appearances are deceptive. In particular, the first trial was not really a clinical trial but a clinical experiment, and the subsequent two trials were more like different stages of a single trial than independent trials. In other words, apart from the final comparative trials, the planning system differed substantially from the phase system that replaced it.

The Eastern cooperative group conceived of their trials as part of a process that began with data concerning new agents, usually generated by the NCI's animal screen, and that led ultimately to the use of the agent in standard therapy. Gehan divided the process into planned and unplanned stages. The first consisted of all the animal and human experimental work that led up to and included a first, early trial. In this type of trial, by varying the doses of the compound under investigation in different types of patients, the investigator tested both the effectiveness and toxicity of compounds. While we might be tempted to equate

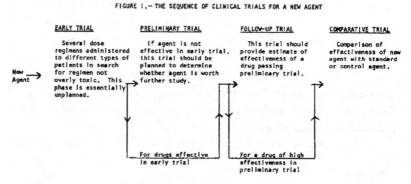

FIGURE 4.3. The sequence of clinical trials for a new agent. Reprinted from E. A. Gehan, "The Determination of the Number of Patients Required in a Preliminary and a Follow-Up Trial of a New Chemotherapeutic Agent," *Journal of Chronic Diseases* 13 (1961): 348. © 1961, with permission from Elsevier.

an early trial with a modern Phase I trial, the mixed finalities—effectiveness *and* toxicity—and their equal importance make such identification impossible. Moreover, unlike a modern Phase I trial, the early trial did not produce data that could be submitted to a statistical test; the early trial resulted in collections of clinical observations and there was consequently no way to evaluate the significance of the data from a mathematical point of view.

Looking at figure 4.3, we see that if an agent proved effective in an early trial, at least in the mind of the clinician-observer, and was not overly toxic, then it skipped the next preliminary trial and moved directly to the third type of trial, the follow-up trial. Good clinical data could, in other words, be used as a substitute for a clinical trial in the early stages. If, on the other hand, an agent proved ineffective, other options opened up. A simple lack of toxicity sufficed to shift the agent to the next stage, the aforementioned preliminary trial (Gehan, 1961). Here researchers examined agents that had not shown clinical effectiveness in a planned study with a view toward minimizing the chance of mistakenly concluding that the drug was not effective when indeed it was or, as Gehan put it, reducing the chance of failing to send an agent to further trial, when it should have been. Recall that a similar set-up had first been instituted in the animal screen.

The first issue Gehan confronted in this respect was how to determine the sample size needed to produce statistically significant numbers in the two sequential trials. As we have already seen, the larger the effect of the agent tested, the smaller the number of patients needed to uncover it. In cancer therapy, expectations were rather low given that no truly effective anticancer compounds had yet been found. Thus as Gehan admitted, the setting of the expected level of effect was "arbitrary" given that:

> the way the plan works depends upon the unknown distribution of percentage effectiveness among the compounds that will be screened. The sampling plans specified here are designed to reject completely ineffective drugs quickly and send on to further trial those that possess some degree of effectiveness. (351)

Still, in order to determine sample size, the clinician and the statistician had to decide what size of therapeutic effect could be reasonably posited: what level of effect would an agent have to express in order to be of interest? This is where the clinician/statistician interaction became crucial. And this is how Gehan imagined the negotiation:

> The usual statement to the consulting statistician is that the agent is of interest "if it's effective at all." The statistician then points out that such a trial is impossible to plan, but that if a drug of 1 per cent effectiveness is being looked for, it will require a very large number of patients. The physician states that this number is impossible and then revises his original statement to something like the following: "I am interested in a new drug that can provide definite benefit to one patient in five or more." The statistician then has something to work with and might specify an initial sample of 11 or 14 patients . . . The point is that the physician might not be at all unhappy if his sampling plan had a high probability of sending on to further trial a drug of 5 per cent effectiveness. (351)

Gehan describes a pragmatic adaptation of the rules of statistical tests in the sense that setting the significance level and/or setting the sample size could both be subordinated to a practical concern—namely: How many patients are truly available for the trial? As regards the status of the statistician in the clinical trial process, two possible roles emerge. First, we might view the statistician as an external consultant as originally prescribed by the NCI. The preliminary trial, like the screening of substances with rodents, was organized in such a way so as to reduce the chances that a useful drug might be rejected, not to compare therapies or to determine the effect of a particular therapy on a specific form of cancer. In this respect the preliminary and follow-up clinical trials functioned *primarily* as tests, and the statistician acted *primarily* as a consultant in a testing procedure. Under this description the statistician did not need any specific knowledge of the disease in question. Setting the sample size, however, amounted to a statistical decision and to a clinical/experimental decision that depended on the state of knowledge in the disease domain in question and that required considerable interaction between the statistician and clinician. In this case the statistician acted as a collaborator with regard to the interpretation of the meaning of the data.

Given the mixed objectives of seeking to determine both toxicity and effect, we cannot say that the trials described by Gehan would today fall under Phase I, Phase II, or even Phase I/II designations. Given the loose terminology, even at the time it was not always clear what kind of trial investigators conducted. In 1962, for example, ECOG reported a "preliminary trial" also labeled as Phase I with forty-five patients. As the number of patients shows, the trial did not conform either to Gehan's specifications of a preliminary trial or to present-day Phase I standards. The trial examined four different dose schedules of a compound that had shown some promise but also expressed some toxicity, testing them across twelve different solid tumor sites. The group recorded objective and subjective responses. Statistics were clearly powerless in this unplanned trial in the sense that none of the categories generated numbers large enough to be subjected to significance testing or error measurement or even the calculation of an average. In the two cases of cancer of the esophagus studied, for example, there was but one objective and one subjective response.[9]

Trials like this became relatively rare in the years that followed as stricter techniques and divisions such as the phases/planning stages became standard. By 1963, the year the FDA adopted the phase terminology, ECOG summarized the results of its activities in terms of the phases using an added category known as "Preliminary Studies." Given the experimental and exploratory nature of the group's work it should be no surprise that in 1963, over 50% of patients treated by the group were treated in the context of preliminary and Phase I studies. Likewise, from 1963 to 1968, an average of 55% of patients fell under similar categories.[10]

A number of protocols set up by different cooperative groups further illustrate the informal drift from a Phase I into a Phase II study and thus the emergent nature of those distinctions. When the Central-Western cooperative group in the mid-1960s set out to study a steroid compound in eighty-one patients with eight different tumor types, more than half of the patients suffered from malignant melanoma. Why? According to the authors of the protocol: "The large number of patients with melanoma in the series is a reflection of the enthusiasm of investigators to increase the value of this Phase I study by pursuing evidence of objective tumor response (Phase II) during the study" (Johnson et al.,

9. "Preliminary Trial of Methyl-glyoxal-bisguanyl-hydrazone Intramuscularly in Solid Tumor Patients (Phase I)," Overall Group Research Plan and Progress Report, May 1958 to May 1963, for the Eastern Cooperative Group in Solid Tumor Chemotherapy, Frontier Science and Technology Research Foundation Archives (Boston).

10. Eastern Cooperative Oncology Group, *Chairman's Report, 1963–1968*, 144–45, Frontier Science and Technology Research Foundation Archives (Boston).

1966). In other words, what had begun as a Phase I toxicology study became, based on data that emerged in the course of the study, a Phase II efficacy study targeting a specific type of cancer. Another protocol from the 1960s had the stated intention of seeking to "define the qualitative toxicity of Hydroxyurea and determine the approximate dose level."[11] To do so the protocol randomized 110 patients according to four different dose sizes. By modern standards that number of patients would have categorized the study as Phase II. Why so many patients? Once again, the protocol tested not only for toxicity but also for efficacy by varying tumor types. Within each dose size, the varying tumor types included chronic myeloid leukemia, colon cancer, breast cancer, lymphoma, and kidney tumors. While the study was in principle categorized as Phase I, the authors openly acknowledged that they were just a step away from screening:

> This kind of study easily merges into a broad screening program against a wide variety of human tumors through minor alterations in the protocol by adding patients at the dose level(s) and schedule(s) that is (are) tolerated and associated with some biological activity. (Carbone et al. 1965: 24)

Conclusion

Unlike Phase I trials, Phase II trials, in our present understanding, have an explicit curative intent insofar as they are designed to assess the activity of the agent or procedure. In order to have curative intent, however, one must have a curative agent. The early years of the NCI's cooperative program generated numerous agents, but not many that by themselves could be considered curative. The situation changed radically by the early 1960s with the emergence of therapeutic combinations of agents (both chemotherapeutic and otherwise) or regimens. Phase I trials increasingly fell under the purview of individual institutions and thus individual Institutional Review Boards. Within the cooperative groups, Phase II and Phase III trials consequently moved to the fore, and by the mid-1970s groups like ECOG no longer performed Phase I studies. The

11. Overall Group Research Plan and Progress Report, May 1958 to May 1963, for the Eastern Cooperative Group in Solid Tumor Chemotherapy, n.p., Frontier Science and Technology Research Foundation Archives (Boston).

problems associated with such trials consequently disappeared (temporarily) as a concern within the groups.[12]

While present-day definitions of Phase II/III trials prescribe therapeutic intent as the primary purpose of the trial, that intention can be framed in a number of ways that introduce major varieties of the Phase II/III distinction. Briefly put, like Phase I trials, Phase II trials can be viewed as a form of screening. The point of a Phase II study was defined at the time as "identifying significant antitumor activity in any type of human tumor."[13] While there is no doubt a difference between testing the efficacy of a curative agent and looking for such an agent in some sort of screening process, absent any curative agent the difference is moot. In other words, at the beginning of the chemotherapy program, tests for an effective agent constituted a search for an agent and thus a human screen that followed up on positive results obtained in an animal screen. While human screening would be ethically unacceptable today, that is only because effective agents now exist and thus create a background against which one may carry out a test of effectiveness. Once again, with the emergence of effective *combinations*, the research-therapy mix involved in Phase II trials moved significantly toward the therapeutic end of the spectrum. As we have seen with the VAMP trial, the decisive move in this direction came in the mid-1960s when clinicians switched from testing substances to evaluating regimens.

12. Phase I trials returned to the cooperative group research agenda in the 1990s with the proliferation of new biotherapeutic agents.

13. "Research Plan," Overall Group Research Plan and Progress Report, May 1958 to May 1963, for the Eastern Cooperative Group in Solid Tumor Chemotherapy, 2, Frontier Science and Technology Research Foundation Archives (Boston).

CHAPTER FIVE

Criticism and the Redefinition of Clinical Cancer Trials as an Autonomous Form of Research

B y pivoting attention away from substances to the regimens deployed within trial protocols, the VAMP trial exemplified a new way of doing clinical trials as drug screening gave way to the testing of therapeutic regimens within clinically defined subspecies of cancer. The term *regimen* refers to dispositions of substances specified not only according to the quantities administered and the modes and schedules of administration but also to the type and number of substances used. In this regard two different regimens may deploy the same substance while using different routes of admission (e.g., oral versus intravenous) or dose size. Similarly, two different regimens may use different combinations of substances. Although regimens are disease specific, clinical trials constantly redefine the disease stage or subtype against which a particular regimen has been devised. The redefinition of disease through clinical trials may sometimes lead to the production of new nosological categories, as when a given cancer is shown to consist of two distinct pathobiological entities that respond differently to similar treatments. Furthermore, as we saw, Protocol No. 3, the forerunner of VAMP, innovated in disease definition by treating different phases of the disease in an individual as independent events, thus allowing multiple interventions to be made on the same person with ostensibly a single disease.

Although relatively unique in that it eschewed common drug screening practices for an all-out assault on a single species of cancer, the VAMP approach became the norm in the years that followed, a transformation that coincided with a reformation of the epistemic status of cancer trials. Beginning with the task forces set up in 1963 (discussed in chapter 2), a series of reforms initiated shortly after VAMP culminated in the mid-1960s with a complete overhaul of the CCNSC. Human trials ceased to be seen as a mere extension of the CCNSC and became an autonomous form of clinical research endowed with their own grant committees. Below we describe how the chemotherapy program adjusted to this new reality. First we discuss the criticisms that dogged the program from the outset.

Clinical Cancer Research under Fire

As much energy was expended in the scientific community to stop cancer drug development as was exerted to get it off the ground. —DeVita, 1984: 1

As we saw in chapter 3, the open-ended nature of cooperative research initially led trialists to adopt learning by doing: "the goals of individual studies in [the] cooperative plan were selected as much to achieve familiarity with this kind of clinical investigation as for the clinical investigation itself" (Ravdin, 1960: 6). Critics of the program questioned the value and rationale of the whole endeavor. As a result the chemotherapy program was quickly caught up in a web of arguments that confronted opposing values and approaches: scientific creativity versus bureaucratic procedures, individualism versus collective or quasi-parliamentary decision-making, and mindless empiricism versus rational research design. As a result the program had a rough start. Sidney Farber (1960: 294) observed: "when this program began, there was a period of hard work, anxiety and concern over the use of new conceptions in research in the field of cancer," and Kenneth Endicott (1958: 171) similarly noted that "even now, two years after the program operations began, the program is questioned from time to time by segments of the academic world, the pharmaceutical industry, and the medical profession." Criticism continued unabated in the subsequent years, leaving program supporters on the defensive. Summing up the situation in 1959, the head of the clinical panel admitted that "there are those who have stated that . . . the chemotherapy program . . . is an irrational approach to the major problems of malignant diseases" (Ravdin, 1960: 2).

An analogous situation prevailed in Europe where the establishment of GECA failed to garner universal applause. Confronted by those who held that chemotherapy did not qualify as a true scientific endeavor and who considered the pursuit, at best, an empirical undertaking tainted by the commercial interests of pharmaceutical companies, GECA countered that not only did chemotherapy achieve positive results in cases where surgery and radiotherapy failed, but also that the discovery of chemotherapeutic agents increasingly mobilized fundamental research in biochemistry, organic chemistry, and molecular biology and in turn contributed to the development of these sciences.[1] The chair of the CCNSC Clinical Studies Panel had mounted a similar defense of the chemotherapy program by arguing that "further extension of surgical and radiation therapy alone [would] not greatly improve the presently accepted end results" (Ravdin, 1960: 1).

Remarks such as these might lead one to believe that the therapeutic specialties had been engaged in some kind of turf war. There is indeed some evidence to that effect. A French clinician, for example, recalled how the arrival of the chemotherapy "intruders" in the 1960s led surgeons and radiotherapists to abandon their 1950s rivalries and join forces against the newcomers (Tubiana, 1991: 303). Yet the CCNSC Clinical Studies Panel chair was a surgeon, and the fact that the aforementioned French clinician was one of a number of radiotherapists who joined the EORTC leads to a different conclusion: confrontations between specialists were often less a clash of specialties than a matter of different individual trajectories.

In the United States, the oft-criticized empiricism of the chemotherapy program and of the clinical trials conducted by the cooperative groups led to a series of investigations and reports in the late 1950s and early 1960s. The first investigation began in 1959 when the National Advisory Cancer Council, the ultimate advisory body for the NCI, invited the director of the Institute of Cancer Research at Columbia University, Alfred Gellhorn, to appear before it. Gellhorn, it will be recalled, had conducted the original studies that led to the adoption of the murine screening system. Despite his early implication in screening programs, he charged that most of the CCNSC studies were contradictory, trivial, or better carried out by individual investigators. In short, he suggested that the NCI money would be better spent on basic research (Gellhorn, 1959). As an example of the low esteem in which the program results

1. Le Groupe Européen de Chimiothérapie Anticancéreuse (GECA), undated brochure (ca. 1964), EORTC Archives (Brussels).

were held and the lack of cooperation between the cooperative groups, Gellhorn pointed out that even though the Eastern Solid Tumor Cooperative Group (later ECOG) had concluded that 5-fluorouracil was ineffective against human tumors, not only had researchers at the University of Wisconsin found the opposite but also that "despite the pronouncement of the 'Cooperative,' 5-fluorouracil is being tested throughout the length and breadth of the land by independent investigators and by other units of the clinical cooperative program" (7). In 1962, in an article polemically titled "Clinical cancer therapy in a single institution," he noted that in his group "cancer chemotherapy [had] been a part of the overall studies of the tumor-bearing host rather than a primary objective" and restated his "conviction that greater understanding of cancer and ultimately effective therapy will result from study of the effect of neoplastic disease on host physiology and from broad inquiry into regulating mechanisms of normal and abnormal growth *rather than merely from the evaluation of a large variety and great number of potential anticancer compounds*" (Ultmann, Hirschberg, & Gellhorn, 1962: 262, our emphasis).

In their defense, chemotherapists argued that even though tumors had not responded "with sufficient regularity" to the limited number of available anticancer drugs, some compounds had showed "some effect upon . . . previously unaffected tumors" (Ravdin, 1960: 3). More importantly, they insisted, the chemotherapy program had already produced important results both directly and indirectly. The program had spurred work on disease mechanisms; generated "objective criteria" for anticancer drug evaluation; established clinical study groups in hundreds of institutions; and created new support mechanisms linking universities, hospitals, government laboratories, and private enterprise. In addition, and indirectly, the program had promoted interest in cancer research among scientists by providing, for instance, more uniform definitions in the pathology and therapy of cancer via spin-offs such as the National Cooperative Cancer End-Results Evaluation Program whose Uniform Punch Card Code was designed to generate much-needed information on cancer patients in the United States (Cutler & Latourette, 1959). The program had also assisted in the training of scientists and clinicians and developed biometric units in medical institutions.

While the program's elaborate infrastructure—and in particular the fact that it demanded "obedience to rules, democratically achieved and agreed to"—might have irritated those who believed in the unrestricted freedom and creativity of individual basic researchers, its results could not have been achieved by "any other organizational pattern of sponsorship" (Ravdin, 1960: 6). Those results, moreover, belied the existence of a meaningful distinction between basic and applied research in chemotherapy, thus eliminating the dichotomy on which much criticism was based (Ravdin, 1960). In sum, the program had made collaboration between investigators "with diverse talents and interests" possible; provided a "more universal understanding of language, definitions, and mechanisms" in cancer; trained a large number of clinicians and basic scientists; shown the necessity of biometrics for assessing results and their causes; and "last, and by no means least, it [had] brought to light the true morbidity and mortality" of certain therapeutic procedures, "particularly surgery," once again undermining criticism by falsifying its premises (Lee, 1960: 73).

Seven years after the Gellhorn report, the principal architects of the chemotherapy program revisited their critics and the opposition between basic research and empiricism that sustained them. In the official history of the program, they observed:

> Gellhorn's critique reflected the attitude of a section of the scientific community which views empiricism with disdain and even with alarm as a source of dilution of basic research funds and efforts. It has been difficult for such individuals to accept the major role that empiricism has played in the development of almost all therapeutic agents. (Zubrod et al., 1966: 375)

Nonetheless, Gellhorn's criticisms had had an impact when first formulated. Program administrators admitted that there were "inherent difficulties" in the program. The drug screens, for instance, lacked the precision that would make it possible to extrapolate from animal models to humans. While there "existed differences of opinion" as to how to improve the screen and the evaluation of the products of the screen, possible solutions included "a reorientation of mechanisms of choosing projects" (Ravdin, 1960: 3). Following the publication of Gellhorn's report in 1959, the program director, Gordon Zubrod, was "told indirectly that there is a need for a document on reorganization of the CCNSC."[2] Recognizing that although the program had been a success in terms of procuring compounds and in "arresting the interest and attention of many

2. Memorandum from Clinical Director, NCI to Director, NCI, 22 December 1959. "Recommendations on Reorganization of CCNSC," 1, NCI Archives.

scientists and physicians in the problems of cancer chemotherapy," Zubrod admitted that there were several "things which CCNSC performs poorly or not at all." They were: "1. Primary screen. 2. Analysis and planning, utilizing the information obtained from the secondary screen, preclinical pharmacology and clinical trial."[3] In other words, and notwithstanding its "indirect benefits," most, if not all, of its key components were deficient.

A mere four years after its inauguration, Zubrod felt that the program, as initially designed, had become too complex to manage.[4] There can be no doubt that the problems of scale and heterogeneity had come home to roost with a vengeance. In 1960 the program had over 150 new compounds in clinical trials; only thirty of these substances had existed prior to the advent of the program five years earlier. Implicating over 150 hospitals throughout United States and now Western Europe and 16,500 patients, the program managed some two hundred clinical studies. To a certain extent the trials were simply the tip of the iceberg of an extremely successful enterprise that had "expanded at an unprecedented rate." Between 1957 and 1960, the program had screened over 100,000 substances and there appeared to be no end in sight.[5] But the program lacked focus. Cancer was not a single disease; what was true for one type of cancer was not necessarily true for another. The clinical trials conducted over the previous five years had made that clear. Because cancer was "so grossly heterogeneous," Zubrod believed that "further large-scale progress will come about in single types of cancer and I believe that we must plan our attack accordingly."[6] Shortly thereafter, and as mentioned in chapter 2, Zubrod instituted the task forces in order to provide the missing focus. The task forces were but the beginning of a much larger reform.

While the US program had already been underway for several years by the early 1960s, in Europe GECA continued to work on both its modus operandi and its identity vis-à-vis established institutions such as the UICC (that provided financial support) and the Co-ordinating Committee for Human Tumour Investigations, a recently initiated, pan-European

3. Ibid.

4. Ibid., 1–2.

5. Annual Report of Program Activities, 1960, *I*, National Institutes of Health, National Cancer Institute, 12, 16.

6. Memorandum from Clinical Director, NCI to Director, NCI, 22 December 1959, "Recommendations on Reorganization of CCNSC," NCI Archives, 1–2.

project that had established study groups on human tumor biochemistry, cytogenetics, cell proliferation, and the role of hormones in human cancer. A turning point for the organization came in 1968 when GECA, partly in order to procure funding from Euratom, changed its name to EORTC. The name change signaled a broadening of the group's activities to include other treatment modalities (e.g., immunotherapy) and research not immediately related to the screening program, although group members made it clear that their main focus would remain the development and testing of anticancer drugs and regimens. Regarding preclinical screening, several problems had emerged. Group finances remained tight, and the screening system was slow. In turn, pharmaceutical firms likely to provide financial support were just as likely to be put off by extensive delays.[7] As a solution the group developed a simplified preclinical screening system that took into account the chemical structure of compounds and reduced the amount of in vivo testing.[8]

Concerns about the performance of the clinical cooperative groups had also surfaced. Thus in the wake of its name change, the EORTC undertook several reforms. In 1969 it reconstituted its administrative board and its council that now included members from the United Kingdom. More importantly, it established the position of a group coordinator charged with evaluating group activities, submitting proposals for changes of procedures or policies, and establishing criteria for bona fide group membership. In his initial reports the group coordinator made a number of observations that would ultimately lead to important changes in the kinds and the organization of groups supported. He noted, for example, that cooperative groups operated at three distinct levels: discovery work carried out with new, potentially dangerous or difficult drugs; work of average difficulty carried out with known drugs or drugs of moderate toxicity; and largely pedestrian work—that is, the repetition of old work or work with known agents without much risk or interest. The normative nature of the categories entailed consequences: pedestrian groups were deemed "useless."[9]

As just indicated, the end of the 1960s and the early 1970s were turbulent times for the EORTC. Indeed, when asked about his tenure as pres-

7. EORTC/GECA, proceedings of the meeting of 21–22 January 1967, EORTC Archives (Brussels).

8. EORTC/GECA, proceedings of the meeting of 21 March 1961, EORTC Archives (Brussels).

9. *EORTC Newsletter*, no. 2, December 1969; *EORTC Newsletter*, no. 6, August 1970.

ident during this period (1969–75), Dr. van Bekkum (a radiotherapist) averred that his principal accomplishment had been "to keep the group together because it was falling off. People were fighting like hell at that time. They didn't agree on anything."[10] EORTC nonetheless survived. In his 1970 presidential report, after noting that a number of "extremely weak spots" had been brought to his attention, van Bekkum concluded that "although some problems concerned with cooperation between European clinical groups may seem to be very difficult to solve, . . . in spite of diverging viewpoints between some of the cooperative teams, there has never been any disagreement about the fact that many of the problems we are faced with in cancer treatment can only be solved by the combined activities of a number of teams," adding that the major problem for Europeans was a serious shortage of cancer specialists and that the challenge in coming years would be to establish a training program to produce this new kind of specialist.[11] In the meantime, to improve communication among its members, the EORTC launched a newsletter in the fall of 1969 and installed a telex network linking its members' offices. To today's readers both initiatives might appear relatively minor and obvious. That they were not so at the time underscores the novelty of collaborative infrastructures that we now take for granted.

The Reorganization of Screening and Clinical Trials in the United States

Following Gellhorn's criticisms in 1959, the US Cooperative Group Program fared little better in the evaluation of the NIH commissioned by Lyndon Johnson in 1964. Describing the results of the program as "less than were anticipated," the NIH Study Committee (known as the Wooldridge Committee) concluded that not only had the projects been scientifically "less distinguished" than the usual NIH undertaking, but also that they had heretofore escaped peer review, and thus recommended that the program be subjected to that form of scrutiny and control (Wooldridge, 1965: 85–86). The Wooldridge Committee's rec-

---

10. Interview with Dirk van Bekkum, Rotterdam, 15 March 2005.

11. "Annual report of the President for 1970," *EORTC Newsletter*, no. 9, February 1971, 4.

ommendations converged with earlier proposals that had been formulated by the executive secretary of the Clinical Studies Panel, Margaret Sloan. In a report to the director of the NCI toward the end of 1964, Sloan had proposed the formation of a Cancer Therapy Evaluation Center charged, as its name implied, with research on the evaluation of therapies and developing new experimental approaches. At the same time, the center would supervise research carried out by the cooperative groups through the review of research grant applications for cooperative studies.[12]

Following Sloan's proposal and the publication of the Wooldridge Report, the director of the NIH appointed a committee chaired by Arthur Richardson to inquire further into the matter.[13] Although the committee members did not examine the clinical collaborative trials in detail, they raised no doubts about their quality. Indeed, they went so far as to suggest that "one might well look on the collaborative chemotherapy groups as exemplary models for establishing sophisticated clinical groups around the country for the evaluation of therapy in other fields of medicine."[14] The Richardson Committee also recognized that the groups had made a number of significant contributions despite the fact that they had been criticized for having been relatively slow in accepting the "principle of the controlled study" and for using protocols that were too "rigid." The committee furthermore noted that the groups had developed the criteria for evaluating drugs and for tracking disease progress as well as making some therapeutic inroads in the case of leukemia ("possibly a cure in a small percentage of cases of acute leukemia in children").[15] As regards the accusation that the program itself amounted to little more than crass empiricism, the committee simply recommended that "in the continuing search for chemotherapeutic agents greater emphasis should be given to the acquisition of new information through basic research. Additional funds should be directed to individual investigators who can provide more rational bases for the continuing search."[16]

12. Margaret Sloan, "Proposal for the establishment of a Cancer Therapy Evaluation center as a part of Collaborative Research," 9 December 1964, NCI Archives.

13. Arthur P. Richardson (Chairman), Report of the Cancer Chemotherapy Collaborative Program Review Committee, June 1966, NCI Archives.

14. Ibid., 58.

15. Ibid., 27.

16. Ibid., 22.

The cooperative clinical trials differed from the other components of the CCNSC in that they were supported by research grants. As "programmed" research grants, however, they were not subject to peer review.[17] The Richardson Committee solved the problem of how to fund the trials through peer review by splitting the difference between the Wooldridge Committee (which had suggested that the clinical trials be submitted to normal peer review) and the Sloan proposal (which had suggested that the trials be controlled by in-house, NCI expertise). Because of the mix of program components—routine and basic—in the funded projects, the committee suggested that a special clinical trial review committee be set up to peer review the projects.[18] In other words, peer review, but with clinical peers.

While the Richardson report was not published until 1984, it tempered criticisms made at the time by observing that the program had improved the care of cancer patients. Endicott received the report with enthusiasm and endorsed the central recommendation of the committee; namely, that the peer review mechanism be adapted to support the groups.[19] Responsibility for the cooperative group program was transferred from the Division of Grants and Training to the Cancer Therapy Evaluation Branch of the Chemotherapy Division.[20] When the dust settled in 1966, the clinical trials program had become autonomous.[21] On the one hand they were formally separated from the screening program, consecrating their status as an independent enterprise. On the other hand they were not subjected to the same kind of peer review as more basic cancer research. The new system of peer review would allow the chairman of ECOG to justifiably claim more than a decade after its institution in 1966 that "the molding and reshaping of the Cooperative

17. Other programmed research in the 1960s included the "Special Virus Leukemia Program" (1964), the "Special Virus Cancer Program" (1969), and the "Breast Cancer Task Force" (1966). See Baker (1977).

18. Richardson, Report of the Cancer Chemotherapy Collaborative Program Review Committee, 10.

19. Memorandum, Dr. Kenneth Endicott (Director, NCI) to Dr. James A. Shannon (Director, NIH), "Preliminary Comments on the Report of the Richardson Committee," 17 September 1966, NCI Archives.

20. National Institutes of Health, National Cancer Institute, Cancer Therapy Evaluation Branch (Chemotherapy), Annual Report of Program Activities, 1967–68, vol. 1, 97.

21. Ibid,, 97–141.

Groups has been the result of peer-review" (Carbone, 1979: 4). Clinical cancer trials had thus become clinical research.

In 1966 Gordon Zubrod, now the Scientific Director of Chemotherapy at the National Cancer Institute, and his NCI colleagues summarized the reforms of the chemotherapy program in a detailed flow chart that described the organization as a process that began with the development of new drugs and flowed ineluctably toward clinical trials for the drugs or, in other words, toward the "determination of [the] role of agents in [the] total care of patients" (Zubrod et al., 1966: 537). The flow chart (figure 5.1 a & b) was the product of a new research planning methodology known as a *convergence technique* derived from the network analysis developed just after World War II (not to be confused with today's network analysis) and better known today as a component of operations research. Given the prominence of such techniques in industry, should we conclude contra our previous assertions that observers who have equated the chemotherapy program with industrial research were right after all? Not in the least, since it turns out that the convergence technique used at the NCI was designed specifically as an alternative to the networking techniques used in industry.

To understand how, we must first realize that networking techniques broke down industrial and engineering projects into a set of tasks or jobs, established the logical relations between the tasks, and estimated the time for the completion of each activity. NCI planners explicitly rejected such techniques because the tasks to be fulfilled in order to achieve the programs' objectives escaped simple definition. Nobody knew, for example, if some tasks were feasible, just as it was impossible to estimate their time to completion, the order in which they should be carried out, or even if the results of a given task would have a positive effect on the project as a whole. Instead of concluding that biomedical science could not be planned and thus throw out the baby with the bathwater, the authors of the convergence technique devised a plan that was scientist-oriented rather than job-oriented. They thus minimized "the more quantified and structured elements of network analysis" by formulating "a series of flows and arrays depicting major research program elements in a hierarchy of phases, steps, and individual projects, sequentially ordered on the basis of research logic, and graphically represented by a matrix which relates research performance to resources required" (Carrese & Baker, 1967: B-423).

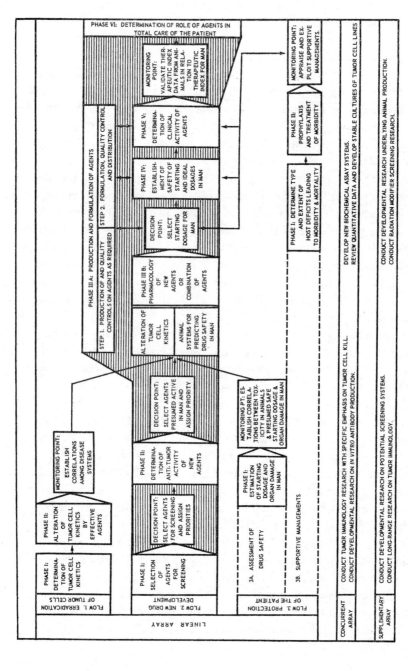

FIGURE 5.1. (a) Overview of the cancer chemotherapy program's linear array; (b) detail of Phase IIIa of the linear array. Reproduced with permission from L. M. Carrese & C. G. Baker, "The Convergence Technique: A Method for the Planning and Programming of Research Efforts, *Management Science* 13, no. 8 (1967): B427–28. © 1967, Institute for Operations Research & the Management Sciences.

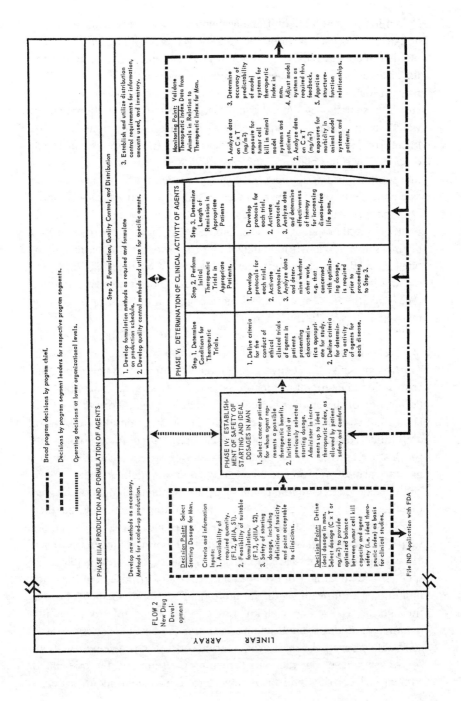

While this general definition has engineering overtones, a closer look at figure 5.1 a (itself a simplified version of the original chart) reveals a number of concessions to the unknowns of scientific research. Below the linear array lie both a concurrent and a supplementary array, set out in recognition of the frequent surprises of the research process. Should the current research program not pan out, for example, the additional arrays would supply alternative approaches. The ongoing examination of the linear array's scientific assumptions was thus part of the system. In addition, while other planning approaches viewed research activities as either fundamental or applied, the convergence approach treated research as a continuum. With separate mechanisms already in place to support both independent and planned research, the convergence technique covered the middle range (Carrese & Baker, 1967). In other words, the NCI planners invented a hybrid managerial form to contain the basic and applied research that characterized the new style of practice. The linear array and its convergence technique could be deployed with considerable reflexivity, the least of which was the recognition of the simple fact that many activities lay beyond its reach.

The linear array consisted of three flows that, according to the planners, reflected the underlying logic of the cancer chemotherapy program. The first and third flow converged into the development of new drugs, the central flow. The first flow focused on cell kinetics as the conceptual underpinning of the eradication of tumor cells and the theory that articulated the two basic arms of the chemotherapy program: drug discovery and clinical trials. Other lines of research such as tumor immunology, natural history, and disease prognosis were relegated to concurrent or supplementary arrays. Zubrod and colleagues (1966: 485) defined them as "not now considered so essential to the successful accomplishment of the goal that they should be in the linear array," maintaining nonetheless that they were essential to the "linear array" insofar as successful concurrent or supplementary projects would contribute to the overall flow in the array: "In each instance, these are projects that, if successful, will increase the speed of the linear flow or even become part of it." Hindsight tells us that immunology, for example, did far more than accelerate the flow of drugs through the NCI screen. Cell kinetics, on the other hand, while initially helpful as a model, soon dried up as a source of questions for clinical trials that had developed their own, internal sources of problem formation.

We saw above that the drug screening program and the associated clinical cancer trials had been subject to almost continuous criticism since their inception on the grounds that they were based on crude empiricism. Zubrod's language of logical arrays, flows, and kinetics was in part a response to these criticisms. We have also seen that clinical trials freed themselves both from NCI's screening program and from laboratory animal studies and created an autonomous space of representation (Rheinberger, 1997) for their conduct. Herein lay yet another response to criticism. By creating their own questions and setting their own timetable, clinical trials shifted from an empirical test of drugs to an inquiry into disease. As important as cell kinetics may have been, it was not indispensable for the emergence of a cure for acute leukemia. Retrospectively, the animal studies reinforced the biological insights gained in the clinic and made them robust. It was not a case of biological understanding being applied in the clinic, but the emergence of a kind of clinical trial that created a new biological and pathological understanding of the disease process; a kind of "biomedicine."

Surgery and Radiotherapy in the Cooperative Groups

The chairman of the Clinical Studies Panel of the CCNSC, who oversaw a program primarily devoted to chemotherapy, was a surgeon. His appointment seems to have been in part justified by the fact that the original mandate of the CCNSC included, in addition to research in chemotherapy, research in surgery and radiotherapy. As one of the principal architects of the statistical methodology of the screening program put it: "The aim of this large group is the development of a chemical, or perhaps joint chemical-surgical, or chemical-surgical-radiological 'cure' for cancer" (Schneiderman, 1961: 232). This medical pluralism, however, did not prevent the former chief of the radiation branch of the NCI from noting, somewhat tartly, that chemotherapy had been the favored son of the NCI in the 1950s and 1960s: "During my years at the National Cancer Institute, the emphasis given to research on cancer chemotherapy, while representing a milestone in the relief of cancer, allowed little attention to, recognition of, or support for radiotherapy" (Andrews, 1978: 2). Before ending part 1 of this book, devoted to the emergence of cooperative cancer research, we will take a brief look at how surgery and radiotherapy participated in clinical cancer research. This will provide us with a better understanding of the place of chemotherapy in the field of cancer medicine in the 1950s and 1960s.

Surgery

Throughout the present period, randomized clinical trials entered cancer surgery slowly and inconsistently even though surgical experimentation with the chemotherapeutic ur-substance, nitrogen mustard, goes back to the very discovery of the compound. In the 1950s, for example, a number of surgeons used nitrogen mustard preoperatively in an attempt to forestall the spread of cancer cells to other organs following surgery (Schwarz, 1961). Upon formation of the CCNSC, organizers instituted a cooperative group that undertook a number of clinical trials combining surgery and chemotherapy, an endeavor termed "adjuvant chemotherapy." Prominent early studies sought to see if the postoperative use of mustard gas and thioTEPA, a recently discovered agent with effects similar to nitrogen mustard, had any benefit in a number of anatomic sites (Sykes et al., 1953). Other studies analyzed the results of performing radical mastectomy with and without chemotherapy, concluding that chemotherapy added little to the original surgery (Noer, 1961). Similar findings emerged in trials that combined chemotherapeutic agents and surgery in gastric cancer (Longmire, Kuzma & Dixon, 1968). Despite the poverty of the early results, chemotherapeutically minded surgeons believed that they had "several advantages over [their] internist colleagues," since chemotherapy used postoperatively had only to "destroy microscopic amounts of cancer tissue" and thus "less than perfect agents [would] still have clinical value" (Moore, 1962a: 121). Surgeons, moreover, sought the short-term destruction of cancerous cells immediately following surgery. Fast-acting compounds that would be useless for chemotherapists seeking palliation would be perfectly acceptable to surgeons. Chemotherapy, in other words, could be construed as a handmaiden of surgery, rather than a competitor.

Despite the initial optimism, seven years of surgical adjuvant studies carried out by dedicated cooperative groups delivered less than encouraging findings. At an NCI-sponsored conference held in 1962, the Surgery Adjuvant Lung Group, which comprised over one thousand patients in twenty-two different institutions, reported the results of a comparison of surgery followed by or without chemotherapy. Not only did survival rates not differ, but also the number of postoperative complications stood significantly higher in the chemotherapy group (Curreri, 1962). Two other groups reported similar negative results (Longmire, 1962; Holden & Dixon, 1962) and all three disbanded shortly after the

1962 conference.[22] Notwithstanding the negative outcome, the exercise was not judged a complete failure since once again the possibility of carrying out this kind of investigation was as much at stake as the actual results. As one participant concluded at the end of the conference: "The greatest accomplishment of the surgical cooperative clinical studies has been the demonstration that such studies can be done" (Moore, 1962b: 591).

There was, moreover, some ground for optimism: the Surgical Adjuvant Chemotherapy Breast Group and the Veterans Administration (VA) Surgical Adjuvant Group had by the early 1960s managed to generate some positive data, although only the former lived up to its original promise, as we will see in part 2 of this book (Higgins et al., 1962). A later review of more complete data by the VA group leader concluded that there was no evidence of improved survival in the lung, stomach, and colorectal cancer patients who had received chemotherapy (Higgins & White, 1968). Undaunted, the VA group pursued a number of studies throughout the 1970s only to see their grants terminated at the beginning of the 1980s (Higgins, 1984). Thus disappointing results and persistent ambivalence in the realm of adjuvant therapy in the 1950s left surgical oncology in a relatively marginal position with regard to clinical trials. Most of the early groups devoted to clinical cancer trials implicating surgery were shut down, with one very important exception, the Surgical Adjuvant Chemotherapy Breast Group, renamed National Surgical Adjuvant Breast and Bowel Project (see chapter 6).

Radiotherapy

While progress in surgery was incremental, radiotherapy was in the throes of a revolutionary retooling. Before 1950 both diagnostic and therapeutic radiologists used low-energy sources of radiation that lay in the general range of 200 to 400 kilovolts. Then cobalt and subsequently linear accelerators pushed voltages into the million-volt range, resulting in what is generally termed the supervoltage (or megavoltage) revolution in radiotherapy and the functional separation of diagnostic and

22. See H. L. Davis, J. R. Durant, and J. F. Holland, "An Overview of the Cooperative Groups and Their Interrelationships with the National Cancer Institute and Other Governmental Agencies," *Cooperative Groups Clinical Trials Review*, March 1979, NCI, on p. 27, James Holland personal papers.

therapeutic radiology (Schultz, 1975; The technical history of radiology, 1989). Use of the new equipment resulted in the doubling of the five-year survival rates in most cancers treated by radiotherapy in the period 1955–70 (Kramer et al., 1980). More importantly for our present purpose, the equipment revolution also served as a continual distraction from clinical trial research. Equipment standardization took precedence over clinical research, as the former became a substitute for the latter. In particular, the onslaught of higher voltage equipment generated constantly changing protocols and problems of standardization without creating an immediate need for the production of clinical research protocols or interinstitutional clinical trials.

Moreover, US radiotherapists in particular showed little enthusiasm for clinical trials. In 1955, when the NCI invited "outstanding radiologists" to a meeting devoted to the project of organizing clinical trials in their field, it quickly became obvious they were not yet ready for such an undertaking (Brady et al., 2001: 2). So instead of incorporating radiotherapists directly into the clinical trials program, the NCI set up a Radiation Study Section that instituted a number of cooperative group studies of tumor classifications. Absent clinical trials, it was believed that radiotherapists could at least be induced to report their results following collective standards. If practices could be normalized, then routine practice would follow a standard protocol. In turn, questions of how much and in what circumstances radiation should be administered could be answered and clinical trials would be unnecessary.

Rather than simply unnecessary, radiotherapists saw clinical trials as unequal to the task of evaluating radiotherapeutic practices. In a 1960 presentation to the Clinical Studies Panel of the NCI, one of the most prominent radiotherapists in the United States, the Paris-trained Juan A. del Regato (1909–1999),[23] raised a series of obstacles to the conduct of randomized clinical trials in radiotherapy that effectively precluded any such effort. The list of impediments started with an argument routinely advanced by opponents of randomized clinical trials in surgery; namely, that the uneven distribution of skills among both practitioners and institutions excluded meaningful comparisons. Perhaps somewhat surprisingly for external observers who understood radiotherapy as largely constrained by commercial equipment and quantitative physical variables, del Regato maintained furthermore that "practiced

23. At http://www.juanadelregatofoundation.org/Biography/Biography.html.

techniques of radiotherapy are much more varied than are surgical procedures" (1960: 48). Radiotherapists could, for example, modulate radiation in the course of any single treatment so that dose size could vary from case to case; they could also control the therapeutic ratio—that is, the ratio of the treatment response over the toxicity response. Radiotherapy textbooks in the 1950s and 1960s concurred: "Treatment must be tailored to meet the particular manifestations of the disease in each patient. Both the local and the general signs and symptoms must be evaluated. Subsequent conduct of therapy is guided entirely by the patient's response" (Moss, 1959: 313).

Del Regato's (1960) list contained two further barriers to clinical trials. The first was the common ethical objection that a randomized trial would inevitably result in some patients being unjustly deprived of treatment, in particular since "radiotherapy [was], unlike chemotherapy, curative in many of its indications and consequently relied upon, whether rightly or wrongly, as an asset not to be denied to the patient." The second objection raised the possibility that there was somehow a practical difficulty in following the long-term consequences of therapy in a clinical trial and that this problem was particularly acute in radiotherapy, where "experiments are not easily judged on a short-term basis and their benefits may not be amenable to biometric analysis" (48). In light of these reservations, del Regato proposed a series of clinical trials using radiotherapy, but only in conjunction with other therapies: chemotherapy as an adjuvant for radiotherapy, comparisons of chemotherapy and radiotherapy as adjuvants for surgery, and the use of chemotherapeutic substances as radiotherapy-enhancing agents.

Del Regato's proposals followed on the heels of the creation of the Committee for Radiation Therapy in 1959 "to assist in the development of clinical trials in radiation therapy" (Kramer, 1973: 26). The committee's business advanced at a snail's pace. One year after its formation, "protocols [had] been discussed [by the committee] but experiments [were] not yet afoot" (del Regato, 1960: 49). Finally, in 1961 the committee launched its first cooperative clinical trial, a study of radiotherapy as adjuvant in lung cancer surgery. Then things went even slower. Reconvened in 1963 as the Committee for Radiation Therapy Studies (CRTS) and having secured funding for regular meetings, they decided to reorient their efforts from clinical trials to radiotherapists and radiotherapeutic institutions. Focusing on training, manpower, and the role of radiotherapy in the emerging cancer centers promoted by cancer commissions worldwide since the mid-1960s,[24] the CRTS all but abandoned clinical trials, organizing a mere three studies in the following four years. All told, in almost ten years of existence, the committee had conducted four clinical trials (Kramer, 1973; Cox, 1995).

In the pursuit of standards, the CRTS produced a number of reports known as Blue Books that served as unofficial guidelines for radiotherapy and the development of radiotherapy facilities. The first was submitted to the NCI in 1968. Titled "A Prospect for Radiation Therapy in the United States," in all more than 14,000 copies of the document were distributed (Kramer, 1973). While not a practice guideline, the earliest Blue Book set the background and standards for practice by laying out the conditions for the establishment and maintenance of radiotherapy facilities. They defined, for example, the ratio of therapists to patients, equipment required for a given number of patients, and training standards.[25]

The Blue Books that followed resulted in the collection of statistics on all sorts of practice variables including equipment, patient loads, and personnel. One can thus follow the rise and decline of types of equipment—in 1975, there were 407 linear accelerators in the United States; in 1994 the number had climbed to 2,777; in the same period, cobalt units declined from 970 to 314—and the steady rise in radiation oncology personnel—radiation oncologists doubled their number (from 1,400 to 2,800) in the same twenty-year period (Owen, Coia & Hanks, 1997). In the early 1970s, building on the Blue Books and as a natural extension thereof, radiotherapists launched a self-audit program titled the Patterns of Care Study that documented patient loads and facilities for the entire country (Kramer, Hanks & Diamond, 1982). It also presented the "best management programs for specific cancer entities" (Rubin, 1985: 1259; see Kramer, 1977).

Now, as useful as these studies no doubt were, they were not clinical trials or research protocols. In fact, in 1968, aside from the CRTS clinical trials, the Cooperative Oncology Groups carried out just about the only clinical trials in radiotherapy funded by the NCI. Even here, however, radiotherapy researchers were thin on the ground: of the many hundreds of protocols then in operation, only three involved radiother-

24. See, e.g., U.S. President's Commission on Heart Disease, Cancer and Stroke (1964) and Cancer Treatment (1966).

25. Committee for Radiation Oncology Studies, A Prospect for Radiation Therapy in the United States. Final Report. October 24, 1968. Bethesda, MD, NIH. Cited in Rubin (1985).

apy, and those three implicated a mere thirty-five radiotherapy institutions of the many hundreds then in existence.[26] To remedy the situation, the NCI replaced the CRTS with a bona fide cooperative group devoted to radiation therapy studies known as the Radiation Therapy Oncology Group (RTOG) (Brady et al., 2001). More than any of the Blue Books, the RTOG has spearheaded research in the domain and ultimately set the standards of practice.

When the NCI set up the RTOG, they realized that in order to carry out clinical trials in radiation therapy they needed a different type of standardization. The problem was this: unlike chemotherapy, where relatively pure compounds emanating from a central source assured an initial standard for comparability, each radiation facility depended on internal calibration to set the standards for doses of radiation. Dose size, in other words, had to be standardized at the site of administration. In order to create the conditions for comparability, the NCI set up an institution devoted to the standardization of radiation doses, or so-called dosimetry. Established the same year as the RTOG, the Radiologic Physics Center (RPC) quickly set about calibrating the apparatus used in radiotherapy centers that participated in clinical trials. In the first ten years of operation, members of the center visited almost two hundred different institutions. As an NCI review of the center pointed out in 1977, the "massive expansion" of the operation was "coincident with increasing multidisciplinary efforts of the various cooperative groups" who had introduced multimodal protocols in 1975 in an attempt to attract radiotherapists and surgeons to their groups.[27]

Radiotherapy trials initially fared slightly better in Europe in the mid-1960s. In 1964 radiotherapists in GECA established a Clinical Radiobiology cooperative group that included two of GECA's founders, a chemotherapist and a radiotherapist.[28] The cooperative group quickly launched a multicenter trial that compared radiotherapy alone to radiotherapy and chemotherapy in Hodgkin's disease. Seven years later, the group, now called the Radiotherapy-Chemotherapy group, had grown

26. At http://rpc.mdanderson.org/rpc/htm/AboutUs_htm/History.htm.

27. National Institute of Health, National Cancer Institute, Division of Cancer Treatment, Cancer Therapy Evaluation Program, Clinical Investigations Branch. Cooperative Group Program Review, November 1977, *I*, 50, NCI Archives.

28. A. Laugier, "H1 story, or 'The first seven years of a clinical trial,'" *EORTC Newsletter*, no. 14, September 1971, 1–5; GECA, "Compte-rendu de la reunion du 5 avril 1964," EORTC Archives (Brussels).

from two dozen to over forty members who, as noted by a participant, fell into three categories: the clinicians performing the trial, the scientists conducting ancillary studies (e.g., pathologists, radiophysicists, statisticians), and "those who would like to participate actively but cannot yet master a local situation where spontaneous investigations are carried out with prescientific methods," a thinly veiled reference to the obstacles the pioneers of cancer clinical trials continued to face.

The group chose to study Hodgkin's for a number reasons: its biopathological interest as both a systemic and localized form of cancer, the recent introduction of a new technique for staging the disease, the availability of an anticancer drug that showed specificity for some forms of the disease, and the existence of a concurrent US clinical trial that compared two purely radiotherapeutic approaches. Notice how even at this early stage, individual trials had become part of a mutually referential system; notice, moreover, the heterogeneous nature of the elements that contributed to the design of a trial and the choice of its target.

A 1971 interim report on the trial provides us with a sense of the issues confronting trialists during this early period. Between 1964 and 1971 the group met fifteen times, and the problems they discussed were often the by-product of the sheer length of the trial. More often than not, the understanding of the disease (in particular clinical and pathological staging) and its management (implicating the use of new techniques such as laparotomy) had evolved significantly since the inception of the trial. To deal with the evolution of knowledge in the domain as well as other unexpected contingencies, the trialists were forced to make a number of amendments to the protocol. Moreover, given that the number of cooperating centers had increased from the original three to ten, the group also felt constrained to undertake a cooperative evaluation of clinical material to ensure quality control and reduce discrepancies from center to center. Not to be underestimated, finally, was the motivational aspect. Few researchers appreciated the mundane yet time-consuming tasks such as entering patients in the trial and filling out the required paperwork all the while refraining from public disclosure of the protocol and its results for years on end.

Although the interim report lacked definitive numbers, there were at least two ways in which the trial could already be counted as a success. First, the trial partook in the new style of practice insofar as the groups had begun planning follow-up trials even before the end of the original trial. In a depiction of the process reminiscent of Foucault's (1963) account of the threefold spatialization of clinical knowledge, a member of the group described how, by 1971, the group had expanded its study of Hodgkin's disease in three directions: (i) geographically: twenty English centers had joined the study, (ii) anatomically: studies now took into account abdominal involvement in the disease; and (iii) epistemically: stage III of the disease now fell within the purview of group studies. Second, in what would become a common argument in support of clinical trials, the trial had changed clinical practice: "in the [hospital] departments which participate in a supra-national study, patients are more carefully evaluated and more precisely treated even if they are not individually included in a trial. This side effect of randomization may perhaps be the greatest achievement of our cooperative activities."[29]

To sum up, with a few notable exceptions, such as a Hodgkin's disease radiotherapy trial started in Stanford in 1962 and that grew into a well-known multicentric trial in the mid-1960s, US surgeons and radiotherapists initially showed little interest in clinical trials. Within the principal US cooperative groups, as in Europe, interest focused mainly on adjuvant studies. A second wave of interest in clinical trials in radiotherapy and surgery would roll over the cooperative groups in the mid-1970s, when, as noted above, groups such as ECOG, CALGB, and the Southwest Oncology Group (SWOG) became multimodal and began to conduct trials that combined chemotherapy with radiotherapy and surgery. But that is the subject of part 2 of this book.

29. Laugier, "HI story," 1–5.

An Avalanche of Numbers from the New Style of Practice (1965–89)

CHAPTER SIX

A Web of Trials

Breast Cancer and Its Trials

In 1976 the Eastern Cooperative Oncology Group (ECOG) published the results of a clinical trial launched in 1971 by a team led by George Canellos, chief of medicine at the Sidney Farber Cancer Center in Boston (Canellos et al., 1976). The Phase III trial compared two regimens in cases of metastatic breast cancer: CMF (a combination of three drugs: cyclophosphamide, methotrexate, and 5-fluorouracil) versus a single drug known as L-PAM (L-phenylalanine mustard). Named ECOG 0971—the ninth trial of 1971—the investigation drew patients from twenty hospitals in three countries on two continents. ECOG's statistical center at the State University of New York at Buffalo analyzed the research data, and in conformity with ECOG's publication rules, the protocol statistician, Stuart Pocock, occupied second place in the list of authors of the resulting *Cancer* publication.

This thumbnail description of ECOG 0971 highlights several characteristics of the period that we examine in this part of our book; namely, the maturing of the institutions—the cooperative groups—devoted to the routine performance of large, multicenter trials, the enforcement of the rules governing trial performance from design to publication, and the central and sometimes controversial part played by biostatisticians as clinical coinvestigators. Like the VAMP trial, ECOG 0971 was not an isolated event: it was embedded in a nexus of trials that targeted different kinds of breast cancer. The trial did not simply recruit breast cancer patients, but rather, patients suffering from *metastatic* breast cancer— that is, an advanced stage of the disease characterized by the spread of tumor cells beyond their initial location to other parts of the body. As such, ECOG 0971 differed from trials targeting *primary* tumors—that is, those in which cancer cells are still confined to their initial site of production.

The specific configuration of cancer targeted by the trial represents a continuation of a process initiated early on in the clinical research program. In chapter 2 we discussed a trial—Protocol No. 3—that was among the first to intervene in different phases of the treatment history of the disease: clinicians treated remission and relapse as independent events (though they implicated the same patient) and therefore, for all practical purposes, as two different subspecies of the same disease. The dual meaning of the verb "to treat" neatly captures the situation: clinicians regarded those events as two different diseases and acted on this difference by prescribing different treatments. In a continuation of this production of variety, the differential sensitivity of treated tumors to chemotherapeutic agents allowed clinicians to conceptualize the different reactions to treatment as the result of underlying biological differences.

Thus subsequent protocols like ECOG 0971 (and B-05 that we will discuss below) selected patients suffering from *either* metastatic *or* primary forms of a given cancer. In other words, the distinction between early and advanced forms of cancer became the basis for separate protocols. Those protocols, in turn, generated further differences by deploying a host of additional bioclinical distinctions concerning the life-cycle status (e.g., pre- or postmenopausal) of patients and the biological status of their tumors as defined by an increasingly important number of biological markers (e.g., hormone receptors on cancer cells) in their design. Here again the differential reaction of these patients to treatment, their *clinical* status as evidenced by trial results, served to entrench those differences. Briefly, then, using a mix of bioclinical criteria for the constitution of populations and therapy as a form of artificial selection, clinical trials multiplied the number and kinds of patient subpopulations that had to be taken into consideration when stratifying clinical trials, and of the subspecies of diseases for which subsequent trials would be organized.

Pierre Band, the young Canadian oncologist we met at the beginning of chapter 1 and the last author of the 1976 article reporting the results of ECOG 0971, had been involved in a previous breast cancer trial whose results had appeared the year before (Fisher et al., 1975). The trial, conducted jointly by ECOG and another cooperative group known as the National Surgical Adjuvant Breast and Bowel Project (NSABP), had focused on *primary* tumor patients. Standard treatment for these patients was surgery, and the trial was designed to test whether adding drugs to surgery increased the disease-free period by destroying the invisible metastases that lay beyond the surgeon's reach (a strategy called *adjuvant* chemotherapy, as we saw in chapter 5). The protocol had two arms: surgery followed by L-PAM versus surgery alone or, seen from the chemotherapeutic vantage point, L-PAM versus placebo.

This protocol (NSABP B-05) was a variation of a previous adjuvant protocol drafted by Dr. Band that compared cyclophosphamide (a nitrogen mustard derivative) and a placebo. Originally submitted to the National Cancer Institute of Canada (NCIC) as part of a clinical fellowship application, the protocol had found no takers. After working in the United States and becoming a member of ECOG, Dr. Band learned in 1970 that his NCIC protocol had attracted the attention of the chairman of NSABP, Dr. Bernard Fisher. An exchange of letters between Fisher and Paul Carbone, ECOG's chairman, ensued. The resultant B-05 protocol, finalized at a meeting held in the Canadian Rocky Mountains in Jasper, Alberta, replaced cyclophosphamide with L-PAM (another mustard gas derivative) because Fisher feared cyclophosphamide's severe adverse effects and, by the same token, its capacity to jeopardize the future of adjuvant chemotherapy.[1] The B-05 trial was ultimately launched in 1972 and was carried out in thirty-seven institutions in the United States and Canada, implicating twenty-eight members of NSABP, six members of ECOG, and three members of the soon-to-be-disbanded Central Oncology Group (Tormey, 1975). By the end of 1974, 269 women had been admitted and randomized to the two arms of the trial (figure 6.1).

Meanwhile, in 1971 Band had become chair of ECOG's breast cancer committee and had joined the team planning ECOG 0971, a chemotherapy trial, it is worth repeating, for women with *metastatic* as opposed to *primary* breast cancer. ECOG 0971 was construed as an extension of a small (twenty-five patients), single institution, nonrandomized Phase II study of the effect of a four-drug combination (CMF and prednisone) on advanced metastatic breast cancer patients (Canellos et al., 1974). While the final results of this trial—conducted by George Canellos and colleagues at the NCI Clinical Center—did not appear in print until 1974, investigators knew of the initial favorable results well before that date.

1. Interview with Pierre Band, Montreal, 18 October 2007; phone interview, 27 July 2009.

FIGURE 6.1. Dr. Fisher at the magnetic board used to follow patients for the B-05 protocol. Notice the material infrastructure used at the time: computers have today replaced magnetic boards. Adapted and reprinted by permission from the American Association for Cancer Research, V. T. DeVita, Jr. & E. Chu, "A History of Cancer Chemotherapy," *Cancer Research* 68 (2008): 8650.

Based on these interim results, Canellos and Band decided to undertake a larger, multi-institution Phase III trial comparing CMF and L-PAM, using the same dose size for L-PAM as the B-05 adjuvant therapy trial. The protocol, in other words, had a dual purpose: to test a treatment for advanced cancer and, should CMF prove superior, provide the basis for the next generation of adjuvant combination therapy.

ECOG 0971 enrolled 231 women with metastatic breast cancer who had previously undergone surgery, hormonal therapy, radiation therapy, or some combination thereof. The treatment protocol specified dose size and frequency, adopting the cyclical approach used in the Phase II trial. L-PAM was given orally every day for five days once every six weeks. Patients with a complete response—defined as no measurable presence of disease—went through twelve of these six-week cycles: almost a year and a half of treatment. The CMF patients cycled through a similar number of twenty-eight-day treatments that consisted of fourteen days of rest followed by fourteen days of daily oral doses of cyclophosphamide and intravenous doses of 5-FU and methotrexate on days one and eight. Figure 1.10 (chapter 1) summarizes the *treatment* protocol and illustrates the increasingly sophisticated architecture of the competing regimens.

As readers will have by now surmised, ECOG 0971 participated in a complex choreography of trials: clinicians, using different combinations of a limited number of drugs, juggled metastatic and adjuvant trials without waiting for the end of each trial and used preliminary results and sometimes only the protocols of previous trials to design new ones. This nonlinear sequence of interconnected trials, as we will see below, was even more extensive: several other trials preceded and informed the aforementioned protocols, which, in turn, led to further protocols, including a widely cited CMF versus placebo *adjuvant* trial carried out in Italy with NCI funding (Bonadonna, 1985). Clinical trial publications bear witness to the feverish circulation of findings: the concluding paragraphs of the ECOG 0971 article discussed the ongoing Italian trial and even reported a few of its initial results. A group of European "young oncologists" who met in 1978 in Brussels proposed that this flurry of activity become an explicit strategy: "Phase 3 trials should if possible be designed to ask sequential questions which may run consecutively without waiting for the results of prolonged follow-up of earlier studies."[2] Moreover, like their more traditional scientific peers examining the "state of the art" in a given field, trialists resorted to review articles to reconstitute (from a given point of view) the sequence of trials, their mutual relations, cumulative contributions, and to reply to critics (Fisher, 1991; see also Lerner, 2001).

Unlike most of the Phase I/II trials mentioned in part 1 of this book that tracked responses such as tumor reduction by canvassing many different cancers in a small number of patients, ECOG 0971 measured a different and, in some respects, a somewhat less tangible variable known as survival time. Survival-time studies produce data referred to as *censored data*. The term calls attention to the fact that once a trial ends, in contrast to the *known* survival times of patients who passed away in the course of the trial, for patients who are still alive at the time of analysis the date of death is unknown. For these patients the duration of survival (or of any other end point; Bland & Altman, 1998) is *censored* at the last date the patient was known to be alive. Survival-time data differ from tu-

2. "Report on the meeting of young oncologists, Brussels 1978," EORTC Archives (Brussels).

mor response data in a second regard. Since the recruitment of patients to such trials was spaced out over many years (an increasingly common occurrence that resulted from the need to increase sample size in order to obtain significant measurements of small differences between regimens), not all patients entered a trial at the same time. Thus the survival times of patients recruited at the beginning of a trial had to be compared with those of patients enrolled months or even years later.

Statisticians developed different techniques to analyze censored data and estimate the survival times for all patients (Sylvester, Machin & Staquet, 1978). The oldest in this class of techniques known as *survival analysis*, the "life table," emerged in the early 1950s (Berkson & Gage, 1950), but by far the most successful was the "Kaplan-Meier estimator," named after the two American statisticians who developed the technique in the late 1950s. The results of the Kaplan-Meier calculations are expressed in a Kaplan-Meier plot.[3] The article in which Kaplan and Meier introduced their approach enjoyed an exceptional career, receiving over thirty thousand citations. Figure 6.2 shows the remarkably sustained growth in the number of citations and compares this growth to the parallel development of the number of publications reporting cancer clinical trials. This congruence expresses the connection between the growth of large Phase III trials that took off during the second half of the 1970s and the use of survival times as a clinical trial end point and offers a quantitative view of the generation of a new representational space governed by the telltale, stair-step Kaplan-Meier diagrams. Of the over five thousand articles that cited Kaplan-Meier between 1970 and 1989, 69% came from the field of cancer, with the remainder emanating largely from statistics journals (that made up the bulk of the pre-1970 citations) and a cluster of cardiology publications.[4]

Figure 6.3 shows the diagram used in the ECOG 0971 publication to present the survival-time results of their protocol (produced using the life table method). While it clearly conveys the idea that the CMF regimen beat L-PAM, such a plot may deceive the untrained eye. It might suggest, for example, that the statisticians who plotted the data simply began collecting data on 100% of the patients on day one of the trial and

3. The original article (Kaplan & Meier, 1958) had no graphs. On the conditions surrounding its publication, see Kaplan (1983) and Marks (2004).

4. We calculated this percentage by searching the Web of Science for cancer-related articles using a filter kindly provided by Dr. Grant Lewison (Evaluametrics Ltd.) and by combining the results with a cited-reference search.

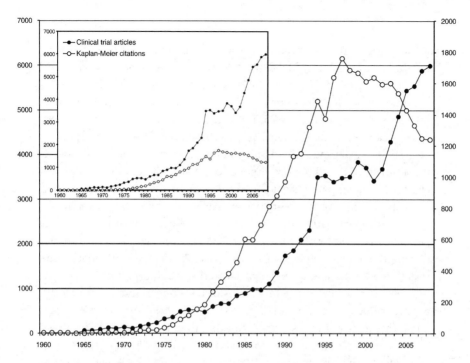

FIGURE 6.2. Comparison of the growth of English-language articles on cancer clinical trials (primary *y* axis) and the number of articles citing the 1958 Kaplan-Meier article (secondary *y* axis). The insert shows the same data using a single *y* axis. Data sources: PubMed and Web of Science.

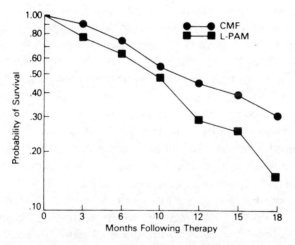

FIGURE 6.3. Overall survival of the two treatment groups in ECOG 0971. The original legend read in part: "The CMF groups had a significantly . . . improved survival beyond the first year of follow-up." Reprinted with kind permission from G. P. Canellos et al., "Combination Chemotherapy for Metastatic Breast Carcinoma: Prospective Comparison of Multi-Drug Therapy with l-phenylalanine mustard," *Cancer* 38 (1976): 1885. © 1976, John Wiley and Sons.

as the patients died subtracted the percentage that died from the total number of participants. They then apparently plotted the resultant numbers against time on a graph. But this is *not* what they did. The zero on the time axis ("Months Following Therapy") is not a calendar date denoting the beginning of the trial. It is an abstract mathematical point where time equals zero, and the subsequent subdivisions of that axis represent abstract physical time divided into months. In turn, the y axis ("Probability of Survival") does not correspond to the absolute number or the percentage of actual patients who were still alive at a given date but, rather, to the probability of surviving beyond some time t which is given on the x axis. To repeat, rather than the visual representation of raw data, the plot describes the results of a series of calculations based on the product of conditional probabilities. In spite of its sophisticated mathematics, the resulting estimate has a simple, visually effective interpretation.

There is a second sense in which survival-time diagrams can be deceptive: while figure 6.3 makes a case for the superiority of the CMF regimen, this does not necessarily mean that the obvious visual difference between the CMF and L-PAM curves is *statistically significant*. Further calculations are needed to determine this fact. Once again, several techniques emerged and one eventually dominated (figure 6.4). A cooperative group statistician whom we have already met, Edmund Gehan, developed one of the first such techniques known as Gehan's modified Wilcoxon test (Gehan, 1965). A year later, Nathan Mantel, a NIH statistician, introduced the more widely used logrank test (Mantel, 1966; see Gail, 1997). Finally, as we will see in chapter 9, in 1972, using data from the VAMP trial supplied by Gehan as test data, David Cox developed the proportional hazards technique (Cox, 1972). The classic paper in which Cox introduced his technique enjoyed a success comparable to that of the Kaplan-Meier estimator paper (over twenty thousand citations).

Part of the success of the Cox model was due to the fact that it opened up trial data to a number of analytic possibilities unavailable to other forms of survival analysis. In particular, using the Cox model, in the case of a trial in which patients had been randomized to two treatments, in addition to the primary analyses (the comparison of the two treatments), trialists would be able to conduct so-called secondary or supportive analyses that would adjust treatment results, taking into account prognostic factors (such as pre- or postmenopausal status) that had been recorded in the patients' charts but had not been included in the randomization scheme (Sylvester et al., 2002).

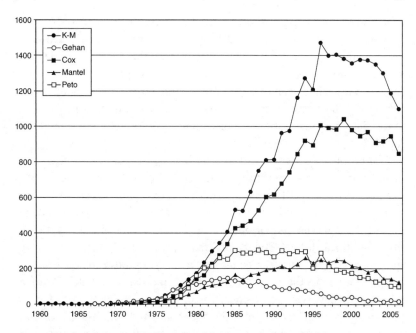

FIGURE 6.4. Comparison of citations (English-language articles only) to publications introducing different methods to calculate the significance of survival analysis curves. Data source: Web of Science. K-M = Kaplan & Meier (1958); Gehan = Gehan (1965); Cox = Cox (1972); Mantel = Mantel (1966); Peto = Peto et al. (1977).

The Lessons of the Breast Cancer Trials

The Kaplan-Meier estimator and the life table techniques were developed during the 1950s, and techniques such as Gehan's and Mantel's in the 1960s. Why were they picked up by trialists only a decade or two later? The answer is that the clinical trial system we will discuss in part 2 changed significantly from the one we discussed in part 1. Focus switched from the clinical screening of substances on a relatively limited number of patients to the performance of increasingly large-scale clinical research: a new kind of experimental work that compared regimens on the basis of biological and pharmacological hypotheses and that simultaneously provided a window on the diseases against which those regimens had been designed. In turn the increasing statistical sophistication of the new set of trials and the complex *dispositif* of clinicians, statisticians, data managers, and the like who carried out the trials, lent itself to renewed rhetorical use of the technique of clinical trials by trialists such as NSABP's Bernard Fisher. Fisher promoted clinical trials findings as a superior form of evidence that would ultimately replace individual clinical insights and reshape therapeutic practices. To be sure, some of these changes have their roots in the period covered in part 1 of this book, which saw the initial recognition of clinical research as an autonomous enterprise. But the transformations then set in motion had by the mid-1970s become the norm.

Unlike the VAMP trial, ECOG 0971 might appear somewhat mundane. Rather than a breakthrough trial bristling with innovation, it was a standard Phase III trial comparing two competing regimens. In this sense, then, we might have chosen any number of Phase III trials conducted throughout this period. We nonetheless believe that 0971's position within a dense network of breast cancer trials offers a fruitful starting point for our discussion. ECOG 0971 stands out not so much as a novelty as in the way it incorporated and expressed standard features of the new style of practice. It can only be understood as part of the larger research strategy in which it participated, and its performance was predicated upon the existence of a number of interconnected institutions such as data centers and protocol review committees that streamlined the work of cooperative groups; professionals such as biostatisticians and data managers who, with clinicians, manned such institutions; and a mounting body of evidence about the effects of drug combinations on tumors emerging from a *relational system* of trials.

Despite this underlying continuity, VAMP and ECOG 0971 also differ insofar as VAMP treated leukemia and 0971 treated breast cancer. We have chosen trials devoted to two different diseases for a reason. During the early period of chemotherapy, leukemia acted as a model disease. As its name suggests, the first cooperative group, ALGB, concentrated on leukemia and while a second cooperative group devoted to "solid tumors"—a term that according to several early trialists unduly conflated different diseases (Morrison, 1960: 78)—was soon established, leukemia continued to occupy center stage for several years. Similarly, when the NCI decided to establish task forces, it first did so for leukemia, and only later for breast cancer. Since incidence rates of leukemia were far lower than those for breast cancer or other solid cancers, the explanation for this initial pattern of institution building has to be sought elsewhere. Because leukemia is by definition a *systemic* disease beyond the reach of surgeons and radiotherapists, it was an obvious target for a *systemic* approach such as chemotherapy. Unlike some operable tumors such as breast cancer, leukemia quickly and, in the absence of even partially effective therapies, inevitably led to death. It was, moreover, relatively easy to measure the effect of drugs on the disease, since blood counts and smears were far simpler to perform than the measurement of solid tumor masses. In other words, researchers saw leukemia as an ideal choice for a "proof of principle" of cancer chemotherapy. From this point of view, subsequent work on breast cancer can be interpreted as an attempt to replicate the success obtained with combination chemotherapy in leukemia.

Things were nonetheless more complex. Solid tumors begin as a localized pathology and subsequently disseminate beyond their initial location. While chemotherapists could hardly pretend to replace surgeons and/or radiotherapists in the treatment of primary solid tumors, their intervention at the disseminated stage remained a legitimate option. To advance the chemotherapy agenda, trialists pursued two complementary strategies: the first was to collaborate with surgeons to design adjuvant trials (see chapter 5), and the second sought to begin chemotherapy with disseminated stages of cancer and move upstream to the primary stage (Carter & Soper, 1974: 3).

Researchers thus began working against the flow in an effort to shift chemotherapy to the front of the therapeutic line (figure 8.9, chapter 8). This relatively straightforward strategy amounted to testing drugs and regimens in late-stage patients and moving the successful ones forward until they reached the front line. The standard treatment flow for breast cancer patients, for instance, consisted of four stages corresponding to the disease trajectory: mastectomy and radiotherapy followed by hormonal manipulation (removal of the ovaries or estrogen treatment) for primary recurrent disease, hormonal manipulation and/or chemotherapy for secondary recurrent or advanced disseminated disease, and, finally, chemotherapy for advanced late disease. Introducing chemotherapy into the flow involved testing chemotherapy drugs and regimens in patients from this last stage, determining the optimum drug combination in patients with secondary recurrent or advanced disseminated disease, and using a combined modality that would henceforth include chemotherapy in patients from the first two stages (Carter & Soper, 1974: 4). Consistent with this strategy, the opening paragraph of the 1975 article reporting the early findings of the B-05 adjuvant breast cancer trial argued that since chemotherapy had shown its effectiveness in treating advanced disease, the "highest priority" was now to be given to the evaluation of the combined use of chemotherapy and surgery in women "with a lesser tumor burden" (Fisher et al., 1975: 117).

Meanwhile, events had moved quickly on the surgical front. The NSABP, under the leadership of Dr. Fisher, had launched a crusade against the dominant approach to breast cancer, Halsted's radical mastectomy, an extremely (and as it turned out, needlessly) mutilating surgery that had become the gold standard (Lerner, 2001: 137–40). The B-04 protocol, initiated in 1971, enrolled almost 1,700 women in thirty-four US and Canadian institutions to compare radical mastectomy to a less drastic alternative, a total (simple) mastectomy. Based on the results of B-04, the 1976 B-06 protocol, involving 2,000 women, compared total mastectomy to a breast conserving surgery (lumpectomy), again demonstrating the value of the less radical option. In spite of significant resistance (several institutions, including major cancer centers, refused to join B-04), the NSABP data eventually carried the day.

In this instance, the issue was not so much the amount of evidence (conservative surgeons also had evidence supporting the radical approach) as the kinds of evidence. The retrospective data or, as Fisher would have it, "anecdotalism," paled in comparison to the data generated by randomized clinical trials (Lerner, 2001: 140). Fisher's insistence on the scientific aspect of the NSABP endeavor drew not only on clinical trial methodology but also on his claim that his trials explored a scientific, biological hypothesis antithetical to the anatomic-mechanistic understanding of breast cancer entertained by traditional surgeons. In short, Fisher (1991) felt entirely vindicated in his claim that breast cancer was a systemic disease and that local therapy was unlikely to have decisive effects. This explains why the B-05 adjuvant trial was sandwiched between the two surgery trials: if, invisible to the eyes of the surgeon, micrometastases were arguably already in circulation, systemic therapy would have to complement local surgery. Chemotherapy, in other words, by taking a multimodal turn, could abandon its status of "last resort therapy" and reposition itself on the front line of therapeutics.

In order to continue the discussion of the web of trials in which ECOG 0971 was embedded, we now need to have a quick look at the substances used for the trial and their combination in previous and concurrent trials.

The Emergence of Combination Chemotherapy

Unlike VAMP's relatively novel substances, those used in the 0971's competing regimens had been around for almost two decades. Let us begin with 5-fluorouracil (5-FU), an inhibitor of DNA synthesis, since its early testing contrasts sharply with subsequent trials exemplified by 0971. Synthesized in the late 1950s by Hoffman-La Roche at the request of Charles Heidelberger (1920–1983), an organic chemist working in the cancer research laboratory at the University of Wisconsin Medical School, by the early 1960s 5-FU had been characterized as a "useful palliative drug for the treatment of solid tumors, particularly of the breast and gastrointestinal tract" (Heidelberger & Ansfield, 1963: 1240). Fred Ansfield, a future member of the cooperative breast cancer group (disbanded in 1975) and one of the founders of the American Society of Clinical Oncology, had conducted much of the early clinical work at the University of Wisconsin. In 1958 he and Heidelberger administered 5-FU to over one hundred patients with twenty-seven different types of cancer in a screening exercise similar to those conducted by the cooperative groups (Curreri et al., 1958). In the years that followed, Hoffman-La Roche farmed out 5-FU to a number of individual researchers who, like Ansfield, collected case series that sought not only to specify dose sizes and toxicity limits but also, by targeting large numbers of different cancers, to screen the substance in humans.

Meanwhile, Ansfield and his colleagues had moved beyond dose exploration and screening and in 1959 had instituted a sequential trial to test for efficacy by comparing patients with localized lung cancer treated with radiotherapy alone or radiotherapy and 5-FU. When they reported their results in 1962, thirteen pairs of patients had been studied. All of the radiotherapy-only patients had died and four of the radiotherapy plus 5-FU patients were still alive. This difference was statistically insignificant, but it did not prevent the team from concluding that "the crucial question as to the influence of 5-FU as a radiotherapeutic adjuvant on the therapeutic ratio remains to be answered" (Gollin et al., 1962: 1215). It is thus not surprising that in the early 1960s the NCI statistician Marvin Schneiderman (1962) criticized early studies on 5-FU as "clinical excursions"—that is, investigations lacking precise questions, end points, and study populations. After ten years of experimental use, Ansfield and his colleagues had treated almost seven hundred advanced breast cancer patients, and the results remained equivocal. In a brief note in 1969,

Ansfield's group confessed that having determined optimal dosage and the size of effect that one might expect from a course of therapy, they had made no attempt "to prove that 5-FU is superior to any other treatment in disseminated breast cancer" (Ansfield et al., 1969: 1066). Having conducted nearly all of their trials without controls, rather than decisive conclusions they were left with nebulous observations.

As we have just seen, before treating breast cancer specifically, 5-FU had been screened in a number of cancers. At the time, other substances had followed similar trajectories, including the remaining components of the CMF regimen, cyclophosphamide and methotrexate. The former, another nitrogen mustard derivative, had been synthesized in 1957 in Germany at Asta-Werke (Arnold, Bourseaux & Brock, 1958; Brock, 1989) and tested across a spectrum of cancers in Europe in the late 1950s (Hoest & Nissen-Meyer, 1960: 47–50). Unlike other nitrogen mustards, cyclophosphamide showed remarkable activity in the NCI's L1210 mouse system. Clinical evaluations carried out in the early 1960s in the United States with breast cancer patients gave responses running from 10% to 60% (Carter, 1972). Like cyclophosphamide and 5-flourouracil, methotrexate also went through the usual round of human screening in a search for the most reactive cancer. Finally, as we saw in part 1, the NCI submitted all three compounds implicated in the 0971 regimen to more systematic testing at the beginning of the 1960s when, in addition to the cooperative group program, the CCNSC initiated the short-lived Clinical Drug Evaluation Program (CDEP).

How did these substances end up as part of a combination therapy? Researchers had tried several combinations of chemotherapeutic substances in the treatment of advanced breast cancer prior to ECOG 0971. The first came out in the early 1960s in a protocol devised by the head of the Chemotherapy Clinic at the Mount Sinai Hospital in New York, Ezra Greenspan (1919–2004). In collaboration with the laboratory researcher Abraham Goldin (1912–1988) at the NCI, he had shown in 1952 that combinations of chemotherapy drugs had a synergistic effect in mice with leukemia (Goldin, Greenspan & Schoenbach, 1952). When he returned to his native New York in the early 1950s, he began clinical work at the Mount Sinai Hospital, and by the late 1950s he had tested his first combination of chemotherapy agents on patients with ovarian and breast cancer in accordance with the synergistic principles he had learned in mice.

The cancer community largely ignored the initial reports of his work that appeared in the early 1960s in the *Journal of the Mount Sinai Hos-*

pital. And yet Greenspan's results were quite astonishing. One of Green-span's colleagues later recalled how "earlier submissions to other jour-nals had been rejected because some reviewers thought that photographs of tumors before and after treatment documenting responses had actu-ally been reversed!" (Ezra Greenspan . . . , 2005: 2; see also Muggia, 2005). Claiming "responses" (tumor regression in this case) of 60% to 80% and maintenance of the response for six months in 40% of all cases treated, it may be wondered why clinical researchers overlooked Green-span's work. The answer clearly lies in the amount of clinical expertise required to reproduce Greenspan's results and the ultimate impossibil-ity of standardizing the treatment regimens he proposed. Greenspan felt that his therapy required extensive clinical know-how, and the drug dos-ages he used had to be varied according to the individual patient:

> The *sine qua non* for the safe use of cytotoxic drug combinations is a wide empirical clinical experience with the toxicity and pharmacology of each in-dividual drug. Modification of dosage and frequent observation in accor-dance with the clinical condition of each patient was paramount in avoiding rapid lethal drug toxicity, especially from the anti-metabolites, methotrexate or 5-FU. (1968: 782)

Greenspan had tailored dose sizes individually through "oral clini-cal titration" (i.e., "give until there is a response"), as it had been im-possible "to employ methotrexate safely on a standard protocolized ar-bitrary or predetermined drug-body weight basis" (1968: 782). The fact that they also varied according to Greenspan's fluid prognostic catego-ries of "good-risk" or "poor-risk" further complicated the personalized regimens. Greenspan did not explain how to classify patients into these categories other than to say that it depended on "a detailed current med-ical status, the type of preceding therapy, and the extent and pattern of organ metastases" (1968: 782). Such an approach did little to inspire comparisons or imitations and obviously contrasted sharply with the ap-proach subsequently taken by trials such as ECOG 0971.

In 1969 a watershed event occurred during the 60th Annual Meet-ing of the American Association for Cancer Research (AACR). Rich-ard Cooper of the Buffalo General Hospital presented the results of a five-drug combination in breast cancer patients using four different classes of reagents. Subsequently titled CMFVP (for cyclophosphamide, methotrexate, 5-flourouracil, vincristine, and prednisone), the regimen involved a mix of oral and IV routes of administration and specific dosages following daily and weekly schedules. It had been applied to a group of sixty hormone-resistant advanced breast cancer patients beginning in 1967. At the meeting, Cooper reported a surprising 90% "complete remission" rate (1969: 15). Such numbers had never been seen before and have never been seen since, not even by Cooper himself.

It was not for lack of trying. Shortly after obtaining his initial results in 1968, Cooper set out to test his regimen on women with a less disseminated form of breast cancer, hoping for even more spectacular results. As in the original study, there was no randomization, and although he selected patients according to anatomic dissemination of disease, he did not control for previous and concurrent therapy, meaning that a significant subgroup of patients studied had also received radiotherapy. When he and his colleagues finally reported their results almost a decade later, they omitted the remission rate; numerous subsequent studies had shown that relapses at twelve months sometimes ran as high as 90% rendering their initial 90% remission rate somewhat meaningless. Instead, they reported a more significant statistic, the remission duration. Here the numbers, though less eye-popping—68% survival at five years (Cooper, Holland & Glidewell, 1979: 793)—were substantially superior to the three-year survival rate of less than 40% previously reported by the breast cancer group (Goldberg et al., 1973: 662). By the time results appeared in print, however, they had been superseded by an entirely new series of trials and regimens that Cooper's initial results had provoked.

The Multiple Meanings of Breast Cancer Trials

When presented at the AACR meeting, Cooper's regimen created genuine excitement. As we saw, though, it was not a thunderbolt in a cloudless sky: Greenspan's "proof in principle" that combination therapy (albeit in a nonstandardizable form) had effects on solid tumors, and the subsequent success of the standardized combination of the VAMP protocol (albeit in a different form of cancer), had set the table. Almost immediately, six different cooperative oncology groups set out to test variations of the Cooper regimen. Among these we find ECOG 0971. Figure 6.5 summarizes the trials underway or completed by 1975. As can be seen, alternative regimens abounded, and numerous variants were made possible

Group	Regimen 1	Regimen 2
Acute Leukemia Group B #6982	1. 5-FU 12 mg/kg/wk IV 2. MTX 0.75 mg/kg/wk IV } × 8 3. VCR 25 μg/kg/wk 4. Cytoxan 2 mg/kg/d PO 5. Prednisone 0.75 mg/kg/d PO × 21, then taper	1. 5-FU 2. VCR } as in Regimen 1 3. Prednisone
Eastern Co-operative Oncology Group #0971	1. 5-FU 600 mg/m² d 1 and 8 2. MTX 60 mg/m² d 1 and 8 3. Cytoxan 100 mg/m²/d × 14 PO, cycles repeated at 28-day intervals	1. Phenylalanine mustard, 6 mg/m²/d × 5 PO every 6 wks
Central Oncology Group #7500	1. 5-FU 500 mg/wk IV 2. MTX 25 mg/wk IV 3. VCR 1 mg/wk IV 4. Cytoxan 100 mg/d PO 5. Prednisone 45 mg/d × 14 then taper	1. 5-FU 2. MTX } as in Regimen 1 3. Cytoxan 4. Prednisone
Southeastern Cancer Study Group #339	1. 5-FU 400 mg/m²/wk IV 2. MTX 20 mg/m²/wk PO } × 8 3. VCR 1 mg/m²/wk IV 4. Cytoxan 100 mg/m²/d PO 5. Prednisone 45 mg/d and taper vs 5-FU 600 mg/m²/wk × 8 ↓progressive disease MTX 20 mg/m² twice weekly PO prog. dis. →	1. 5-FU 400 mg/m² d 0 and 7 2. MTX 30 mg/m² d 0 and 7 3. VCR 1 mg/m² d 0 and 7 4. Cytoxan 400 mg/m² d 0 IV 5. Prednisone 20 mg q. i. d. × 7 VCR 1 mg/m²/wk ↑progressive disease Cytoxan 100 mg/m²/d
Western Co-operative Group #115	1. 5-FU 15 mg/kg IV every 2 wk 2. MTX 0.5 mg/kg IV every 2 wk 3. Cytoxan 2 mg/kg/d PO 4. Prednisone 0.5 mg/kg/d × 14, taper 5. Triiodothyronine 50 μg/d	Sequential use of 1. 5-FU 15 mg/kg/wk ↓ 2. Cytoxan 2mg/kg/d PO ↓ 3. Prednisone + Triiodothryonine ↓ 4. MTX 0.5 mg/kg/wk
Southwestern Group #450	1. 5-FU 300 mg/m²/wk 2. MTX 15 mg/m²/wk } × 8 3. VCR 0.625 mg/m²/wk 4. Cytoxan 60 mg/m²/d 5. Prednisone 30 mg/m²/d × 14, taper	1. 5-FU 180 mg/m²/d × 5 2. MTX 4 mg/m²/d × 5 3. Cytoxan 120 mg/m²/d × 5 4. VCR 0.625 mg/m² d 1 and 5 5. Prednisone 40 mg/m²/d × 5 Repeat courses every 28 d

vs
New Agent
(Adriamycin 60 mg/m² every 3 wks)

FIGURE 6.5. Therapeutic regimens evaluated by cooperative groups on the basis of the "Cooper regimen." Reprinted from S. K. Carter & R. B. Livingston, "Cyclophosphamide in Solid Tumors," *Cancer Treatment Reviews* 2 (1975): 300. © 1975, with permission from Elsevier.

by adding and subtracting substances and by devising different dose schedules.

As already mentioned, the interpretation of trial results takes place in relation to past, present, and future trials. To return to figure 6.5, as its authors suggested, the primary object of ECOG's protocol 0971 was to provide another test of the Cooper regimen. This was certainly the context.[5] It did not, however, supply the entire rationale for the enterprise. According to the sponsors, 0971 did more than simply embroider the Cooper regimen. Canellos and his colleagues had designed the trial, it will be recalled, to compare two regimens: L-PAM versus a three-drug combination chemotherapy (CMF). Consequently, over and above its character as a permutation of the Cooper regimen, 0971 also sought to test a more general principle when launched: compare single and combination chemotherapy regimens in advanced breast cancer.

Whatever might be accomplished in advanced breast cancer did not automatically translate into adjuvant therapy, and so the L-PAM versus placebo (B-05) trial sponsors had set out on a different path. Following initial results that showed a 50% reduction in recurrence in the L-PAM group, the trial team drafted a second protocol that followed directly from the first—B-07—according to which the placebo group was to be transferred to L-PAM and the new experimental arm was to receive L-PAM plus 5-FU. B-07, in other words, transformed B-05 from a single-agent versus placebo into a combination versus single-agent study, in both cases preserving the multimodality (surgical adjuvant) nature of the protocol (Fisher et al., 1977). Interviewed by a reporter from *Science* in 1976 about the trial, Fisher explained: "We are going about this in a very orderly manner. First we tested a drug versus nothing, then one drug versus two. Now, we'll look at other combinations. The point is to find the minimal treatment that will do the job with minimal toxicity" (Culliton, 1976: 1030). The concern with toxicity, while common in any chemotherapy, had been further heightened by general surgical hostility to chemotherapy (recall that Fisher had replaced cyclophosphamide with L-PAM for exactly this reason) and the volatile context within which adjuvant trials were being performed.

The B-05 study continues to be cited as the one of the first "in principle" proofs of the utility of adjuvant chemotherapy in breast cancer (Levine & Whelan, 2006), but its results were quickly overshadowed—in spite of parallel citation curves (figure 6.6) illustrating once again the epistemic and institutional entanglement of these largely concurrent

5. Dr. Canellos, in an e-mail message dated 4 September 2009, denied any direct connection between 0971 and the Cooper regimen. Regardless, however, of the immediate relation between the two trials, the general perception of observers at the time was that the Cooper regimen had set the stage for 0971 and other similar trials.

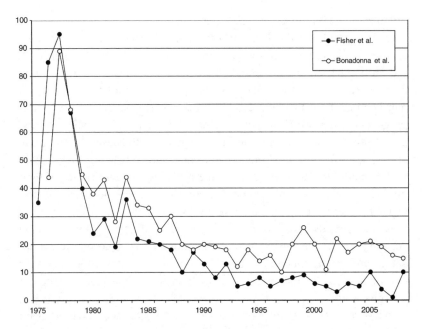

FIGURE 6.6. Citation curves of the publications reporting the initial results of B-05 and of the Milan trial. Data source: Web of Science.

trials—by those of an Italian adjuvant study that began a year later. Funded by the US NCI in spite of its foreign location, the trial, carried out by Gianni Bonadonna and his collaborators at Milan's National Institute of Tumors, compared CMF versus placebo in essentially the same population of patients as the B-05 study (Ribatti, 2007). In an article published barely a year after the publication of the B-05 initial results and in the same journal, the Italian team—albeit stressing the preliminary nature of their results—claimed that the CMF combination had reduced recurrences of breast cancer from 24% to 5%—that is, twice the reduction of the ECOG/NSABP study (Bonadonna et al., 1977). These results, their reliability, and their therapeutic consequences ignited a controversy that quickly reverberated outside clinical circles.

Hailed for its "monumental importance" (Holland, 1976b), popular weeklies (*Time, Newsweek*) broadcast the Italian trial results under headlines such as "Spectacular Hope" and "Breakthrough." As a harbinger of things to come, an article in *Science* titled "Reports of new therapy are greatly exaggerated" quickly reined in the enthusiasm. NCI officials interviewed for the *Science* article agreed that although the study was good, it did not rise to the status of "monumental," an opinion that gained ground in the years that followed (Culliton, 1976: 1029; see Holland [1976a] for a reaction to the exaggeration charge). Commenting in 1978 on the original B-05 and Milan publications, the former head of the CTEP and director of the Northern California Cancer program at Stanford University concluded that clinical investigators and practicing clinicians, caught up in the general enthusiasm for the treatments, had overestimated their preliminary nature (Carter, 1978a: 2). In 1983 Bonadonna and colleagues, while restating the importance of their findings, concurred and acknowledged that "disappointment [could] in part be explained by excessive initial expectation, primarily from those not directly involved in prospective clinical trials" (Bonadonna et al., 1983: 1157). The results of the Bonadonna trial continue to be updated (and to fuel debate) at regular intervals decades after they were first presented (Bonadonna et al., 1995; 2005; Veronesi, Bonadonna & Valagussa, 2008).

Alert readers will now ask: why did the NCI-funded CMF adjuvant trial take place in Italy, since all the other trials leading up to it were North American? Bonadonna, a member of the EORTC council in the 1970s, but who carried out the trial independently of the European organization, has offered the following account: while visiting Paul Carbone in May of 1972 at the National Cancer Institute, the ECOG chairman and NCI Associate Director for Medical Oncology briefed him on recent and ongoing breast cancer investigations carried out by US groups. In the process, Carbone gave him access to the protocol of ECOG 0971 and to the first draft of the B-05 protocol, before B-05 had in fact started. Bonadonna instantly came up with the idea of hybridizing the two by using CMF instead of L-PAM in an adjuvant trial. Following discussion with Carbone, who encouraged him to write a protocol along those lines, he set to work and within "the next few hours," he had produced a draft for the study. As Bonadonna recounts, that same day:

> I got Paul's approval almost immediately—before noon—because the experimental design, including patient selection, stratification parameters, and follow-up evaluation, was almost superimposable on those of the NSABP [B-05] trial. Last but not least, he lent his support to funding the studies through NCI. (Bonadonna, 1985: 22)

Other accounts accord the NCI a more proactive role and suggest that the NCI reached out to Bonadonna to see whether his institute would be willing to carry out the trial, given its large breast cancer patient population (DeVita & Chu, 2008; Culliton, 1976). Regardless of which account is correct, the question still remains: why did Carbone fund an Italian instead of a North American CMF adjuvant trial, especially since its protocol seemed to be the obvious next step down the path laid out by previous trials? It may have been the case that this was merely a way of speeding up the process by simultaneously testing different protocol variations. Two of our informants, however, mentioned special circumstances that made it possible to carry out the trial in Italy but not in the United States. Specifically, given the aforementioned surgeons' hostility to chemotherapy, no single large institution, beyond those already tied up with the B-05 protocol, was apparently willing to host a combination adjuvant trial (Culliton, 1976). Bonadonna's institute in Milan, in contrast, was not only eager to do so, it could also count on a large cancer patient population and special facilities for clinical trials and combined modality treatments, modeled along US lines. Readers will have noticed that Bonadonna's trial was a single-institution, as opposed to a multicenter trial: while this obviously set constraints on the accrual of patients, those constraints, so the Milan trialists argued, were in a sense compensated by improved study accuracy due to a more consistent application of eligibility criteria, treatment administration, and follow-up examinations, which translated into a drastic reduction in protocol violations and follow-up losses (Bonadonna & Valagussa, 1978). As we will see in chapter 9, this implicit criticism of multicenter trials figured prominently in a controversy about the value of large randomized trials in general.

The rush of results from overlapping trials had generated other problems that add further interest to the choice of Milan for the trial. As we know, the B-05 organizers had rapidly switched their placebo arm patients to the treatment regimen and created a new experimental arm. According to the NCI director, once even preliminary data had shown that L-PAM was superior to placebo, it had become unethical to withhold it from patients (Culliton, 1976). How, then, could a CMF versus placebo be carried out in Italy? Or conversely, and in the words of a US reporter, "could such a study [have been] conducted in [the US] where the present [1976] climate is one in which the ethics of human experimentation are foremost in everyone's mind?" (Culliton, 1976: 1031; see also Jones,

1978: 56). There was no easy answer to that question. As noted almost a decade later by two University of Virginia oncologists, the two initial adjuvant trials did make it difficult to design a trial with untreated controls, and yet such a trial was sorely needed since in spite of the Bonadonna results the jury was still out on the value of the CMF regimen (Goldstein & Webb, 1983). In other words, while it may have been "unethical" to withhold treatment from patients, it might have been equally "unethical" to give them inadequately validated treatments (Bonadonna et al., 1983: 1157). A Dutch observer of adjuvant therapies espoused a different rationale, arguing that the US funding system made "it desirable, on one hand to follow each new and popular breakthrough as a nominal background for the [next] studies and, on the other hand, to draw conclusions from a study early enough to support the next grant application."[6]

The Bonadonna trial raised yet another related issue. In the flurry of trials conducted at the time, many decisions had been made on interim and partial results, unconcluded trials, and trials not yet underway. When was an early result too early to be used and who then should control the flow of information from one trial to another? As we will see in part 3 of this book, this dilemma received an institutional "solution" with the establishment of data monitoring committees charged with policing the circulation of data. For now, we note that problems such as the one just described emerged from a series of interlocking protocols and would increasingly become the norm. And their solution, no matter how temporary, required an equivalent application of sociotechnical ingenuity.

The debate over adjuvant therapy did not remain confined to trialists and surgeons. Patient activists entered the fray. Rose Kushner, a prominent activist who had supported the crusade against radical mastectomy (Lerner, 2001), began in the early 1980s to suggest that adjuvant chemotherapy had become the new orthodoxy in the United States, the equivalent of the Halsted radical mastectomy and equally as pernicious for women (Kushner, 1984). In a heated exchange with two NCI trialists (Feedback: Adjuvant chemotherapy . . . , 1985: 184–91), she urged caution, counseling the administration of adjuvant therapy only under controlled clinical trial conditions, and instanced the emergence of new, far

---

6. L. M. van Putten, "Paris, June 1978 EORTC plenary meeting: Adjuvant therapies and markers of postsurgical residual disease. A retrospective evaluation." *EORTC Newsletter*, no. 67, September 1978, 1–2.

less toxic treatments such as tamoxifen that we will discuss later. Much to the dismay of Bernard Fisher (1991: 103), the National Institute of Health Consensus Statement on adjuvant breast cancer therapy (1985) adopted a position that shared several of Kushner's misgivings. In spite of what may appear to be a setback, it is important to observe that it is not so much chemotherapy that was in dispute from this date on but results of specific clinical trials. Debates about chemotherapy henceforth took place *within* the field of clinical cancer trials implicitly endorsing the need to compare protocol results and competing regimens.

Conclusion

What is the lesson for us here? Clearly any clinical trial admits of a number of interpretations, and here we refer not simply to the statistical results but also to the significance of the trial within the totality of trials deemed relevant from a variety of research and therapeutic perspectives. If taken in isolation from concurrent trials, and adopting a narrow focus on the substances tested, ECOG 0971 compared L-PAM to CMF and found the latter superior. This, however, was clearly not the whole story. Over and above the specific substances involved, the trial compared a single-substance to a combination therapy. Thus from the central perspective of the NCI and its clinical research vantage point, ECOG 0971 constituted a test of combination chemotherapy for solid tumors and built on previous success in leukemia. Viewed from within the cooperative groups themselves, although 0971 concerned primarily advanced breast cancer patients, the two arms of the study were prospectively related to adjuvant trials that combined surgery and chemotherapy and thus launched the multimodal era. In this regard the trial fit comfortably into the groups' ongoing effort to integrate the different modalities into their research programs. Finally, from our point of view, 0971 is a typical case of all of the above; in other words, it is emblematic of a new style of practice that among other things tested regimens within an evolving network provided by interacting substances, practices, and diseases that clearly defy reduction to chemical structures or methodological pronouncements.

CHAPTER SEVEN

Statisticians, Data Centers, and the Organization of Large-Scale Clinical Trials

With time flowing from top to bottom, figure 7.1 is a blow-up of the left-hand side of figure 2.9—a map of the articles co-cited in the papers published by US cooperative groups between 1955 and 1975—that we used in chapter 2 to visualize the network of publications documenting ALGB's contributions to leukemia research. Figure 7.1 displays the co-citation pattern of a number of breast cancer articles discussed in chapter 6, but more tellingly, it illustrates the centrality of biostatistics to group publications. These contributions went well beyond conceptual tools. Embodied in computer algorithms and a swarm of daily routines that formed the basis of distinct institutions known as data centers, statisticians, data managers, and a host of ancillary personnel increasingly took over the management of clinical trials and their results.

Centers of Calculation: US Statisticians and the Data Center

As protocol statistician and hence as second signatory to the 1976 *Cancer* publication presenting the results of ECOG 0971, Stuart Pocock symbolized the special position accorded statisticians as full members and collaborators within the cooperative oncology groups. It had not always been so. As we saw, clinical trial statisticians were originally consultants to, and occasionally policemen for, the clinical trials conducted by the cooperative groups. As such, statisticians had worked primarily out of

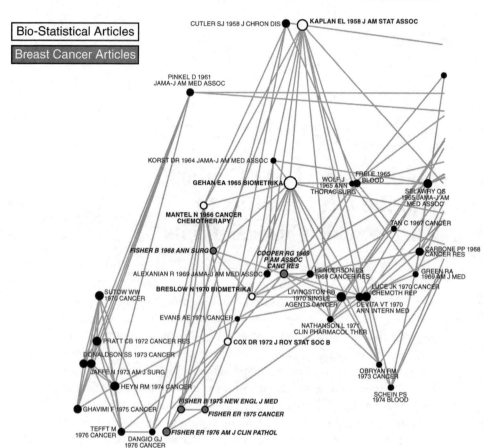

FIGURE 7.1. Co-citation map of cooperative group publications, showing the key role of biostatistical articles. Data source: Web of Science.

the central statistical office of the NIH. Toward the end of the 1960s the NCI decided to provide the groups with funds to hire their own statisticians. Edmund Gehan, for example, left the NCI to become the statistician for the Southwest Oncology Group. Other groups either hired members of the NCI staff or went looking for their own statisticians. At the same time (1967–68), some groups such as the Eastern Cooperative Oncology Group (ECOG) acquired the newly emerging computer technology. The combination of the two events led to some initial ambiguity about the role of a group statistician. It is perhaps hard to believe that after almost fifteen years of cooperative clinical cancer trials there could be some mystery as to what a statistician might do, but nonetheless, there it was, the question: What function would a full-time clinical-trial statistician fulfill in a cooperative oncology group?

In 1968 ECOG, for example, projected that the statistician and the computer would function jointly as a data dredging team to "provide a mechanism for obtaining otherwise inaccessible but valuable information with respect to both epidemiological data and the natural history of disease."[1] The group clearly viewed the statistician's work as tied to the computer and initially believed that the function of the group statistician would be more that of a computer programmer than a researcher. The ambiguity persisted for almost a decade, judging from a 1978 comment according to which "[a] common misunderstanding of the statistician's role [in cooperative groups] concerns his relation to the computer" (Breslow, 1978: 329). In the event, ECOG management placed their newly hired statistician in the operations office and made him responsible for what we would now call data manager operations. He lasted barely more than a year for, as the group chairman reported in 1969:

> The major requirements of our program are for data collection, collation, processing and programming, and the need for statistical manipulation is minimal. As time progressed it became increasingly evident that [the statistician] did not have an interest in our primary requirement, nor was he qualified to handle the problems involved therein.[2]

Moreover, even though the group statistician was supposed to approve all protocols along with the group chairman, his function in the analysis of results and as a collaborator remained more potential than actual.

After the statistician's early departure, the group hired a "programmer-statistician" who was also expected to act more as a programmer than a statistician. Statistical analysis and support for group studies was at the time deemed optional and thus was provided only "upon request." Moreover, the few services that the statistician did provide proved temporary since, according to the *Report*, the group forecasted that "basic

1. Eastern Cooperative Oncology Group, Chairman's Report, June 1963–April 1968, 20, Frontier Science and Technology Research Foundation Archives (Boston).

2. Eastern Cooperative Oncology Group, Group Chairman's Report in Support of Continuation Grant Requests, May 1968–December 1969, 5, Frontier Science and Technology Research Foundation Archives (Boston).

statistical services can be 'computerized.'"[3] The clouds that hung over the group statistician cleared in 1970 when Paul P. Carbone became the chairman of ECOG. In addition to reforming statistical services, Carbone undertook reforms on a number of related fronts.

During the first three years of his leadership, ECOG membership more than doubled and patient accrual tripled. Prefiguring future changes at the NCI Division of Cancer Treatment (DCT), Carbone further reoriented the type of studies carried out by ECOG, changing them from "a predominantly drug oriented broad spectrum Phase II or III trial to disease specific protocols."[4] The explicit goals of the group consequently shifted somewhat seismically from "testing antitumor drugs in humans"[5] to attempting to "improve therapy for patients with cancer."[6] On top of his clinical research expertise, Carbone had previously worked at the NCI with the statisticians who had, up until 1967, provided statistical support for the group. He thus hired his former NCI colleague, Marvin Zelen, as group statistician in order to establish a statistical support center separate from the central operations office. Unlike their predecessors, Carbone and Zelen saw a vastly more important role for the "Statistical Office" within the group.[7] Within a relatively short time, the statistical office became a going concern: by 1974, it included nine statisticians "with advanced degrees," three computer programmers, two keypunchers, three administrators, and a new category of practitioner created by Zelen, the data manager, of whom there were twelve. ECOG regarded statisticians other than Zelen as "protocol statisticians" in recognition of their status as collaborators in the process of protocol production and management, the essential business of the group. Equally as important was the alignment of the statisticians along disease lines in accordance with the internal organization of the group.

To gain some appreciation of the depth of this transformation, it is sufficient to note that prior to Carbone's chairmanship, clinical cancer

3. Ibid.

4. Eastern Cooperative Oncology Group, Group Progress Report, May 1971–May 1974, 4, Frontier Science and Technology Research Foundation Archives (Boston).

5. Eastern Cooperative Oncology Group, Chairman's Progress Report, January 1967–December 1967, 1, Frontier Science and Technology Research Foundation Archives (Boston).

6. Eastern Cooperative Oncology Group, Group Progress Report, May 1971–May 1974, 4, Frontier Science and Technology Research Foundation Archives (Boston).

7. Ibid.

trial protocols concerned the entire group. Thus once the group chair and the disease-oriented committee responsible for the study had reviewed a protocol initiated by a member of the group, it was submitted to the group as a whole and accepted or rejected by majority vote.[8] In keeping with the burgeoning knowledge of cancer and cancer therapeutics and the transfer of interest from drugs to diseases, however, by 1970 Carbone's ECOG no longer thought it possible or, more importantly, scientifically valid to create protocols that cut across all cancers or to submit cancer protocols for approval to the entire group. From the 1970s on, therefore, protocol initiation lay in the hands of the disease committees and although submitted to the group for comment, the disease committee retained control of the protocol, including the final decision as to whether or not to proceed with the study.[9]

Carbone's reforms established the disease committees as one of the central nodes of group activity. Thus as 0971 abundantly illustrates, research continuity was no longer tied to the substance to be tested but to the disease under study. What had begun with VAMP as an innovation in a single disease protocol, now structured the entire enterprise. The change had in fact been prepared several years in advance. As Carbone had pointed out in 1966 in a report that he had prepared as "an attempt at trying to orient more of the group studies to include more disease orientation," he saw "no reason why all Phase II and III studies could not be designed to study remission induction rates, remission maintenance with and without the active drug" and thus "get an idea of how therapy affects the natural history of [the patient's] *disease*."[10] Given the subsequent restructuring of the group, protocol statisticians were accordingly assigned to specific disease groups, and the transformation was more or less complete: from consultants in the quasi-industrial testing of substances, group statisticians became collaborators in the experimental study of human therapy and disease. As Zelen noted in the statistical

8. Eastern Cooperative Oncology Group, Chairman's Report, January 1967–December 1967, Attachment II, Procedures for Processing of Protocols, 1, Frontier Science and Technology Research Foundation Archives (Boston).

9. Eastern Cooperative Oncology Group, Group Progress Report, May 1971–May 1974, 32–33, Frontier Science and Technology Research Foundation Archives (Boston).

10. P. Carbone, M. Schneiderman & J. Lynch, "Report of the Subcommittee on Response," Enclosure I in Eastern Cooperative Oncology Group, Chairman's Progress Report, January 1967–December 1967, 10, Frontier Science and Technology Research Foundation Archives (Boston).

center report for 1974: "Protocol Statisticians are responsible for scientific aspects of the studies. Generally, they concentrate in specific disease areas and work closely with Chairmen of Disease Oriented Committees as well as Drug Study Chairmen."[11] To "work closely with," however, was not to "work for." As collaborators, protocol statisticians not only attended all disease committee meetings, they also maintained independent research activities. The statistical center was consequently referred to both as a *Center* and as a *Statistical Laboratory*. There, Zelen and his collaborators developed data analysis techniques beginning with "new concepts in randomization, analysis of survival data and contingency tables," and a variety of other numerical, graphical, and computer tools for the conduct of clinical trials.[12] In the years that followed, they made numerous contributions to statistical science that included both the application of statistical techniques developed in other fields and the statistical solution of problems generated by clinical trials.

Applications of statistics in clinical cancer trials were more than trivial extensions of methods from one domain to another. They entailed adaptations and, hence, significant modifications of the techniques in question. Consider sequential methods. ECOG began developing sequential methods for Phase III clinical trials in the early 1980s. As we have seen, such methods had originally been created for industry and, subsequently, Phase I and II medical trials. Their extension to Phase III trials required further transformation. Unlike Phase II trials, for example, sequential methods for Phase III trials had to account for discontinuous analysis of the data (the groups met only once every six months for interim analysis) and different kinds of end points—namely, survival data rather than specific clinical effects such as remission or reduction of tumor size (Pocock, 1982). They also had to take into account the fact that in large multicenter trials, patients did not arrive in two's as had been the case in early applications of sequential methods.

When designing multicenter trials, trialists soon realized that they had to balance patient treatment among participating institutions in order to avoid the possibility that in a trial comparing two treatments, centralized, randomized treatment allocation might assign patients in a given institution to only one of the two competing treatments. Such an

11. Eastern Cooperative Oncology Group, Group Progress Report, May 1971–May 1974, 20, Frontier Science and Technology Research Foundation Archives (Boston).
12. Ibid., 21.

outcome could introduce significant bias given the variation in practices from one institution to another and transform a comparison of treatments into a comparison of institutions. To circumvent this problem, cooperative group statisticians devised *stratified randomization* techniques to balance institutions. They moreover developed and modified these techniques as the design and logistics of clinical trials evolved from small, single-institution to large, multi-institution trials.

One of these techniques was known as "block randomization," but it created the following double-bind situation: if carried out centrally in a multi-institutional study, then it might balance the overall treatment allocation among institutions but not necessarily within individual institutions; if performed within a single institution it might allow clinicians to guess the treatment allocation of many patients. Thus cooperative group statisticians developed a technique known as "balanced block randomization" that was carried out centrally and corrected for imbalances within institutions (Zelen, 1974). Block randomization techniques were "static" techniques, and so yet another stage in the delicate balancing act between statistical considerations and practical constraints was the development of a "dynamic" method known as "minimization technique" that balanced institutions and prognostic factors (see below) without having to consider all possible combinations of these factors (Sylvester et al., 2002). In a further attempt to facilitate the trialists' work through augmented patient recruitment, Zelen proposed a Randomized Consent Design that randomized patients *before* asking for their informed consent, thus increasing the willingness of clinicians to enter their patients in a trial. Zelen's proposal caused a minicontroversy that mingled statistical and ethical arguments (Altman et al., 1995); we will not examine it here, except to note, once again, the constant entanglement of statistical and clinical research innovations.

In addition to stratifying treatment randomization by institutions, trialists began to stratify prognostic factors—that is, pretreatment factors, such as the stage or extent of disease, known to influence response to treatment regimens. Depending on the number of factors deployed, the resultant number of strata or categories created by this strategy multiplied quickly and could easily lead to an unequal distribution of patients within categories. In one breast cancer chemotherapy trial, for example, patients were stratified according to four different factors, leading to twenty-four different strata. After eighty patients had entered the trial, twelve of the categories were either empty or had only one patient (Pocock, 1979). Even worse, an ECOG trial comparing three regimens and using five prognostic factors led to seventy-two strata, forty-nine of which were completely empty after one year of patient recruitment; moreover, patients came from nineteen institutions, but these institutions could not be incorporated into the design since this would have led to 1,368 strata (Pocock, 1975: 113). Computer models developed in the 1970s at the NCI showed that only a small number of factors should be examined, lest the whole exercise become self-defeating (Pocock & Simon, 1975).

The extent to which stratification was desirable became subject to debate in the 1970s. An influential publication described stratification as an unnecessary complication and argued that imbalances created by prognostic factors could be corrected during the analysis of results (Peto et al., 1976; see also Meier, 1981). Noting that some researchers tended toward the other extreme and had become "excessively concerned about balancing treatment groups across a whole series of prognostic factors," Pocock promoted an intermediate solution, limiting randomized stratification to one or two prognostic factors. The trial's size also played a role: probably unnecessary in large trials, stratification could be of vital importance in small trials (Pocock, 1979). Even more pragmatically, trialists appeared to use stratified allocations less for statistical efficiency than to increase a trial's credibility, allowing researchers to claim that they had indeed compared comparable groups (Pocock & Lagakos, 1982). Stratifying trial populations caused statistical techniques such as the logrank test, mentioned in chapter 6 and used to compare survival curves, to fail. In the 1980s, ECOG statisticians produced a general solution to this problem (Schoenfeld & Tsiatis, 1987).

In 1982 Pocock and an American colleague carried out an international survey of "the methods by which randomization [was] actually performed" in cancer trials (Pocock & Lagakos, 1982: 368). Canvassing fifteen major cancer centers and cooperative groups in the United States, the United Kingdom, and continental Europe, the survey reached several interesting conclusions. With regard to the mechanics of treatment assignment, the majority of centers resorted to central randomization by telephone and some form of institution balancing. Prognostic stratification, however, divided American and European institutions: the former overwhelmingly endorsed the practice, whereas European centers showed some reserve. Not only did a number of them not even use prognostic stratification, but also those that did used fewer stratifying factors than their American counterparts. As to which factors should be stratified, while the survey showed that in the case of primary breast cancer there was some consistency in the choice of some factors (pre- and postmenopausal status, presence or absence of cancer cells in the lymph nodes), other factors remained controversial; for instance, should one favor "technical" factors such as the presence or absence of hormone receptors or histological classification, as clinicians often did, or opt for more "patient-oriented" factors such as weight loss or performance status? While the patient-oriented factors had less bioclinical interest, statisticians had shown that they did affect treatment response. The selection of factors often turned, pragmatically, on considerations such as the number of strata that a center considered "appropriate or manageable" (Pocock & Lagakos, 1982: 372).

The impact of innovations such as those promoted by Zelen made the ECOG "laboratory" a model for other groups from the outset.[13] The advent of computerized techniques offered a novel conduit for ECOG's influence: the computer program. Zelen recounts, for example, that shortly after the introduction of Cox's proportional hazards analysis (see chapter 6), statisticians at ECOG developed the first software to use this approach. The importance of the computerization of this technique should not be underestimated. Even though Cox's original paper counted among the most highly cited papers in the medical literature, few clinicians and researchers could actually follow the paper's highly technical, mathematical arguments. Even today, although they may cite the original paper, their contact with the technique has most likely been mediated by some off-the-shelf statistical package, since Cox's technique "could only be implemented by means of a computer program" (Machin, 2004: 524). ECOG's contribution thus consisted of a veritable reduction to practice of a mathematical innovation. Much prized by other groups, ECOG's software became an important means for the diffusion of proportional hazards analysis to the point where virtually all cancer clinical trials now use the technique or some variation thereof in the analysis of data.[14]

Work on software applications had begun quite early. Showing remarkable foresight, in the early 1960s the NCI had funded three research centers: MIT (for hardware), Tulane (for clinical research), and UCLA

---

13. Eastern Cooperative Oncology Group, Group Progress Report, May 1971–May 1974, 8, Frontier Science and Technology Research Foundation Archives (Boston).

14. Interview with Marvin Zelen, Boston, 6 December 2004.

(focused on "bringing the computer to the forefront of very general use within the research community"). Work on a software package known as BMD (or "Biomeds," later BMDP) began in 1961, and in 1962 the UCLA developers shipped free tapes of the original source code to anyone who requested it. For the first fifteen years, news of BMDP circulated widely by word of mouth in the United States and abroad as the software incorporated an increasing variety of statistical analyses, including survival analysis.[15] Because of federal government support, the software also ran on a variety of competing platforms. In 1975 BMDP left the public domain as the NIH gradually withdrew financial support, and while it continued to circulate for nominal fees for a number of years, a UCLA spin-off company eventually took over production of the package in 1983.

Statisticians in the United States were not alone in using software for the dissemination of their approach. In the mid-1970s, the well-known British clinical trialist Richard Peto and his collaborators published two widely cited papers (1976; 1977) in the *British Journal of Cancer* on the design and analysis of clinical trials. The publications laid out a template for clinical cancer trials in the years that followed. A great deal of the impact of the publications flowed from the production of a computer program, known as the Oxford program, that automated the methods advocated in the two articles and whose circulation allowed workers in data centers to translate those ideas into routine practice (Machin, 2004).

The Mechanics of Data Production

In addition to their mathematical contributions, statisticians in the cooperative groups introduced a number of organizational innovations. Often overlooked because they seem, at first sight, more managerial than scientific, they deserve notice because of the close interaction between the two in a domain characterized by the high premium placed on organization. In what follows we describe some of the organizational innovations introduced by ECOG statisticians and the statistical office (which includes more than statisticians) who trained an inordinate number of

15. Luanne Johnson (interviewer), oral history of Wilfrid J. (Wil) Dixon and Linda Glassner, recorded 27 March 1986, Los Angeles, California), Computer History Museum, 2007 (CHM reference number: X4365.2008).

clinical trial statisticians and whose practices, as previously noted, became widely recognized as models for other groups.

One of ECOG statisticians' first organizational innovations was the institution of centralized randomization (Kalish & Begg, 1985). Prior to centralization, doctors randomized patients during a clinical trial at the local (hospital) level, a solution that created numerous opportunities for bias. The situation improved with the first stage of central randomization. At this juncture ECOG sent closed envelopes to physicians containing random assignments to one of the two treatments under study. Physicians would receive a batch of envelopes and with each succeeding patient would open one to determine which of the two treatments to use. This system, however, was less than perfect. As previously noted, clinicians would occasionally hold the envelope over a light bulb to determine which of the two treatments was prescribed and thus circumvent the central randomization by conducting their own local randomization. To overcome such local variations, Zelen instituted central randomization by telephone, thus forcing physicians to conform to randomization at the group level.[16] Not all groups immediately followed suit. In the late 1970s CALGB continued to use "sealed, opaque envelopes" because "telephone entry [was] impractical for foreign institutions."[17]

In addition to eliminating local variation, the centralization of randomization schedules allowed the statistical center to deal with the increasingly complex nature of randomization. Whereas early cooperative group studies randomized undifferentiated patient populations with regard to treatments or dose sizes, the subdivision of disease categories and the rising interest in the study of prognostic factors created the need to randomize a far more stratified patient population. To maintain randomization in an increasingly fragmented population, the center instituted computer simulation of randomization schedules to ensure that randomization would indeed produce a random distribution.[18] Finally, in the mid-1980s ECOG computerized the entire registration and randomization systems (Lange & MacIntyre, 1985).

16. Interview with Marvin Zelen, Boston, 6 December 2004; interview with James Holland, New York, 24 July 2003.

17. Cancer and Leukemia Group B (CALGB), Group Progress Report, 1976–1977, 12–13. James Holland, personal papers.

18. Eastern Cooperative Oncology Group, Progress Report, June 1980–June 1984, 119, Frontier Science and Technology Research Foundation Archives (Boston).

In addition to randomization, clinical trials required data forms to control and contain the potential tsunami of data that flowed in from participating hospitals and community cancer centers. To understand the constant growth in data, take the simplest instance: the entry of a patient onto a clinical trial. Not all patients with leukemia, for instance, are suitable for a clinical trial testing an antileukemic substance. Patients must first of all be subdivided into the different disease subcategories (childhood leukemia, acute leukemia, etc.). Patients who have already been treated differ from those who have not. Likewise, patients near the end of the disease differ from those who have only recently been diagnosed. Clinical trial protocols thus created criteria that attempted to generate homogeneous patient populations with regard to a constantly growing number of significant variables concerning response to therapy and the evolution of the disease under study. These criteria could be extensive. Consequently, information relating to these criteria had to be collected in order to assure commonalities (and to catch potentially significant differences) within the patient population. Patient-entry data were merely the beginning of the process; there were also forms to determine study end points, forms to collect data on prognostic variables, flow and measurement sheets, interim forms, summary and evaluation forms, modality specific treatment forms, and so on. In sum, however mundane it may appear, "an essential factor in capturing and reproducing high quality data [was] a well designed form."[19]

In the first three years of the statistical center, ECOG statisticians and data managers had to design, by hand—computerized designs were not possible until the mid-1970s—over fifty different forms. Designing forms entailed, of course, the translation of the vaguely discursive data that appeared on hospital forms—for example, "The patient appears to have some anemia"—into semiquantitative and qualitative data, for example, "Does the patient have anemia? 1 = yes, 2 = no, 3 = questionable, 9 = unknown."[20] That translation has required the continual construction and refinement of study forms that allow investigators to enter the required information in a self-coding manner. The proliferation of studies and disease types has been accompanied by a proliferation of forms. By 1980 their number had almost tripled, and in an average year

19. Ibid., 115.

20. Eastern Cooperative Oncology Group, Group Progress Report, May 1974–June 1977, 71, Frontier Science and Technology Research Foundation Archives (Boston).

ECOG added more than thirty new forms to the database. The workload created a new division of labor, and as of the mid-1980s ECOG turned over the form design process to data managers, the personnel closest to the process.[21]

ECOG also introduced a series of quality control measures to ensure that data forms were filled with consistent data. Data managers combed through the returns throughout the course of the clinical trial checking for blanks and inconsistencies and then followed up with requests for supplementary information. In 1976, for example, data mangers sent out over 1,500 letters to institutions asking for additional information on data form submissions.[22] This somewhat tiresome activity was automated shortly thereafter so that a computerized request for missing data could be sent out on a monthly basis by ECOG computers. In 1992 some 6,000 requests for information were sent out.

Noncoercive measures such as written and computer-generated inquiries had well-defined limits. So in 1977, with the realization that upwards of 20% of data forms were outstanding, the group took further action. Institutions with more than 10% of their forms outstanding—calculated by the computer on a semiannual basis—henceforth lost randomization privileges and were thus prevented from entering patients on protocol.[23] Delinquency quickly fell to less than 4% in the next three years.[24] ECOG statisticians and data managers also launched workshops for clinicians and data managers (the data manager position had become somewhat universal following its creation in the early 1970s). Programs, seminars, and training sessions proliferated to the point that ECOG reports began to speak of the group as a "university without walls."[25]

ECOG augmented its education and delinquency program in 1979 with the establishment of an audit system that randomly audited institutions. The need for the addition of the audit had emerged as the limits of the computerized surveillance system became apparent. Specifically, although the computer audited the timely submission of data and their

21. Eastern Cooperative Oncology Group, Progress Report, June 1980–June 1984, 115, Frontier Science and Technology Research Foundation Archives (Boston).

22. Eastern Cooperative Oncology Group, Group Progress Report, May 1974–June 1977, 62, Frontier Science and Technology Research Foundation Archives (Boston).

23. Eastern Cooperative Oncology Group, Progress Report, June 1977–June 1980, 52, Frontier Science and Technology Research Foundation Archives (Boston).

24. Ibid., 6.

25. Ibid., 10.

quality, there was no way of knowing the degree to which the submitted data reflected the data contained in the original patient hospital record. To be able to make that comparison and thus to maintain true quality control—or at least a quality control that followed the "chain of custody" back to the original case files (including, for example, x-rays)—inspectors had to go on site to compare selected submitted data sheets and case files. On-site inspections also enabled ECOG personnel to correct deficiencies and undertake the educational efforts that computerized mail-outs clearly could not provide.[26]

As one of the offices in charge of quality control, the Statistics Office became an obligatory point of passage in the management of the credit and discredit of participating institutions. The Statistics Office not only evaluated participating hospitals in terms of protocol and data-form compliance, they also assessed the institutions in terms of the number of patients put on protocol—the more patients, the better. Crediting institutions for patients on protocol was more than a matter of distributing brownie points for good behavior; group membership crucially depended on participation. When, for example, budget cuts in the late 1970s created demands for greater efficiency, ECOG pruned its member institutions principally according to patient-accrual numbers and dropped almost 25% of its poorest-performing hospitals. As a result, in less than two years the number of member institutions fell from forty-one to twenty-eight even though patient accrual remained steady.[27] Patient accrual became such a crucial aspect of maintaining member status (and funding) that when ECOG began using protocols that required longer follow-up and greater work to maintain the same patient numbers, the Statistics Center sought ways to provide "extra credit" for physicians and institutions that accrued patients to such long-term protocols.[28]

Meanwhile in Europe, Take Two

In its early years the EORTC experienced a paucity of resources due to the fact that it received very limited and only occasional governmental support. All the EORTC could offer its members, beyond a platform for

26. Ibid., 52.
27. Ibid., 2.
28. Ibid., 18.

interfacing their activities, was its name. Local institutions and national agencies provided the funding and infrastructure for the members' preclinical and clinical work. As a consequence the European groups lacked central oversight. At the end of the 1960s the growing EORTC succeeded in securing funds from the NCI for a central structure. In turn its image enhanced through NCI backing, the EORTC secured increased support from European sources and established a fund-raising foundation.[29] Growth and centralization, however, created their own problems. As the number of clinical cooperative groups increased, so did their heterogeneity.[30] As a result, in 1969 the EORTC decided to reorganize the groups' activities and to this end established the position of special coordinator. Maurice Staquet, a Belgian physician who had specialized in biostatistics at Columbia University, was chosen for the position, which he held until 1990. His initial tasks included the analysis of the reports, protocols, and results of each clinical cooperative group as well as site visits. On this basis he submitted proposals for reorganizing group procedures and policies to EORTC's council.[31] According to the EORTC president, the "weak spots" that Staquet had brought to the council's attention had in most cases little to do with administrative or organizational matters such as the regional nature of some groups and the problems of communication created by the plurality of languages in Europe.[32] Rather, whereas the United States had a plethora of well-trained young

29. In 1972, the United Kingdom, Denmark, and Ireland joined the Common Market, stimulating the interest of European oncologists to work with the EORTC, a process further enhanced by the ripple effects of the US "war on cancer" that was launched in 1971 by President Nixon (Annual Report of the President for 1972, *EORTC Newsletter*, no. 25, February 1973).

30. US cooperative groups such as ECOG are large organizations that include teams working on different types of cancer; the EORTC, from a US perspective, is a cooperative group comparable to ECOG. The EORTC, however, uses the term "cooperative groups" to refer to subgroups focusing on a given type of cancer or activity; in short, what US groups call disease committees. While this might confuse some readers, we have decided to keep the actors' terminology.

31. Annual Report for 1969 by the President of the Council, *EORTC Newsletter*, no. 3, January 1970. For Dr. Staquet's appointment and tasks, see *EORTC Newsletter*, no. 2, December 1969.

32. Analysis of EORTC by the Past President, *EORTC Newsletter*, no. 38, February 1975; Report on the Meeting of the Committee for Clinical Cooperative Groups, the Data Center and Certain Task Forces (November 22, 1977), *EORTC Newsletter*, no. 62, February 1978/B.

oncologists, most European countries lacked formal training in medical oncology in the mid-1970s.[33]

While waiting for the emergence of a critical mass of well-trained European oncologists, more intense collaboration with US cooperative groups offered a temporary solution. Still, other problems remained: in spite of a growing number of projects, due partly to the large number of trials abandoned before completion, the European cooperative groups published only four papers in 1971. Even when completed the clinical trial results were sometimes less than solid, as statistical rules were "not always carefully followed."[34] A report on papers presented as late as 1978 at a Paris meeting on adjuvant therapies noted that many studies had not obtained statistically significant results because of insufficient patient numbers or observation period, or had simply defied interpretation because of the absence of a randomized control population.[35] To remedy the situation, the EORTC organized courses and workshops on the design of trials (Staquet, 1972), developed "minimum requirements for EORTC clinical protocols" ratified by the groups' chairmen and secretaries,[36] and above all established a data center in Brussels that became the heart of the organization.

Opened in 1973, the data center relied initially on NCI funding. Modeled on "statistical units and laboratories serving several of the largest cooperative clinical groups in the United States,"[37] visiting American statisticians from academic centers and the NCI played an important role in its early organization.[38] In addition to the data center, the EORTC established a central Protocol Review Committee whose approval became a mandatory requirement for trial sponsorship by the EORTC and for the use of the EORTC name in publications. Together, the Protocol Review

33. Annual Report of the President for 1970, *EORTC Newsletter*, no. 9, February 1971; van Bekkum, "Report on a Study of Comprehensive Cancer Centers in the USA," *EORTC Newsletter*, no. 43, November 1975.

34. Annual Report of the Coordinator for 1971, *EORTC Newsletter*, no. 18, February 1972.

35. Van Putten, "EORTC Plenary Meeting: Adjuvant Therapies and Markers of Postsurgical Residual Disease," *EORTC Newsletter*, no. 67, September 1978.

36. Activities of Clinical Cooperative Groups during 1972, *EORTC Newsletter*, no. 25, February 1973.

37. Meeting of the Chairmen and Secretaries of the Cooperative Groups, *EORTC Newsletter*, no. 34, May 1974. See also George (1976).

38. Report of the President of EORTC for the Year 1975, *EORTC Newsletter*, no. 45, February 1976.

Committee, the Coordinating Office, and the data center helped the cooperative groups to formulate ideas, draft protocols, and organize their activities.[39] So as not to step on too many toes, both the Protocol Review Committee and the data center initially drew clear lines between consultation and control, and issued statements to assuage fears of a centralized takeover.[40]

By 1974 the data center consisted of a director and principal statistician (Staquet) supported by a protocol statistician, a computer analyst, a data manager, a data clerk, and a secretary. The statisticians ran the center and coordinated the cooperative groups' trials. The data manager collated and verified the data collected from participating institutions, and the data clerk handled central randomization.[41] By 1977 the staff had increased to eleven, including three statisticians and four data managers,[42] and by 1985 the Center housed twenty-five members including five statisticians and eleven data managers.

The establishment of the data center generated some dissension. Since a number of cooperative groups had their own statisticians—in particular the French who, as we saw in part 1, had developed a prominent statistical center at the Gustave-Roussy Institute—some objected to the centralization, arguing that it might inhibit research in other centers.[43] Coordination between existing biostatistical centers was, in their opinion, a more viable option.[44] While proponents recognized that the "usefulness of the data center [had] not been understood by all cooperative groups"[45] and that excellent biostatistical departments already existed in associated institutes and hospitals, they nonetheless maintained that the data center represented progress insofar as it exclusively served EORTC needs and that as a consequence and in contrast to statistical units in academic centers it focused on the special problems arising from

39. Annual Report of the President for 1973, *EORTC Newsletter*, no. 32, February 1974.

40. EORTC Protocol Review Committee, *EORTC Newsletter*, no. 80, January 1980.

41. Meeting of the Chairmen and Secretaries of the Cooperative Groups, *EORTC Newsletter*, no. 34, May 1974.

42. The EORTC Data Center, *EORTC Newsletter*, no. 54, April 1977, Supplement.

43. Report on the 2nd Meeting of the Protocol Review and Coordinating Committee (May 14–15, 1974), *EORTC Newsletter*, no. 35, July 1974.

44. Proceedings of the Council Meeting (30 June 1974; 30 November 1974), EORTC Archives (Brussels).

45. Ibid.

cooperative clinical trials.[46] While the centralizing approach prevailed, the EORTC also established a statisticians' club to foster exchanges among different biostatistical offices.[47] The organization promoted further interaction by convening regular meetings of data managers,[48] and the development of computer technology in the 1980s led to the creation of computer networks linking a dozen European statistical centers.[49]

Once established, debate over the data center shifted from questions of its necessity to its compulsory use by all group trials. In 1976 members of a cooperative group oversight committee advised that the data center handle data for all new trials and that recourse to the center become obligatory for any group receiving EORTC funds. The council rejected the advice, suggesting instead that "an equivalent quality in handling data" would be sufficient to garner EORTC support, adding, nonetheless, that computer and statistical expenditures incurred outside the data center would not be reimbursed.[50] In practice, within a year of operation, six of the fifteen cooperative groups used the data center for their trials, and although the groups remained free to organize data handling, within three years the center had developed close relationships with fifteen out of twenty-three groups.[51] As figure 7.2 shows, data center activities grew steadily throughout the 1980s, reaching a plateau in the number of protocols (but not patients or participating institutions) at the beginning of the 1990s. Figure 7.3 compares the activities of the data centers of a number of US and European cooperative organizations and shows that by 1990, the EORTC data center had attained a size similar to that of ECOG.

In addition to helping cooperative groups plan, conduct, and analyze clinical trials, the data center also helped standardize trial methodology

46. The EORTC Data Center, *EORTC Newsletter*, no. 54, April 1977, Supplement.

47. Analysis of EORTC by the Past President, *EORTC Newsletter*, no. 38, February 1975.

48. Proceedings of Meeting of the Board, 28 November 1985, EORTC Archives (Brussels).

49. Proceedings of the Board Meeting, 23 August 1986, EORTC Archives (Brussels).

50. Report on the Meeting of the Committee for Clinical Cooperative Groups, the Data Center and Certain Task Forces (November 22, 1977), *EORTC Newsletter*, no. 62, February 1978/B.

51. Analysis of EORTC by the Past President, *EORTC Newsletter*, no. 38, February 1975; The EORTC Data Center, *EORTC Newsletter*, no. 54, April 1977, Supplement.

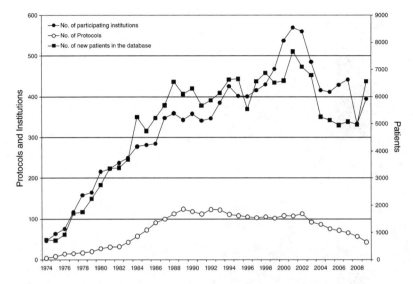

FIGURE 7.2. The activities of the EORTC data center: number of new patients entered in the database (right vertical axis), total number of clinical protocols in progress, and number of participating institutions (left vertical axis). Data kindly provided by the EORTC data center.

and logistics.[52] Trial methodology and logistics remained controversial topics in the 1970s as only a minority—about one-fifth according to a US study (Chalmers, Block & Lee, 1972)—of cancer clinical trials underwent randomization. Beyond the methodological objections to randomization (which we review below) many of the difficulties raised by multicenter randomized trials entailed a mixture of social, technical, and organizational issues and a distinctive moral economy. According to a description of the practice of cooperative trials produced by the EORTC coordinator, one of the primary considerations determining the decision to carry out a clinical trial was the burden created for senior clinicians (Staquet, 1976). Since the EORTC was unable to provide financial rewards in the early 1970s, clinical trial teams had been forced to rely on nonmonetary incentives and local commitments of resources to complete trials. Furthermore, as we saw in America, individual investi-

52. Meeting of the Chairmen and Secretaries of the Cooperative Groups, *EORTC Newsletter*, no. 34, May 1974.

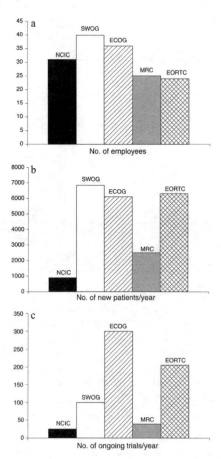

FIGURE 7.3. Comparison between data centers in 1990: (a) number of employees (full-time equivalents); (b) number of new patients/year; (c) number of ongoing trials/year. Source of data: The EORTC data center, March 1992, EORTC Archives (Brussels).

gator motivation was often undermined by the fact that trials resulted in collective publications and thus in diluted scientific credit. Last but not least, effective cooperation required shared interests, skills, and facilities: a cooperative group had to function like a well-oiled machine rather than a learned society where members reported personal experiences.

Motivation, however, was not enough. The design of effective protocols provided yet another challenge: even shared interests among participants who often came from different specialties—medical oncologists, surgeons, radiotherapists, radiologists, pathologists, statisticians, and pharmacologists (Bonadonna & Valagussa, 1978)—could not override the proliferation of diverse ideas about treatment modalities and patient selection. Two partial solutions were at hand: an inclusive compromise that incorporated aspects of the various proposals—a choice that produced "self-destroying" protocols (also known as "octopus protocols") since the intercalation of multiple stratifications of the population and treatments eventually led participants to ignore the most convoluted rules—and a minimalist approach that restricted the protocol to simple questions that could be answered quickly but that discouraged and dismayed the more demanding clinicians. It is important to recall that, as argued in chapter 1, protocols cannot be reduced to a simple treatment scheme or a sequence of tasks; with the development of clinical trial practice, protocols incorporated a growing number of items whose definition was often the object of debate. For instance, criteria such as complete or partial response, progression, disease-free intervals, not to speak of the prognostic factors used in patient stratification and other aspects of statistical design, were as much problems as measurements (Staquet, Sylvester & Jasmin, 1980). While these criteria often became routine, especially when deployed in standard protocols, their use in patient records for assessing patient eligibility or treatment results, for example, ideally required independent review by experts.

The difficulties were not simply conceptual; each problem had a corresponding "managerial" aspect. One of the essential tasks of the cooperative group's protocol committee was to prevent the production of the aforementioned "octopus protocols" and to identify related problems of leadership and feasibility before forwarding the protocol to the central EORTC protocol review committee that constituted a second level of monitoring. The subsequent implementation of a trial predictably created its own logistical difficulties. Geographical distance (and, in Europe, language barriers) disrupted communication, while institutional heterogeneity put the uniform handling of patients at risk. A less obvious difficulty was "chronology bias"; that is, inconsistencies caused by the fact that senior clinicians who lost interest in the trial tended to shift tasks such as patient selection to less experienced investigators. Confronted with problems like varying criteria for patient inclusion, the EORTC data center was forced to make hard choices that, following a necessarily joint assessment by the statistician and the clinician, often resulted in data rejection.

As at ECOG's statistical center, tasks at the EORTC data center included many routine activities such as data collection and editing: the flow of patient information had to be checked for omissions, inconsistencies, and protocol violations; and the valid records had to be computerized. But the data center was also involved in protocol design and the creation of data forms that required the review of new and standard therapies, the selection of prognostic variables for stratification, the determination of sample sizes and the duration of trial, and the preparation of report forms. Fresh problems arose given the nature of the participating institutions. The urological group noted in one of its reports that the fact that the data center allowed hospitals with no academic tradition to participate in projects created special needs. For such institutions, trial design had to be "as uncomplicated as possible and not require facilities and time that were available only in research institutions." Moreover, if too many centers took part in a study (the urology group had recruited almost one hundred centers), the quality of data sometimes suffered, hence the need for heightened control of protocol violations.[53]

Finally, as data flowed in, the data center produced two kinds of reports: administrative and statistical. Administrative reports determined whether the trial had progressed satisfactorily and focused on the performance of participating institutions in order to spot and correct data collection problems. Statistical reports included interim analyses and final analyses of all aspects of a trial and were presented at the biannual meeting of the cooperative group.[54] Interim analyses would soon become controversial, as we will detail in part 3 of this book. Interestingly, this kind of data circulation initially raised few eyebrows. As a former EORTC statistician recalled:

> I think the field has moved a lot in terms of doing interim analysis. For a long time, certainly during all the time I was with the EORTC, there was a policy of doing one interim analysis per year on each randomized study and probably for most non-randomized phase 2 studies as well. . . . This was very naïve when you think of it retrospectively. We would actually show the results of the interim analysis at every group meeting and say "Well, this study has accrued two hundred patients and here are the [interim results]" And people would say "How very interesting; we need to continue." Very informal.

53. The Urological Group: Progress, Possibilities and Problems, *EORTC Newsletter*, no. 66, July 1978.

54. Meeting of the Chairmen and Secretaries of the Cooperative Groups, *EORTC Newsletter*, no. 34, May 1974; The EORTC Data Center, *EORTC Newsletter*, no. 54, April 1977, Supplement.

I think today this would be considered completely heretic but it was okay, I think. It didn't lead to any major disaster at all.[55]

The data center, the protocol review, and the cooperative group over-sight committee established new relations within and among coopera-tive groups and between the groups and the EORTC central office. In order to improve the groups' studies, the data center, as we saw, partic-ipated in group activities from the early planning stages to the final re-porting of results. Moreover, as we also saw, the data center evaluated cooperative groups. As noted by the committee established in 1976 to perform this task, group evaluation—intended to improve trial practices rather than merely discipline them—drew on both quantitative and qual-itative data.[56] Quantitative data included the numbers of patients entered in a trial, the number of trials performed, and such, and were routinely collected by the data center. Qualitative data relied partly on indicators collected by the data center (e.g., trial entry requirements, protocol vio-lations, publications) and partly on more elusive information based on interviews with group representatives: originality of the studies, current objectives, evidence of self-criticism, and such.[57] Given the increasing encroachment of the data center and other committees on group activ-ities and the danger of polarizing central and local institutions, group representatives were asked to join the EORTC council and board, and thus to participate in discussions to reassess and redefine the aims and underlying philosophy as well as the logistics of the operations of the organization.[58]

Back in the USA: Group Statisticians and NCI Statisticians

Group statisticians and group statistical centers did not work alone in the United States. Since the end of the 1970s, clinical cancer protocols

55. Interview with Marc Buyse, Paris, 25 May 2005.

56. Report on the Meeting of the Committee for Clinical Cooperative Groups, the Data Center and Certain Task Forces (November 22, 1977), *EORTC Newsletter*, no. 62, February 1978/B.

57. Criteria for the Evaluation of EORTC Cooperative Clinical Groups, *EORTC Newsletter*, no. 76, August 1979.

58. Analysis of EORTC by the Past President, *EORTC Newsletter*, no. 38, February 1975; President's Report, *EORTC Newsletter*, no. 80, January 1980.

devised by the cooperative groups had been subject to oversight by clinicians and statisticians at the NCI's Cancer Therapy Evaluation Program
(CTEP). Structured along disease lines, CTEP appointed a disease coordinator to review protocols in 1978. The cooperative groups, and in
particular their chairmen, clearly found such oversight a challenge to
their expertise and authority. In the following exchange, Vincent De-
Vita, the director of the Division of Cancer Treatment, confronted the
chairmen shortly after the nomination of the first disease coordinators:

> The chairmen bristled when the discussion moved on to DCT's "disease co-
> ordinators."
>
> "What can he tell us we don't already know?" Hoogstraten [Southwest
> Oncology Group] asked.
>
> "If there are five protocols already. He won't be telling you what to do, but
> hopefully can help reduce duplication," DeVita said.
>
> "I say bring the five protocols, not some junior man's concept of duplica-
> tion," Holland [Cancer and Acute Leukemia B] said.
>
> "Who decides if a protocol is acceptable?" Holland asked.
>
> "Dr. Jacobs," Muggia [Franco Muggia, director, DCT Cancer Therapy
> Evaluation Program] answered. "Ted Jacobs, administrative officer in the
> Clinical Investigations Branch, is in charge of protocol review."
>
> "A one-man decision," Holland said. "We're not totally ignorant of what is
> going on in other institutions."
>
> . . .
>
> Carbone said . . . "We've had a protocol here two months, with no feed back.
> I don't care what kind of protocol it is, if you get enough people looking at
> it, it will be rejected. Give us some flexibility. Okay, if there is gross duplica-
> tion, stop it. But we're the research arm of NCI. Let us have some flexibility."
> (Group chairmen unhappy . . . , 1978: 7–8)

As it turned out, outside of duplication or ethical considerations, the
groups occasionally pursued protocols despite scientific disagreement
between themselves and CTEP staff. As James Holland recounted in a
1978 discussion with members of CTEP:

> CALGB submitted an application for a study and it got mixed reviews from
> staff . . . members," Holland said. "I said we respectfully disagreed and would

go ahead. Ted (Jacobs, Clinical Investigations Branch administrative officer who is in charge of protocol review) wrote back and said there is still some scientific disagreement. Well, that's fine, that's what studies are for. (Ibid.)

CTEP's policy hardened shortly thereafter. In 1980, following a major review of the NCI's clinical trials program (see chapter 9), the Board of Scientific Counselors announced key changes in the organization of clinical trials at the NCI. In particular, in order to forestall scientific disagreements like those cited above, the board decreed that "the NCI project officer will assist a group's protocol design committee and advise with respect to a) duplication of proposed study by other groups or institutions, b) scientific rationale, c) design and implementation, and d) availability of necessary drugs and/or other treatment modalities." Such input would evidently remove most possibilities of protocol rejection by the review committee (DCT board approves . . . , 1980: 5). It did not, however, remove all. For in addition to closer coordination with the cooperative groups, the NCI established a sort of counterexpertise with the formation of a Biometric Research Branch (BRB).

Armed with the mandate to review all clinical trial protocols proposed by cooperative groups, the members of the BRB, formed in 1980, replaced the disease coordinators. The BRB represented the "interest of the NCI to ensure that clinical trials [were] planned, conducted, completed, analyzed and reported in a sound statistical manner."[59] According to its longtime director, Richard Simon, the model for the BRB was Zelen's statistical center at ECOG where statisticians maintained both a research and a practical profile by working on specific projects while also devoting time to methodology developments.[60] Given the expertise developed by the BRB, the unit reviewed group protocols according to both clinical relevance and statistical rigor. Notwithstanding the participation of NCI project officers in the protocol production process, protocols were regularly returned to groups for revision. Even today, approximately 50% of all protocols undergo revision following review at the BRB.

One of the more contentious issues that separated and continues to divide CTEP statisticians and group statisticians was the question of

59. Division of Cancer Treatment, *Annual Report*, 1983–84, Volume 1, 412, NCI Archives.

60. Interview with Richard Simon, Bethesda, Maryland, 23 October 2003, 14.

sample size. At first glance, disputes over sample size may appear rather anomalous. As readers will recall from the First Interlude, once the significance or the error size and power are set for a given trial and once investigators decide upon the size of effect that is being sought, the size of sample follows more or less automatically to the point where it is possible to pick the numbers out of precalculated tables. Indeed, Gehan established the first such tables in the late 1960s. The M. D. Anderson statistician Stephen George created a similar and more widely used set of tables in the mid-1970s (George & Desu, 1974; Machin, 2004). And as we also know, these days computers do the necessary calculations and generate the tables. So wherein lie the seeds of discontent?

The problem, it turns out, lies not in the calculations but in the choice of parameters for calculation. Since the beginning of clinical cancer trials, statisticians have known that choosing an appropriate sample size often requires more art than science. The author of *Intuitive Biostatistics*, for example, cautions readers not to confuse the methods section of scientific papers reporting clinical trial results with the reality of statistical practices. With regard to setting sample size, he points out that whereas "it always sounds like the investigator calculated sample size from [type 1 errors, type 2 errors, and degree of difference sought], often the process went the other way. The investigator chose the sample size and then calculated values of [type 1 errors, type 2 errors, and degree of difference sought] that would justify the sample size." Alternatively, the statistician may have approached the problem in an iterative manner. Having specified "ideal values" for type 1 errors, type 2 errors, and degree of difference sought, the investigator would then calculate the sample size only to be "horrified at the enormous number of subjects required. The investigator then altered those values and recalculated [the sample size]. This process was repeated until [the sample size] sounded 'reasonable'" (Motulsky, 1995: 202).

Those charged with reviewing protocols, like Richard Simon, have long been aware of these practical considerations and have considered them downright wasteful:

> Sample size planning for some trials is treated as a game of starting with the available number of subjects and, having that, determine [the treatment effect]. Unfortunately this practice leads to a proliferation of trials that have inadequate statistical power for detecting effects of realistic size. (Simon, 1999: 282)

Under constant pressure to conduct clinical trials, cooperative groups have played this game, according to Simon (2000: 103), in order to maintain group prestige. Other equally as pressing factors have forced their hand over the years. Funding for major groups has been conditional on running protocols at most major disease sites, so that an underpowered study was better than no study at all. Moreover, the institutions that formed the network feeding patients into the groups depended on the groups to run protocols on a regular basis:

> we mounted studies where there weren't enough patients. . . . But many of the cooperative groups felt that we had to have a protocol for every major disease site because many of the small institutions built their clinical program around these cooperative groups.[61]

Although sample size was the variable that could sink a study, group statisticians merely increased the size of effect looked for in the study in order to justify the reduced sample size. ECOG rarely had its protocols returned for revision on the basis of sample size:

> ZELEN: Instead of looking for a 20 percent change in survival, we would look for a 50% change in survival.
> INTERVIEWER: I see, and then Richard Simon would say that you should be looking for a 20 percent difference.
> ZELEN: Yeah. Well, it was anyone's guess [i.e., the true survival rate could not be known before the trial].[62]

Insiders' Dissent: The Randomization Debate

The consolidation of the new style of practice and of its methodological underpinnings is not a smooth narrative of trouble-free progress. Before ending this chapter, we need to briefly discuss a fierce controversy over one of the central tenets of cancer clinical trials: randomization. We note that the dissent came from *within* the trialists' camp, both in Europe and in the United States. The debate was in part precipitated and in many respects prefigured by a 1972 study of the "natural history of

---

61. Interview with Marvin Zelen, Boston, 5 November 2004.
62. Ibid.

clinical trials in cancer chemotherapy" that described the poor penetration of randomized clinical trials into clinical cancer research. Based on an examination of all abstracts submitted to the American Association for Cancer Research, the study showed that of the more than 2,300 studies reported during a six-year period, 10% concerned human clinical trials and of these a mere 20% merited the designation "controlled" in the sense that patients had been randomized to different treatments in the course of the trial (Chalmers, Block & Lee, 1972).

Switching from the descriptive to the prescriptive, the authors of the study examined (and refuted) common objections to randomized studies. Critics had argued that if a treatment really did cure a disease like cancer, then it would be quite obvious to simple clinical observation and there would be no need for a clinical trial to evaluate its efficacy. Had not history shown that all the great therapeutic breakthroughs (penicillin, insulin, etc.) had been accomplished without randomized clinical trials? The rejoinder, in this case, was that there would be no single cure for cancer. Rather, there would be many cures for the many cancers targeted by surgery, radiotherapy, and chemotherapy. More to the point, those cures would not happen all at once, they would happen by degrees: improvements of as little as 10–20% at a time. Working through such a labyrinth of shifting disease definitions and prognoses with drugs that often worked only in combination with other drugs required strict protocols and constant randomization to clear away the confusion. Critics had further argued that randomization deprived patients of the best care possible. In any randomized clinical trial, at least half of the patients would receive an inferior treatment. The answer to this objection pitted present against future patients: randomizing a study reduced the chances that investigators would be led astray by the data and thus reduced the number of patients that would in the future be subjected to less than adequate or dangerous doses of the substance in question.

In 1974, shortly after the "natural history" study, Georges Mathé — one of the founders and first president of GECA/EORTC (chapter 3) — published a diatribe against randomized comparative trials denying that they were the only scientifically valid means to assess the efficacy of competing treatments. Worse, he claimed, clinical trials had become an end in themselves. Mathé (1974; see also 1973; 1978) based his claim on two arguments; namely, that trials carried out by different groups very often reached different conclusions, and that clinical trial outcomes frequently clashed with results obtained by other, equally as "objective," methods of observation. By adopting randomized trials, oncologists had abandoned any hope of making real therapeutic breakthroughs and had instead opted for the search of small differences that could only be made (statistically) visible by enrolling large patient populations. These large-scale trials were deficient on two counts. By definition, the patient populations were heterogeneous, which thus diminished a trial's reliability. In addition, the performance of large-scale clinical trials was distributed among unequally skilled and equipped institutions, making both collaboration and treatment compromises necessary. According to Mathé, randomized Phase III trials made sense only when carried out in a single institution by a highly skilled team testing promising new treatments on very small, homogeneous patient populations (10–20 in each arm).

In the United States, former ALGB members Edmund Gehan and Emil Freireich similarly disagreed with calls for universal randomization, citing their own experience as seasoned researchers who had undertaken randomized studies and in some cases had found them wanting. From the outset, Gehan and Freireich (1974) maintained that there was no obvious need for randomization in Phase I and II trials. The research objective pursued in Phase I trials was the determination of the toxicity of the treatment, whereas the point of a Phase II trial was to estimate the efficacy of the treatment, and in neither case did the trial require a comparison. Yet while most trialists could agree that there was no obvious reason for randomization in Phase I trials, the same was (and still is) not apparent for Phase II trials. Efficacy in cancer therapy remained a relative concept. As a result, Phase II cancer trials often required some kind of comparison to evaluate the efficacy of a given treatment, so that even by the late 1980s, many Phase II trials conducted by the cooperative groups functioned like mini Phase III trials. Of the seventy-six Phase II trials described in ECOG's 1980–84 progress report, more than two-thirds involved comparisons between regimens—often consisting of combinations of multiple substances rather than a single new agent—and were thus randomized.[63]

Gehan and Freireich agreed that in determining the efficacy of a treatment, any clinical cancer researcher necessarily compared the rate of success of the new treatment with existing practice. But in a Phase II

63. Eastern Cooperative Oncology Group, "Disease Committee Reports," in *Progress Report, 1980–1984*, Frontier Science and Technology Research Foundation Archives (Boston).

trial, they observed, the researcher need only compare success rates of the current trial with success rates from previous trials. To make sure that the comparison was valid, the researcher had, of course, to select appropriate patients for comparison, the most important criterion being patient prognosis both with respect to the evolution of disease and the expected response to therapy (Gehan & Freireich, 1974). This selected patient population was referred to as *historical controls*. The existence of the cooperative groups, rather than being an argument in favor of randomization, became, in Gehan and Freireich's hands, an argument in favor of historical controls insofar as the groups constantly created material for comparison:

> There are over 20 cooperative groups in the United States supported by the National Cancer Institute that proceed directly from one study to the next, have a stable group of clinical investigators, see the same types of patients from year to year, have the same access to supportive-therapy measures and generally use the same criteria of response in successive studies. Using patients from a previous study as historical controls is often feasible for such cooperative groups. (1974: 202)

The positions defended by the two sides in this debate rested on a different understanding of the cooperative groups as either the equivalent of a large, specialized cancer center—a well-integrated machine deploying standard or at least comparable levels of clinical skills—versus a distributed endeavor characterized by internal differences and centrifugal forces that had to be reined in by a central statistical office. This explains how clinicians such as Mathé and Freireich could be perceived as apostates of the very practices they had helped to establish. Given that randomized clinical trials presented themselves as a replacement for the outdated practice of comparing present practice with the recent past, Gehan and Freireich sought to frame their campaign in favor of historical controls not as a return to the past but as the promotion of a new technique against a new orthodoxy. They castigated promoters of randomized clinical trials as the only approach to clinical research, as dogmatists resisting innovation (Freireich & Gehan, 1974: 623). Resistance to historical controls was simply the result of "a new anti-intellectualism . . . based on the notion that the perfect technology [randomized trials] has been achieved [and that] any variation from the standard format represents a deviation from 'true science'" (ibid.).

This criticism apparently made some inroads. The keynote speaker at the 1974 meeting of the Biometric Society reported that "no fewer than three separate investigators, already planning what seemed to me well conceived randomized trials, [had] questioned me on whether it has now been shown that they were misguided" (Byar et al., 1976: 74; see Meier 1975). Citing this incident, a collective of eight prominent NIH/NCI biostatisticians decided to refute the claim that historical controls could usefully replace randomized controls. For historical controls to be a valid alternative, they noted, their selection had to be based on full knowledge of the biopathological characteristics of each group, including knowledge of all prognostic factors affecting members of each group. But such knowledge was "always imperfect" (Byar et al., 1976: 75). Absent such perfect knowledge, only randomization made comparison possible by canceling out the different factors that caused one patient to survive and another to perish even under identical treatment regimes. Moreover, while Gehan and Freireich had stressed the importance of the cooperative groups and their uniformity of practice in creating the conditions for fair historical comparisons, the statisticians pointed out that even under the most homogeneous conditions of comparison, investigators could still be led astray by unknown prognostic variables.

The epistemic issue—How can we know if a comparison is fair?—was not the only issue that Gehan and Freireich had raised. They had also charged that randomized clinical trials effectively precluded the possibility of giving patients the best care possible. Either knowledge to date showed that the new treatment was better than the old, in which case patients could not be ethically randomized to the older treatment, or the data showed that the treatment was not better than the old, in which case there was no need for a clinical trial. But once again, according to the NIH statisticians, Gehan and Freireich's view presumed a knowledge consensus that simply did not exist. Any given clinician, they argued, might have an opinion as to which of two treatments was better, but this opinion remained anecdotal knowledge in the absence of results from a randomized clinical trial.

The debate raged for several years. An overview of a 1978 meeting devoted to the issue noted that the controlled clinical trial had become "an area of controversy" in which "widely divergent views . . . in the arena of public discussion [were] generating emotional debate" (Carter & Mathé, 1978: 11). One could be tempted to ascribe those divergent views to differences in the outlooks of patients, clinicians, and biostatisticians as a

FIGURE 7.4. The "devil" from different points of view. Reproduced with kind permission from A. A. Rimm & M. Bortin, "Clinical Trials as a Religion," *Biomedicine* 28 (Special Issue) (1978): 62–63. © Elsevier Masson (1978).

participant did using a cartoon (figure 7.4) in which the patient's devil corresponds to the randomization scheme that could deprive her of optimal care, the clinician's devil to the statistical requirements prevailing over clinical considerations, and the biostatistician's devil to the anecdotal evidence and the underpowered trials organized by clinicians. Yet as we have seen, there were clinicians and statisticians on both sides of the debate, thus defying simplistic explanations. In any event, the immediate controversy seemed slated to retreat from center stage given that proponents of randomization at the 1978 meeting, like Richard Peto, "appeared to have the edge (at least in terms of supporters) through much of the discussion," even though it became at times "quite heated" (Jones, 1978: 55).

Gehan and Freireich revisited the topic a number of times in subsequent years, generally repeating the original arguments (Gehan & Freireich, 1981; Gehan, 1980; 1982; Freireich, 1990; 1997). The only significant change in their brief against randomized clinical trials concerned the role of the cooperative groups. Whereas the groups had made historical controls possible in the original article, in the writings that followed the cooperative groups became a veritable obstacle to innovation in clinical research. The controls that would have made comparisons possible had now become stifling: "While these organizations have had certain positive effects on cancer clinical studies . . . the sheer size of such organizations and the multiple requirements before any study can be started have tended to limit the development of creative ideas for clinical studies" (Freireich & Gehan, 1979: 299). In addition, reversing their previous argument, Gehan and Freireich now claimed that rather than eliminating institutional differences through standardization, cooperative group protocols merely papered them over. The differences generated in such studies were therefore entirely spurious:

> When a multi-institute study is analyzed, frequently the difference[s] in treatment results . . . are larger among institutions than between treatments under comparison. In one institution, treatment A may be better that treatment B, while in another, treatment B is better than treatment A. When results are summed over institutions, there may be no difference between treatments. Of course, it is possible that there is some real difference between treatments which cannot be established because of the high degree of variability among institutions. (Freireich & Gehan, 1979: 300)

In 1984, summarizing the results of the dispute and setting out the conditions under which historical controls could be justified as a technique, Gehan conceded that absolutist claims of the sort that "all studies must be randomized" or "all studies must be historical control" were equally wrong (315; see also Carter, 1979). Historical controls were justified in limited circumstances:

· When the preliminary data suggest that the new treatment was "quite superior" to results obtained historically (a relatively rare occurrence).
· In cases of controversy where opponents refused to submit to randomization, for instance when comparing competing treatment modalities.
· In the case of investigators in institutions with limited patient enrollment, in which case they could either resort to historical controls . . . or "join a clinical co-operative group" (Gehan, 1984: 315), an implicit admission that an epistemic choice (historical controls versus randomized clinical trial) could also be construed as an organizational choice (to join or not to join a cooperative group).
· The final case consisted of yet another pragmatic argument: suppose costs

militated against a prolonged randomized clinical trial? Would not historical controls be cheaper and more efficient given the reduced number of patients required?

As it is often the case in science, the debate remained without official closure. It simply vanished from the limelight as statistical techniques became entrenched in computer algorithms and as statisticians devoted their attention to more pressing issues, such as the circulation of interim data. Moreover, since much of the debate on randomization focused on Phase II trials, it is interesting to note that in the course of the 1980s Phase II trials ceased to be comparative trials in groups like ECOG. By the end of the 1980s, the percentage of randomized Phase II trials reported by ECOG had fallen to 50% and by the mid-1990s of the seventy-one Phase II trials conducted by the group in that period, only four involved a comparison and were therefore randomized.[64] As ECOG statisticians noted toward the end of the 1990s:

> On the subject of randomized phase II studies, we agree that they should not be used to formally test for differences in efficacy between regimens. Generally, we have serious reservations about using control arms in Phase II studies, except possibly in cases where there is little prior information on which to decide whether the results of an uncontrolled study are "promising."[65]

And yet recent surveys point to movement in the opposite direction in the field as a whole. Overall, the number of *randomized* Phase II cancer trials increased between 1986 and 2002 (Lee & Feng, 2005). Obviously, individual cooperative groups have their own distinctive practices that, as in the present case, do not necessarily conform to the patterns displayed by aggregate surveys of the field. Aggregate numbers must thus be interpreted with caution as they gloss over differences between various segments of the clinical trial domain (Michaelis & Ratain, 2007). Moreover, as we saw in chapter 4, the distinction between the different

64. Eastern Cooperative Oncology Group, *Progress Report, 1984–1988*; Eastern Cooperative Oncology Group, *Competitive Renewal 1999–2004; Progress Report 1993–1997*, Vol. 2, *Disease Committee Reports*, Frontier Science and Technology Research Foundation Archives (Boston).

65. Eastern Cooperative Oncology Group, "Statistical Center," in *Competitive Renewal 2004–2010; Progress Report 1998–2002*, Vol. 1, *Overview and Central Resources*, 70, Frontier Science and Technology Research Foundation Archives (Boston).

phases was not self-evident, but involved several gray zones and shifted over time. In the mid-1970s, at a meeting of the EORTC protocol review committee attended by US participants, the committee commented that a document outlining the planning and design of multi-institutional trials failed to clarify the difference between Phase II and Phase III trials.[66] In fact, Phase II trials operated in a liminal zone, providing trialists with room for considerable experimental maneuvering. Phase I trials could shift seamlessly into Phase II trials—a situation that has become more common in recent years with the emergence of targeted therapies (see part 3)—just as Phase II trials could be performed as small Phase III trials (Riechelmann et al., 2009; Korn et al., 2001; Van Glabbeke, Steward & Armand, 2002).

In sum, the dispute over randomization in clinical trials concerned not so much the use of randomization as its *bon usage*. At what point in the clinical research process did a randomized clinical trial become necessary? Should all Phase II trials be randomized or were there conditions under which randomization became counterproductive? The answer depended, in part, on the role one assigned to Phase II trials in the overall economy of the cancer clinical trial system. As for Phase III trials, maverick positions such as the ones occasionally defended by Mathé remained extremely rare. Here again, a number of the questions raised by this debate would ultimately be answered or, rather, rendered moot, by technical advances in clinical trial methodology. Data monitoring committees, crossover trials, groups sequential methods—all to be discussed in later chapters—went a long way toward addressing the charges of waste and futility originally raised against the randomized clinical trial.

66. Report on the 2nd Meeting of the Protocol Review and Coordinating Committee (May 14–15, 1974), *EORTC Newsletter*, no. 35, July 1974.

CHAPTER EIGHT

A Relational Space of
Substances and Regimens

In chapter 6 we examined the relationships between clinical trials such as ECOG 0971, ECOG/NSABP B-05 and other breast cancer protocols, and showed how a single trial drew much of its rationale and meaning from the surrounding trials. A similar argument applies to the substances used in the chemotherapy trials. They, too, defy static categories of understanding: their attributes change as their environment evolves in at least two important ways. First, a drug's values and properties develop along with the treatment regimens in which it figures. While toxicity may be a defining feature of a drug in one regimen, for example, it may not be so in another. Those values and properties are, moreover, defined in relation to those attributed to the other molecules that compete for attention. Second, the definition of the mechanism of action of a drug and the definition of a disease entity are two sides of the same coin: knowledge of one directly influences knowledge of the other. Knowing how and where a drug acts is based in part on knowledge of the putative causes of the disease. Changing knowledge of a disease mechanism thus continually redefines drug action. Different drugs likewise redefine population groups and disease subgroups by, in the simplest case, creating reactive and nonreactive clinical subgroups of patients. Anticancer drugs evolved not only within human trials but also in the preclinical studies performed on animal models and other assay systems. In addition, following the demise of the CCNSC's brute-force methods discussed in part 1 of this book, the procurement of substances acquired an interactive dimension wherein the introduction of active (but often quite toxic) substances spurred the development of less toxic *analogues* with the same or greater therapeutic action. Adriamycin (doxorubicin), for instance, one of the first and most successful of a new category of anticancer agents known as anthracyclines had by the early 1990s spurred the production of over two thousand analogues (Weiss, 1992).

In what follows we invert the usual chronological sequence of the journey from the procurement of a substance to animal model testing and to human clinical trials. We do so mainly for expository reasons but also because the lifeline of anticancer substances is not necessarily linear: we saw, for example, how the chemotherapeutic substances used in ECOG 0971 and related protocols meandered through an assortment of clinical research programs in the 1950s and 1960s before settling into an effective combination in the 1970s. The next section provides an overview of this tangle of substances and trials from the 1970s to the 1990s, a period dominated by large Phase III clinical trials that compared numerous combinations of a small number of drugs.

The Clinical Evaluation of Substances

We have seen how assorted clinical criteria entered into the decision of whether or not to experiment with a given substance in a clinical trial. If we wish to understand the nature of chemotherapeutic compounds, we must take into account their clinical use and not just their (bio)chemical and physical structure. Their properties, in other words, are the outcome of their changing relations with diseases and substances. The movement of therapeutic substances from the laboratory, through the screen, and into the clinic should thus be viewed not as the trajectory of well-defined compounds but more as a learning process, a process of progressive informational enrichment of the substances (Barry, 2005). More precisely, we should consider therapeutic substances not as single compounds but as members of groups, clusters, and families, in short, as inhabitants of a space defined, first, by the relation between their emerging properties and those of the biochemical and clinical models used in deriving those properties, and, second, by the symbiotic relations between sets of substances in the clinical environment in which they operate and to whose definition they contribute.

Our claim is consistent with the fact that clinicians and researchers rarely deal with a single substance but rather with regimens of therapeutic substances that are often combined with surgery or radiotherapy and that may be preceded or followed by treatment with other regimens. This idea is not new to the reader. In particular, we have seen in our preceding discussions that clinical cancer trials operate on *regimens* deployed within *protocols*, and that those trials coproduce the disease events (resistance, recurrence, stage, etc.) against which a particular regimen has been devised. With regard to the use of disease categories in clinical trials, recall that one of the innovations of Protocol No. 3 (see chapter 2) consisted of treating different phases of the disease—the direct creation of the protocol itself—as independent events upon which therapy might intervene, a process pursued in subsequent protocols that selectively focused on primary or advanced cancers and subsequently on even finer pathological subspecies (see chapter 6). By replacing the natural history of a disease with its treated history in their conception and design, protocols incorporated an evolving understanding of the biology of the disease, thus opening up new targets for therapeutic substances and therefore further modifying their place within the scientific and clinical space. Contemporary clinical cancer protocols continue to define their target population (or target disease) on the basis of former treatments with other substances (and the presence or absence of prognostic factors similarly defined by previous trial experience). Article titles such as "Tamoxifen-treated, node-negative breast cancer" (Paik et al., 2004) or "Dasatinib in imatinib-resistant, Philadelphia chromosome-positive leukemias" (Talpaz et al., 2006) bear witness to this phenomenon.

In what follows we resort to network analysis techniques to represent interactions among biomedical entities on two-dimensional maps. Using publication data obtained from PubMed/Medline and the Web of Science, we will attempt to visualize, at least in part, the complex set of relations engendered by cancer clinical trials. The period covered will include both the previous section and the next section of the book as the results of these semiquantitative displays provide a useful synopsis of the development of the field as a whole.

Figure 8.1 shows the number of articles listed in PubMed that reported the results of clinical trials using a number of anticancer drugs. The constant growth in the number of papers referring to a given substance over a period of approximately forty years is obviously not due to an increase in the number of clinical trials devoted exclusively to that substance but to the fact that a large proportion of Phase III clinical trials compared and explored the effects of regimens. These regimens vary not only according to the quantities administered and the modes

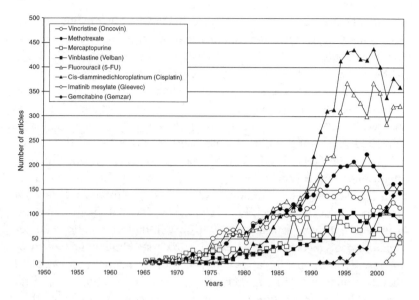

FIGURE 8.1. Number of articles listed in PubMed reporting the results of clinical trials with a selection of antineoplastic agents.

and schedules of administration but also to the type and number of substances they deploy. Two different regimens may use the same substance with different routes of admission (e.g., oral versus intravenous) or dose size. Similarly, two different regimens may use different combinations of substances, often testing slight variations of preexisting regimens, as we saw in chapter 6.

While figure 8.1 displays growth curves for individual substances, the substances act within an evolving context defined by their relation to other anticancer drugs, other modalities such as surgery and radiotherapy, and changing nosologies. For instance, in addition to the previously discussed first-line (standard) chemotherapy and adjuvant chemotherapy, regimens can target other forms of chemotherapy such as neoadjuvant (before surgery), induction (remission-inducing), consolidation (post-remission), maintenance (remission-prolonging), or second-line (post-relapse) chemotherapies. And while regulatory bodies such as the FDA or its Canadian and European equivalents authorize manufacturers to market individual molecules, those drugs are used according to specific therapeutic claims; that is, with regard to a given disease and disease stage that, as we have seen, are constituted within and through historically situated regimens. Figure 8.2 shows the number of new anticancer molecules and the number of claims for those molecules approved by the FDA. That number is clearly quite small compared to the number of cancer clinical trials carried out by cooperative groups over the last four decades. From the mid-1980s on, the number of claims (diseases and disease stages for which the molecule has been approved) quickly outstrips the number of new molecules. Figure 8.3 shows the distribution of substances in the approximately 1,200 protocols listed in the MeSH database (the controlled vocabulary used for indexing articles for Medline/PubMed). The plot fits a power-law pattern, indicating a small number of substances used in a large number of protocols: in the late 1980s, five of the approximately thirty commercially available anticancer drugs accounted for about 70% of sales (Laker, 1989), and a decade later, 80% of the standard antineoplastic treatments used only fifteen of the roughly sixty major anticancer drugs then available

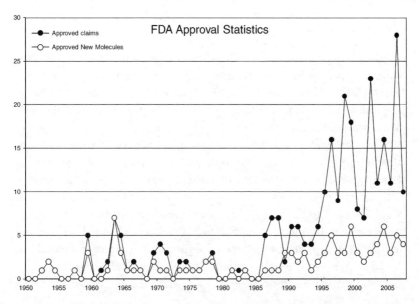

FIGURE 8.2. Number of new molecules and claims for antineoplastic agents approved by the FDA. Data source: FDA website, with supplemental data from http://www.nytimes.com/imagepages/2009/09/16/health/research/16cancergraphic.ready.html.

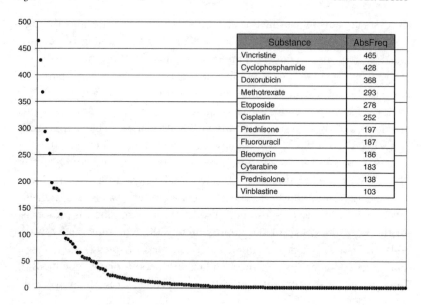

Substance	AbsFreq
Vincristine	465
Cyclophosphamide	428
Doxorubicin	368
Methotrexate	293
Etoposide	278
Cisplatin	252
Prednisone	197
Fluorouracil	187
Bleomycin	186
Cytarabine	183
Prednisolone	138
Vinblastine	103

FIGURE 8.3. Distribution of substances in the approximately 1,200 antineoplastic protocols listed in the MeSH database (2006).

(Cvitkovic-Bertheault, 2000).[1] Thus the "same" substance was simultaneously employed in a variety of trials, from which it emerged with different qualities and qualifiers: its therapeutic space, as defined by these properties, is therefore a function of past and future trials and the regimens those trials test.

While we wish to emphasize the interdependence of trials and substances and to examine the multiple processes that define a given substance's properties, single compounds can make a difference. Oncologists, for example, often foreground a single substance and claims like "cisplatin has become the gold standard treatment against cervical cancer in combination with radiotherapy" (Pasetto et al., 2006: 67) are not unusual. Yet even here where the importance of cisplatin is undeniable, there is a significant simplification. Having set up cisplatin as the principal actor, the same authors went on to describe the qualities of an entire

---

1. Following the targeted therapy turn (see part 3), hundreds of compounds, however, are waiting in the pipeline (Chabner & Roberts, 2005: 70; DiMasi & Grabowski, 2007; Schwartsmann et al., 2002).

spectrum of cisplatin analogues and their combination with other drugs. In other words, trialists may talk substances but they practice regimens. Given these variables, the number of possible regimens and cancer clinical trials is quite large, and this alone accounts for much of the pattern of growth displayed in figure 8.1.

Subtle shifts in a substance's relations with other substances, diseases, and regimes transform the clinical niches within which a substance evolves. A telling illustration of the forgoing lies in the history of the aforementioned cisplatin. First administered to humans in 1971, evidence of a substantial antitumor effect emerged in 1974. The FDA subsequently approved cisplatin in 1978 for the treatment of genito-urinary tumors. The drug went on to display major effects on ovarian cancer and became a mainstay for a number of other cancers, including bladder, cervix, head and neck, and esophageal cancers. By the end of the 1990s, a further twenty-eight platinum compounds had been tested in clinical trials, and three had been approved in the Western world (cisplatin, carboplatin, and oxaliplatin). This linear view obscures a number of major obstacles and detours.

Cisplatin produced serious side effects, including severe kidney toxicity. As a result, clinicians initially argued that it was too toxic for general use. Then in 1976 researchers found a work-around. By keeping patients properly hydrated in the course of therapy and by administering large doses of diuretics, they were able to shelter the kidneys from major damage. Other substantial side effects included severe nausea and vomiting, loss of hair, and loss of hearing. To solve these problems, clinicians deployed a set of "protective compounds" such as antiemetic substances and drugs countering neurotoxicity. Bob Pinedo, a Dutch chemotherapy pioneer who, after transiting through the NCI, created the first Division of Medical Oncology in the Netherlands, recalls that when he returned to the Netherlands with cisplatin in his pockets and tried it out for the first time in 1976 on a testicular cancer patient, he faced stiff criticism from his colleagues who argued that he would make his patient terribly sick, which was indeed the case although the patient was eventually cured. He subsequently set up trials "to learn how to properly use the drug," by which he meant:

> The scheduling, the infusion, the diuretics; all the things around it and the antiemetics. Because those patients were very sick; they were vomiting a lot. Nowadays we don't see vomiting anymore. . . . But you still need the infusion

and there are other side-effects that we cannot solve, like the deafness; it's a problem.[2]

On the other hand, since cisplatin caused only mild myelosuppression (reduction of platelets and white blood cells) it made a good candidate for combination chemotherapy with drugs whose myelosuppressive action was more pronounced (Pil & Lippard, 2002; Lebwohl & Canetta, 1998).

Figure 8.4 shows cisplatin's evolving affiliation with different diseases.[3] At the center of the map we see a group of major cancers (e.g., lung, breast, and ovarian cancer) that have been repeatedly connected to cisplatin, while the most specific subcategories (e.g., metastatic adenocarcinoma of the lung, locally advanced breast cancer, etc.) targeted by regimens containing cisplatin are displayed on the outer layer of the map and are linked to the year(s) during which those studies were published. Figure 8.5 shows a map of the different anticancer drugs more specifically associated with cisplatin between 1980 and 2006. While relational maps such as figure 8.5 allow us to visualize the relations that substances entertain with one another, they do not display the more subtle dynamics of clinical pharmacology research. An example of clinical pharmacology research is the notion of a *therapeutic index* or the ratio between the toxic dose (the maximum tolerated dose) and the therapeutic dose of a drug. This ratio is used as a measure of the relative safety of the drug for a particular treatment. For instance, behind the synthesis and testing of thousands of cisplatin analogues lies the attempt to enhance the therapeutic index, to reduce the degree of cross-resistance, and to optimize a number of other parameters (Pasetto et al., 2006; Lebwohl & Canetta, 1998). The therapeutic index is clearly a relational parameter. Over the years the maximum tolerated dose has changed due to the simultaneous use of other substances such as growth factors and antibiotics (that can rescue a substance abandoned because of its toxic side effects) and to the general adoption of more aggressive therapies—that is, a willingness to tolerate harsher side effects in exchange for increased efficacy. Cross-resistance and synergy are also relational parameters since they obviously refer to the other components of a regimen (Cvitkovic-Bertheault, 2000).

2. Interview with Bob Pinedo, Amsterdam, 18 February 2007. On Dr. Pinedo, see Wagstaff (2004).

3. This and the following maps were created using the software package ReseauLu X2 (http://www.aguidel.com). For more details, see Cambrosio et al. (2006; 2010).

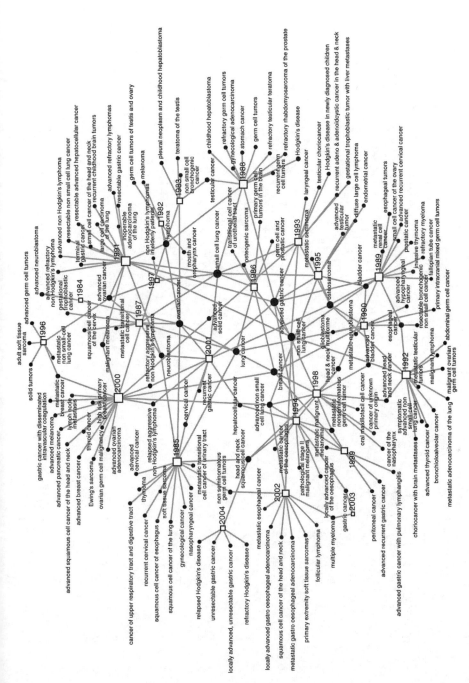

FIGURE 8.4. Diseases associated with cisplatin in cancer clinical trial articles listed in PubMed (1980–2006).

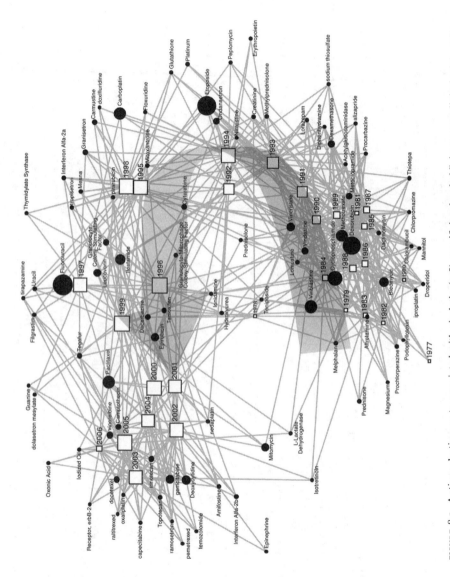

FIGURE 8.5. Antineoplastic agents associated with cisplatin (40% specificity threshold) in cancer clinical trial articles listed in PubMed (1980–2006).

Let us now return to our overview of the development of chemotherapeutic agents. Figure 8.6 shows the cooccurrence between the more common anticancer drugs—those that appeared in over 40,000 cancer clinical trial articles retrieved from PubMed—and the publication years during which these substances were the subjects of clinical investigation. The map displays a neat chronological sequence, a curving trajectory beginning with the late 1960s on the bottom left and ending with the new century on the bottom right. While the most recent period accents the emergence of a new set of "targeted" substances that will be discussed in part 3 of the book, the center of the map demonstrates the preeminence—the size of the nodes being an indication of the number of articles devoted to a given agent—of a limited set of substances with which the reader should by now be familiar—fluorouracil, vincristine, methotrexate, cyclophosphamide, and cisplatin. These drugs function as workhorses for a majority of the therapeutic regimens developed since the 1980s, in particular during the period covered in this part of the book and characterized, as previously noted, by Phase III trials testing manifold combinations of a relatively small number of substances.

If we now shift from substances to regimens, we can again take advantage of the list of anticancer protocols and regimens (the two terms are used interchangeably) recorded in the MeSH database to create a corpus of English-language, clinical trial articles (approximately 2,500 articles between 1980 and 1999). Using this corpus we have created a map showing the more specific relations between the 150 most frequent regimens and the publication years of the articles in which they were discussed. Figure 8.7 shows the result: we have two distinctive clusters to the left and the right of the map, corresponding to the 1980s and 1990s, linked by several transitional years. The map is meant to provide readers with a sense of the intense activity fueled by regimens. The regimens are bona fide research objects and as such they are the focus of numerous publications extending over several years. Since the sizes of the nodes are related to the number of articles, several regimens stand out. These include, in particular, the CMF regimen discussed in chapter 6 as well as other landmark regimens, such as MOPP against Hodgkin's disease, developed in 1964 at the NCI, later largely replaced by the ABVD regimen developed in Milan in the mid-1970s by the Bonadonna group (DeVita, 2003), and CHOP against non-Hodgkin's lymphomas, developed in the mid-1970s at the NCI. Like substances, regimens also evolve in a relational space. Regimens and slight variations thereof are constantly com-

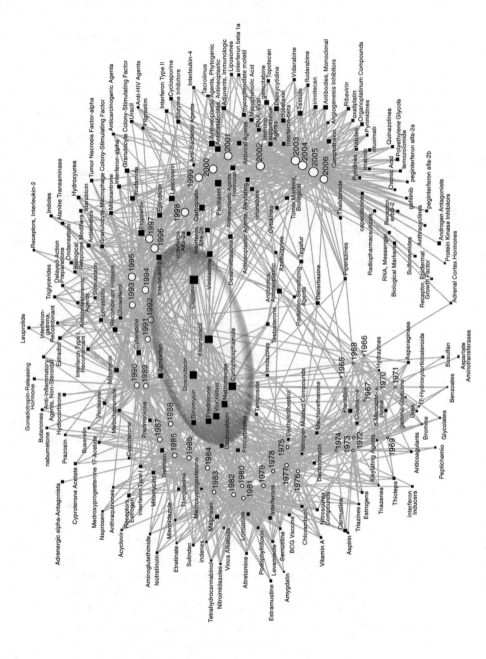

FIGURE 8.6. Cooccurrence between antineoplastic agents (substances and categories) and years in cancer clinical trial articles listed in PubMed.

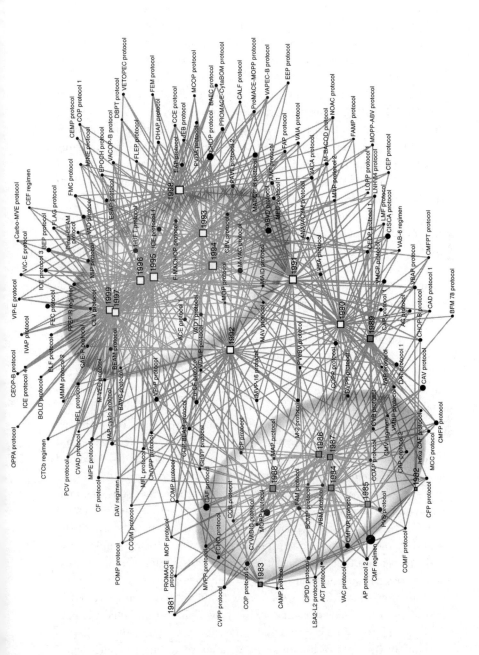

FIGURE 8.7. Cooccurrence between the top 150 regimens and the publication years of the articles discussing those regimens (1980–99).

pared, and, in particular, regimens that have shown activity in a given cancer are tested against other forms of cancer and contrasted with standard therapies. Figure 8.8 maps the cooccurrence of regimens in articles from the same corpus. The map displays two main clusters. At the bottom we see a cluster centered on Hodgkin's and non-Hodgkin's lymphomas: MOPP and ABVD are, as expected, closely connected, other non-Hodgkin's regimens (in particular MACOP-B) are linked to them, and these treatments in turn relate to CHOP. The cluster at the top is more heterogeneous: it consists of a breast cancer subcluster dominated by CMF (see chapter 6) and CAF with links to other regimens for bladder (CISCA), lung (CAV, VAC), head and neck (COB), and testicular (PVB, BEP) cancers.

Like the substances of which they are composed, regimens circulate in the chemotherapeutic space, defining its parameters and boundaries. Determining whether or not to deploy a given regimen, at which stage of a disease—first occurrence, relapse, metastasis—and in conjunction with what other modalities—surgery and/or radiotherapy—involves a complex set of tests, experiments, and comparisons or, more generally, "trials of strength" (Latour, 1987). But how do substances enter regimens?

The establishment of a number of standard chemotherapeutic substances in the 1960s created new constraints for the evaluation of substances in the 1970s. Prior to entering a regimen, substances had henceforth to undergo a primary *information* test: could they be deemed an analogue of a substance already in use? If so, clinical evaluation differed from the evaluation of other substances insofar as "the analogue must always be compared in some way to its parent *clinical* [note that it doesn't say: *chemical*] structure while the new drug stands alone and need only search for its role in competition with currently available *regimens*" (Carter, 1978b: 69, our emphasis). In such comparisons, the analogue might show a greater or wider spectrum of activity or diminished acute or chronic toxicity. Yet none of these end points could normally be tested by themselves: most substances in use were already employed in a given regimen that included other compounds so that activity, for example, was not absolute but relative to a given combination. Since every combination was different, "these new combinations then [became] in many ways like a new drug again requiring a phase I and phase II process which [was] often covered under the rubric of pilot study" (70).

Thus in the case of analogues, the standard testing sequence inaugurated in the 1950s required modification in terms of planning and

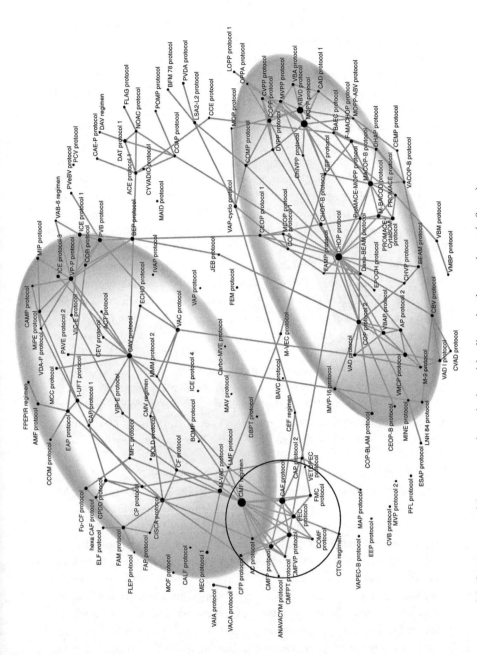

FIGURE 8.8. Cooccurrence of the top 150 regimens in the articles discussing those regimens (1980–99).

assessment. While the move from a Phase I study to a Phase II trial usually sought "to see if a reasonable activity level [could] be demonstrated," if one examined an analogue in order to find a broader spectrum of activity than the parent drug, then the Phase II trial would focus on "unresponsive tumors to parent structure to see if any level of activity is seen," and the subsequent Phase III trial would be used "to establish a level of activity." A different and more complex strategy emerged in the case of analogues designed to diminish acute or chronic toxicity. Here the Phase II trial sought "to establish a reasonable expectation of activity and to get an early feeling for chronic toxicity in responding patients given maintenance therapy on reasonable follow-up," while the Phase III trial focused on establishing "comparative activity as against comparative chronic toxicity" (Carter, 1978b: 71). It goes without saying that terms such as *reasonable* or *early feel* allowed trialists a certain degree of interpretative flexibility.

Obviously, these strategies idealized complex trajectories. Consider, for instance, the life of CCNU (lomustine) in the first half of the 1970s. An analogue of BCNU (also known as carmustine, and yet another mustard gas derivative, today used primarily against brain tumors), CCNU was synthesized as part of the program to produce analogues of successful substances. Because of this status, researchers assumed for several years that CCNU would act like its parent substance. It turned out that the drug's action varied according to the disease encountered. In other words, while structure indicated activity, it did not define it, and so clinicians were forced to conclude in 1974 that "the active form and mechanism of action of CCNU are unknown" (Wasserman, Slavik & Carter, 1974: 131).

When Stephen Carter, the deputy director of the Division of Cancer Treatment, reviewed the numerous clinical trials conducted using CCNU between 1971 and 1974, he divided them into the following categories: (a) alone against many tumors: Phase II; (b) alone against a single tumor type: Phase II and III; (c) in combination against many tumors: Phase I and II; and (d) in combination against a single tumor type: Phase I and II. Two features of this classification are particularly striking. The first is that the unit category was not the substance but the regimen. The second is that from the point of view of the evaluation of a substance, the Phase I–III trajectory was subsumed under a broader classification (one/many tumors; alone/in combination) that cut across the phase categories. The phase distinctions, in other words and somewhat counterintuitively, sometimes described the trajectory of a compound synchronically, not diachronically.

Broad Phase I–II trials of combination chemotherapy screened regimens in humans using patients with advanced disease that had failed to respond to conventional therapies. They gathered together many diseases within a single protocol. For instance, the summary table of a typical study of CCNU in combination with Adriamycin, published in 1973, listed a series of tumors (breast, melanoma, etc.) and their breakdown by the type of response (complete or partial remission, no change, progression) elicited by the combination treatment. A similar table presented an overall evaluation of CCNU as a single agent. The next step in this process assessed the comparative activity of CCNU vis-à-vis other anticancer drugs in different subspecies of a given type of cancer. Finally came a list of "promising drug combinations" of CCNU and other agents, and of the "target tumors" against which each of these combinations should be tested (Wasserman, Slavik & Carter, 1974: 149). Such was the therapeutic space of a compound like CCNU in the 1970s.

In chapter 6 we discussed the emergence and development of *adjuvant* chemotherapeutic protocols, as trialists sought to articulate solid tumor chemotherapy with radiotherapy and surgery in order to move systemic therapy from a last-resort option for advanced-late disease to first-line treatment for primary tumors. Figure 8.9 summarizes this strategy that eventually managed to reverse the treatment flow. An idea of the size of the project facing chemotherapists in the mid-1970s can be gathered from a "cross-reference chart of drug-tumor interactions" whose rows and columns confronted two parameters: the twenty-nine substances that had so far shown activity in solid tumors and the sixteen most common solid tumors. The resulting matrix provided at-a-glance visual proof that a lot of work remained to be done, for 263 (57%) of the matrix's 464 cells contained the acronym NE ("Not Evaluated") and a further 37 (8%) the symbol for "inadequately tested" (Carter & Soper, 1974: 5). A few tumors, such as lung, colon, and breast cancer, had received considerable attention, while others (e.g., pancreas, stomach, prostate) had suffered comparative neglect. Substances could thus be ranked in relation to the number of tumors in which they had been "adequately" or "inadequately" tested (7). The notions of adequate and inadequate obviously opened the door to a number of interpretations of what any given trial signified within the evolving therapeutic space of substances and their regimens.

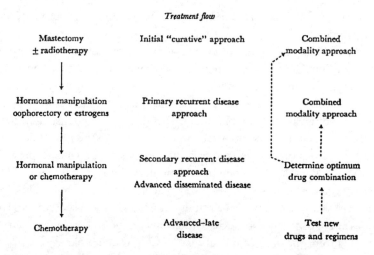

FIGURE 8.9. Reversing the treatment flow. Reprinted from S. K. Carter & W. T. Soper, "Integration of Chemotherapy into Combined Modality Treatment of Solid Tumors. 1. The Overall Strategy," *Cancer Treatment Reviews* 1 (1974): 4. © 1974, with permission from Elsevier.

Once again, regimens and not substances were the name of the game. Combination chemotherapy raised the following problem: although individual agents may have shown effectiveness in a given human tumor, the effectiveness of their combination was largely an empirical matter. The empiricism, however, was not completely naked: it followed rules of thumb that had been shown to hold for the most successful combination regimens; namely, that each drug in the combination should be active when used alone against the tumor, that the drugs should have different mechanisms of action, and that their toxic effects should not overlap, so that each could be given near its maximum tolerated dose (Carter & Soper, 1974). Having ranked substances on the basis of these considerations, one could now integrate these rankings with data about the incidence of a given type of tumor and of the success of traditional therapies (surgery and radiotherapy) in order to produce a priority list of solid tumors for combined modality clinical trials. In planning their trials, cooperative group clinicians did not conform to any precise blueprint derived from the aforementioned hybrid, rational-empirical design; nor, given the distributed nature of the cancer trial system, was their work centrally controlled. Yet the proliferation of large-scale Phase III trials that characterized the 1970s and 1980s did, in an aggregate sense, follow this logic, albeit with all the qualifications derived from the contingencies characterizing the inter- and intragroup dynamics of the cooperative endeavor.

Screening Substances: The Animal Screen

Before entering human trials as part of regimens, candidate compounds had to be screened for anticancer activity. In the United States, the fate of any given compound rested in the hands of the Decision Network Committee (DNC) of the NCI Division of Cancer Treatment (DCT) that between 1966 and 2000 made most of the go/no-go decisions concerning drug development. Initially composed of approximately thirty scientists, the DNC made the decisions at critical junctures as they followed individual substances through the procurement and evaluation process up to and including Phase II clinical (human) trials. The committee also made higher-order decisions: What kinds of substances should be procured? What kinds of tests should be performed? While most of the members of the committee were intramural staff from the DCT, the committee also included industry and university representatives. Once a compound passed the initial set of tests, the DNC issued a priority rating and assigned the substance to a promoter from either within or without the NCI. Committee members thus functioned as part of an advocate system widely used in the pharmaceutical industry at the time (DeVita et al., 1979).

Despite the representation of the process in the form of a *linear* array (see chapter 5), substances did not always move through the DNC in a straight line. For instance, following a series of heated meetings in the fall of 1978, Vincent DeVita sent a memo to members of the DNC titled "The Definition of the Role of the Decision Network Committee," in which he explained that the committee's members were "sometimes too concerned with the precise definition of the decision point rather than the concept of dropping a compound, holding it static to develop more data, or moving it along." Members, he suggested, should recall that they were "here to facilitate drug flow through the system, and when we have a clear interest, forward motion through several decision points is quite appropriate, particularly for unusual drugs. *Some compounds may actu-*

ally be at several stages at the same time.[4] Compounds usually worked their way through the linear array at a fairly steady pace. As a general principle, they were more often eliminated than selected since screening "eliminates the majority of the materials that you don't want, and brings through a mixture of materials that have an enrichment factor, such that a higher percentage will be active, clinically."[5] The criteria for retention were, on the other hand, quite open-ended:

> Things were pretty liberal at that stage. If anybody had any reason that made any sense at all of why you should go forward with that compound, it went forward because we had a big throughput screen. . . . The major decision was made in the decision network at Decision Point 2A. . . . This is where you brought in the chemist, the biologist, the pharmacologist, toxicologist and the clinical people to look at a compound and see if there was something wrong with the compound. The basic concept was that unless it was stopped by toxicology, production problems or whatever, it was going under clinical trial at that point.[6]

With the passage of time, the linear array clogged up and movement slowed. By the late 1970s, the director of the DCT felt obliged to impose even more drastic measures and ordered the network committee to speed things up by holding special meetings to prioritize the sixty compounds currently working their way through the system. Moving with considerable dispatch, the committee eliminated almost 25% of the substances under review in a single year.[7]

There were no ironclad rules for accepting or rejecting any given substance. Rather, it was often a question of choosing between compounds rather than choosing a compound based on preset criteria. Moreover, given the variety of clinical, chemical, and biological considerations entering into the decision-making process, the outcome of the voting procedures could be quite close even though some cases were clear-cut. Take, for example, a meeting held by the DNC in 1979 to decide which of

4. Memo from Director, DCT, NCI to Members, Decision Network Committee, "The Definition of the Role of the Decision Network Committee," 10 October 1978, Abraham Goldin Papers, Acc. 1997–021, National Library of Medicine, our emphasis.

5. Group interview with John M. Venditti, Saul Schepartz, Melinda Hollingshead, Joseph G. Mayo, and Michael C. Alley, Frederick, Maryland, 29 October 2003.

6. Ibid.

7. Division of Cancer Treatment, Annual Report, 1976–77, 249.

twenty-two compounds would graduate to the next level of study. With twenty-five members on the committee, a compound could receive a maximum of twenty-five points and needed thirteen points in order to move on. While more than half of the compounds were either clear winners or clear losers, 35% of the compounds lay somewhere in the gray zone.[8]

The DNC decisions were based in large part on evidence provided by animal models. By the mid-1970s, the pitfalls and limitations of the mouse screen had become increasingly evident. Most cancer researchers had come to consider the use of mice as test subjects as less than optimal even though they realized that an ideal screen, some form of a human system, lay somewhere in the future. In the spirit of compromise, the 1970s and 1980s saw the emergence of a new screen and a new system of screening that lay between the original L1210 mouse leukemia model and the much-sought-after system of human cancer cells in a test tube. The interim test system consisted of a new animal device; namely, *nude mice* that, lacking immune systems, were unable to reject foreign growths and thus allowed transplanted human cancers (xenografts) to grow under their skin (Rygaard, 1973; 1978). Less a bridge from the animal world to the human domain than an entirely new field of play with its own norms and protocols, the mouse-human hybrids provided an essential service by allowing researchers to test substances on human tumor cells in the context of an entire (animal) organism. They thus remained an integral part of the testing system even though the front line of the two-stage, *compound-oriented* screen was subsequently replaced by a set of human cancer cell lines in a test tube as part of a switch to a two-stage, *disease-oriented* approach that we will discuss in part 3 of this book.

Let us have a closer look at the animal model screening process. While often criticized over the years, the use of animal-animal models (animal tumors in animal organisms) for screening chemotherapeutic compounds had become downright controversial by 1975 when Vincent DeVita appeared before an advisory working group reporting to the director of the NCI and suggested: "The whole philosophy of screening needs reevaluation. Perhaps the right questions have not yet been

8. Memorandum from Acting Associate Director, DTP, DCT, NCI to the Director, DCT, NCI, "Ballot Results from August 1st DN Meeting," Abraham Goldin Papers, Acc. 1997–021, National Library of Medicine.

asked."[9] DeVita claimed to have been particularly impressed, indeed, spurred to action, by the fact that although the screen was capable of handling 50,000 agents a year, it had, since its inception, selected only three successful new compounds. The working group responded to De-Vita's criticism by organizing a series of presentations by members of the DCT intended to explain the rationale behind the screening system to group members. The associate director of Drug Research and Development, Saul Schepartz, affirmed that contrary to DeVita's initial and highly provocative remarks that suggested the entire system was "based largely on the effects of compounds on L1210 leukemia in mice,"[10] the specific systems had "changed significantly since 1955 on the basis of retrospective analyses, using primarily drugs of known activity."[11]

Even though other biochemical in vitro systems had become available by the mid-1970s, they, too, showed continued deficiencies.[12] The picture had not changed substantially since the original Gellhorn report twenty years earlier. As John Venditti, the head of the screening program, admitted in the course of his presentation, "at this time, and I want to emphasize 'at this time,' we have no data that disputes the earlier conclusion of Gellhorn and Hirschberg that 'there is no evidence for the existence of any non-tumor system which could replace a tumor system as a screening tool for carcinostatic agents.'"[13] Consequently, Venditti proposed that the screen be widened to include a larger number of in vivo animal models such as those provided by the immunologically deficient mice.[14] The use of these animal models would in turn be preceded by the use of a relatively cheap mouse model such as the P388 (also a leukemia model). The proposed two-stage system thus entailed the existence of a primary or prescreen system that was sufficiently sensitive and, above all, sufficiently cheap to eliminate the bulk of nonactive substances and

9. Molecular Control Working Group, Annual Report for FY 1975, National Archives USA, RG 443, Entry 29, Box 18, p. 3.

10. Ibid.

11. Saul A. Schepartz, "Presentation to Molecular Control Working Group," Minutes of the Sixth Meeting of the Molecular Control Working Group, Appendix 1, 28 May 1975, National Archives USA, RG 443, Entry 29, Box 18, p. 1.

12. John Venditti (Chief, Drug Evaluation Branch, DCT, NCI), "Presentation to Molecular Control Committee," 28 May 1975, National Archives USA, RG 443, Entry 29, Box 18, p. 2.

13. Ibid., 3.

14. Ibid., 4.

the development of a secondary screen system that would subsequently perform an advanced analysis of the substances surviving the primary screen.

Comparisons of responses of tumors in humans and in nude mouse xenografts to chemotherapeutic compounds conducted in the late 1970s had showed that nude mice were clearly superior to existing in vitro systems with regard to predicting clinical response (Fiebig & Berger, 1991: 318). Unfortunately, the closer the xenograft mimicked behavior in the human, the more expensive the entire process became. Subcutaneous tumor implants could take up to four months to develop, and given the immunologically compromised nature of the mice, special breeding facilities were required. Nonetheless, the NCI added nude mice to the screen in 1977 instituting, at the same time, a program for the production of the animals that it contracted out to the Battelle Columbus laboratories. Replete with air-locked rooms attached to each animal test room, the complex was slated to produce one thousand nude mice a week to test approximately eight hundred compounds per year (Ovejera et al., 1973). In the meantime, knowledge of the maintenance of the nude mice evolved into a full-time concern (Committee on Immunologically Compromised Rodents, 1989). On the European side, where the nude mouse had originated, the EORTC followed suit less than a decade later in 1986 when its New Drug Development Office initiated a cooperative program to share tumors and standardize procedures. Here, however, the screen was intended to have a more direct impact on clinical cancer therapy. The principal objective of the Xenograft Group that ran the nude mouse screen was to reduce the number of negative Phase II trials in humans (Boven et al., 1988).

To monitor and evaluate the screen modifications, the NCI instituted a prospective study of "'animal-human' correlations" that ran concurrently with the screening process (Venditti, Wesley & Plowman, 1984). The study employed a panel of eight transplantable tumors including three recent human tumors engrafted onto nude mice. Between 1976 and 1982, the DCT tested over two thousand compounds in the panel and addressed a number of questions, such as: "To what extent do human tumor xenografts and animal tumor screens select the same or different drugs as active?" or "What contribution will the data of the new screening panel make to prediction of clinical effectiveness of new drugs?" (Goldin et al., 1978: 116). Based on the study results, the Drug Evaluation Branch of the Developmental Therapeutics Program proposed and

applied a revised version of the screen. The mouse P388 leukemia henceforth prescreened compounds. Screeners expected that approximately
ten thousand compounds a year could be handled by this part of the
screen that would hopefully winnow down the contenders to about 250
candidates. Given that not all active compounds would be trapped in the
prescreen, the DCT made provisions for a bypass that would allow compounds that escaped the screen to be passed on for further tests if outside information—referred to as "intellectual selection" (Goldin et al.,
1978: 115)—indicated that they were likely candidates. The compounds
that passed the prescreen test, either by expressing the preset level of activity or by reputation, would then be submitted to a pared-down version
of the tumor panel.

So, did the new screen work better than the old screen? Given the
evolving criteria of evaluation, this remained an open question for cancer researchers. Overall, from 1975 to 1984, the program isolated forty-
eight cytotoxic compounds. By 1985, Phase I trials eliminated five of
these because of toxicity. Nine more entered Phase I trials and eleven
more entered Phase II trials (DCT board committee to recommend . . . ,
1985: 3). The primary panel of P388 leukemia and the secondary panel
of murine models and human xenografts were judged unsuccessful by
most accounts as they failed to detect any new structure active against
solid tumors (Marsoni & Wittes, 1984). That judgment remained miti-
gated by the fact that it was not clear where the blame lay: in the P388
mice or the xenografts. Retrospective studies published in the year 2000
yielded the following open-ended conclusion: "One must conclude that
either animal model systems using transplantable tumors do not predict for clinical activity or that the P388 prescreen effectively selected
against compounds specifically active in human solid tumors" (Khleif &
Curt, 2000).

What lessons do we draw from these events? Because it mimicked human tumor reactions to prospective anticancer agents, the hybrid animal-
human system was particularly appealing to trialists as a model *for* human cancer. Our choice of the preposition *for* is not innocent, since
rather than as models *of* the disease, the NCI mass-produced xeno-
grafted nude mice with a specific intent—namely, *for* therapy—that is, as
a screening device for anticancer compounds. Thus when trialists spoke
of animal models they meant that they enacted a specific pathogenic process or event that was also present in humans and thus constituted part

of the disease under investigation. The models never displayed more than a part of the whole disease, since models were defined as "a living organism with an inherited, naturally acquired or induced pathological process that *in one or more respects* closely resembles the same phenomenon occurring in man" (Zwieten, 1984: 206, our emphasis). So it would actually be better to speak of *modeling* rather than *models*. The process, moreover, was a two-way street. Pathological processes first uncovered in animals sometimes prompted a search for similar phenomena in humans. In the specific case of cancer, the spectacular growth of oncology meant that there were increasingly few untreated cancer patients, and this transition from natural histories to treated histories of human cancer shifted the burden of natural history from the object to be modeled to the model itself. This process will become even more evident following the introduction of molecular biology as discussed in part 3, but its roots can be traced back to previous periods. Given the reversible nature of the modeling process, then, it might be more appropriate to speak of disease descriptions as being distributed among humans and animals.

A second lesson instructs us that trialists were not simply aware of but explicitly defined and discussed the problems raised by the use of animal models. Far from naively relying on inadequate models, and from glossing over the problems associated with the relation between the model and the entity or process to be modeled, NCI researchers translated this issue into a research project, instituting the aforementioned prospective study of animal-human correlations to run concurrently with the screening process. Thus rather than broaching the elusive question of the relation between the model and the modeled, the NCI researchers respecified the interrogation, pragmatically, as a comparison between different kinds of models and in so doing introduced a degree of reflexivity in their operations (Lynch, 1982).

Procuring Compounds

By the time ECOG published the results of 0971 in 1976, responsibility for substance procurement for screening at the NCI had devolved from the Cancer Chemotherapy National Service Center (CCNSC) to the Division of Cancer Treatment (DCT). The DCT continued to develop new drugs by contracting out their synthesis, acquiring drugs for the screen from the "scientific-community-at-large," and developing and maintaining a chemical information system.[15] All three components of the DCT program were articulated through an extensive network of contacts in industry, government, and university that constituted a form of "worldwide surveillance." By the mid-1970s, the system contained approximately four thousand suppliers. Industry provided 75% of submissions and made up almost half of the suppliers, while foreign universities provided over 60% of academic submissions. As interest in natural products grew in the mid-1980s, however, input by the synthetic chemistry industry dwindled. By 1990, there were about half as many suppliers (677) as there had been fifteen years before.[16]

In cases where clinical success indicated that an entire class of compounds might prove fruitful for exploration, the DCT contracted out the synthesis of selected substances both domestically and abroad. This structure-function relationship was not the only determining factor, and leads for promising syntheses came from a variety of sources. As the director of drug synthesis at the DCT pointed out: "Syntheses also may originate from the imaginative designs of a scientist based on leads from the screening program, resolution of questions on biochemical mechanisms of action, biochemical or biological rationales and from the literature."[17] The rules of thumb that had guided this minimalist selection of chemical compounds in the 1960s were gradually replaced in the mid-1970s with more formal selection criteria that resulted in a substantial fivefold increase in the number of active compounds detected in the primary screen.[18] The DCT believed that as the matrix of chemical-structure, anticancer-activity relations gradually filled out, the move from random gathering to rational selection would be completed.[19]

Despite the importance of *synthetic* organic compounds in cancer treatment, by the beginning of the twenty-first century almost 60% of drugs approved for cancer treatment were of *natural* origin (Grever, 2001). In the 1970s and 1980s, three different kinds of natural products—fermentation products, plant products, and animal products—entered the NCI screen from multiple industrial and academic sources. The largest source of US compounds came from the NCI's fermentation program

15. Drug Synthesis and Chemistry Branch, DCT, Annual Report, 1976–77, 512.
16. Office of Director, Director's Report, DCT, Annual Report, 1989–90, 15.
17. Drug Synthesis and Chemistry Branch, DCT, Annual Report, 1976–77, 531.
18. Ibid., 513.
19. Ibid., 514.

that had originally been developed as part of the search for antibiotics following World War II (Goldin et al., 1979b). A large proportion of antibiotic-derived anticancer agents, however, were produced by chemists outside the United States: bleomycin was developed by Japanese scientists, daunorubicin was simultaneously developed in France (Rhône-Poulenc) and Italy (Farmitalia), and doxorubicin (Adriamycin) was not only produced in Italy (Farmitalia) but was also clinically tested in Bonadonna's Milan institute where, with bleomycin, it became a component of the standard ABVD regimen against Hodgkin's lymphoma.

Eli Lily's discovery of the anticancer properties of the vinca alkaloids in 1959 in extracts of the Madagascar periwinkle reported in part 1 of this book set off the global search for similar plant products. The NCI was particularly quick off the mark. Barely a year after the Eli Lily discovery, it began collecting plants in order to feed extracts into the screen (Cragg et al., 1996), and in 1962 a tree extract containing taxol, ultimately one of the most successful plant extracts, entered the system, although a pure form of the extract would not reach the clinic until 1977 (Goodman & Walsh, 2001). In the meantime, between 1960 and 1980, the NCI screened approximately 35,000 plants species that produced 115,000 extracts that in turn added fourteen compounds to the list of chemotherapeutic agents.[20] In the intervening years, other companies and institutions followed the NCI into the natural products arena, reducing the need for the NCI to shoulder the entire burden. By the early 1980s, approximately twenty different companies supplied over 30% of the NCI's natural production acquisitions.[21] The search for active animal extracts began in 1975. By 1980 the NCI had examined 16,500 extracts taken from approximately 3,000 animal species to produce seven useful compounds.[22] Targeting principally marine organisms,[23] the program expanded significantly in 1986, prompted in part by the AIDS epidemic, as the NCI put out contracts to collect marine organisms in the Indo-Pacific region. About 1,000 new marine invertebrates a year were fed into the system throughout the 1980s from sites around the world.[24]

As previously noted, the NCI was not only aware of work outside the United States, it actively procured foreign (mainly European)

20. Natural Products Branch, DCT, Annual Report, 1981–82, 206.
21. Ibid., 213.
22. Ibid., 206.
23. At http://www.cancer.gov/cancertopics/factsheet/Therapy/natural-products.
24. Natural Products Branch, DCT, Annual Report, 1993–94.

drugs. Indeed, a number of major cancer drugs—including, in addition to those just mentioned, carboplatin (UK), oxaliplatin, vinorelbine and irinotecan (France), cyclophosphamide (Germany), and etoposide (Switzerland)—came from European countries that had long been world leaders in pharmaceutical chemistry (Wagstaff, 2004). Thus when the NCI provided funds to create the EORTC data center in the early 1970s, it also financed the establishment of a screening laboratory at Jules Bordet in Brussels to test compounds coming from European sources. At the same time, it established an NCI liaison office in Brussels, and a few years later, given the success of the liaison office, it expanded the Brussels screening facilities (McIntosh, 1992). Initially designed to link European chemical and pharmaceutical companies, the EORTC and the NCI campus in Bethesda (Yoder, 1997), the NCI liaison office went on to organize a series of NCI-EORTC symposia on new drugs and to promote the standardization of review procedures and clinical protocols with a view toward facilitating the flow of information and compounds between Europe and North America. The NCI also struck agreements with the United Kingdom, in particular with the Cancer Research Campaign to harness drug development in the United Kingdom, and these initiatives led in 1986 to a joint agreement between the CRC, the EORTC, and the NCI that allowed newly developed anticancer agents to freely enter trials in the United Kingdom, continental Europe, or North America.

Hardly a tale of effortless harmonization, disagreements and obstacles quickly appeared. In the early 1980s, for example, the EORTC board sought "to avoid applying requirements such as those of the FDA" since an "overwhelming majority" of EORTC participants did not "support the scientific basis of [those] requirements." The board felt that "it should avoid even assemblance [sic: a semblance] of modifying its procedures for the benefit of adherence to regulations which are not supported by European scientific opinion." To do so would have been "unacceptable to clinicians and scientists in Europe" and would have paved the way for accusations of "using European patients for studies carried out in the interest of US pharmaceutical companies."[25] The existing disparities between regulatory principles and practices would not be addressed in earnest until the subsequent decade, with the establishment, for instance, of the International Conference on Harmonization of Technical Requirements for Registration of Pharmaceuticals for Human

25. Letter from M. Staquet to P. Hilgard, 20 March 1981, EORTC Archives (Brussels).

Use (ICH) in 1990 and the promulgation of the 2001 European Directive on human experimentation. In the meantime, and despite these frictions, the circulation of substances for experimental testing continued unabated.

The EORTC had initially set up a new drug development committee in Brussels in 1980 to coordinate the increasing number of Phase II trials with new drugs carried out by the organization. Rapidly, and following an agreement with the NCI facilitating the exchange of experimental drugs of mutual interest,[26] the EORTC opted for the creation of a New Drug Development Office (NDDO) in Amsterdam that opened in 1982 under the direction of Dr. Pinedo. The NDDO focused on late preclinical (formulation development, toxicology) and early clinical phases of drug development. The Amsterdam center created a second data center to manage clinical trials (including site visits of each center) carried out by the early clinical studies group of the EORTC, a network that by the end of the 1990s consisted of fifty European oncology centers, leaving the Brussels data center in charge of large Phase III trials.[27] These developments were in part spurred by the fact that while European trialists had previously carried out only confirmatory Phase II trials, during the second half of the 1980s so-called pivotal Phase II trials; that is, trials that could be used to register drugs with the FDA, had become increasingly frequent in Europe. Early Phase II trials of this nature had to be performed under special conditions dictated by Good Clinical Practice regulations, which included the aforementioned site visits, the necessity of a flow of information to the company in addition to the data center, the impossibility of correcting data at the data center, and so on (Gerlis, 1989). In short, a "normal" data center was usually not equipped to handle Phase I and early Phase II pivotal trials.[28] The independent data center provided a sign of the increasing autonomy of drug discovery and

26. EORTC representatives attended the NCI Decision Network meetings and participated in the selection of compounds that were subsequently released for testing in Europe. Letter from Vincent T. DeVita to Laszlo Lajtha, 12 May 1981; letter from M. Staquet to Gisela Hämmerli, 5 January 1982, enclosing a document titled "New Drug Development and Coordinating Committee," EORTC Archives (Brussels).

27. Interview with Franco Cavalli, Bellinzona, Switzerland, 22 December 2006; interview with Bob Pinedo, Amsterdam, 7 December 2006; Omar C. Yoder, NCI European Collaborative Program, draft manuscript, May 1990.

28. Memorandum from Franco Cavalli to the EORTC Board, 17 August 1989, EORTC Archives (Brussels).

Phase I trials vis-à-vis the EORTC and was further accentuated in 1984 when the aforementioned NCI-EORTC new drug symposium, initially held in Brussels, moved to Amsterdam (Krul, 1998). The symposia became major biennial events attracting 1,750 participants from over fifty countries by the late 1990s.[29] Remembering the early days of drug development, the Argentine-Spanish oncologist Hernán Cortés-Funes noted:

> Each centre developed their own phase I trials, receiving drugs from many sources. Then they would say "this looks promising," and everyone would try it. It was a small club. When the big laboratories and pharmaceutical industry started producing drugs and offering them to different people, then came competition between the units. (McIntyre, 2007: 35)

Despite the competition and national and institutional boundaries, oncologists and substances continued to circulate in the clinical-experimental space established by the new style of practice. In the words of the former head of the NCI liaison office, trialists from both sides of the Atlantic formed "one community" (Wittes & Yoder, 1998), but one, we would add, whose growing complexity, internal differentiation, and burgeoning interfaces with a growing number of stakeholders led to new forms of "oncopolitics," as we will see in what follows.

29. At the turn of the century the symposia underwent a further transformation as part of a tripartite agreement with the American Association of Cancer Research and are now organized on a yearly basis alternately in the United States and Europe as meetings on "Molecular Targets and Cancer Therapeutics." As of 2010 the number of participants was restricted to 2,200 to promote exchange and discussion.

CHAPTER NINE

Oncopolitics? Reshaping Collaborative Research

As knowledge evolves, so too does the organization of knowledge production. This is especially true in transdisciplinary and inter-organizational domains. In this chapter we examine the shifting *architectures of knowledge* (Amin & Cohendet, 2004) that accompanied the events discussed in previous chapters. Despite institutional differences such as the absence of a European agency comparable to the NCI, trialists have noted that similar tensions characterized the distribution of author-ity on both sides of the Atlantic (Wittes & Yoder, 1998). These tensions expressed themselves within organizational milieus whose main differ-ences are encapsulated by the notions of *formal organizations* (such as firms or state agencies) and *metaorganizations* (organizations of organi-zations, whose prototype is the association) (Ahrne & Brunsson, 2006). Composed of individual members, formal organizations maintain a cen-ter of control and coordination. Metaorganizations lack central control and have a limited ability to mobilize resources; instead, they facilitate interaction and communication within a field. While an agency like the NCI easily qualifies as a formal organization, the tensions between cen-tralized decision-making at the NCI and the cooperative groups can be seen as a manifestation of the groups' drive to increase their autonomy vis-à-vis the NCI, and in particular the Division of Cancer Treatment that supervised them. In Europe, members of a metaorganization like the EORTC debated how and to what extent their organization should move toward a more formal organizational status. We have already had a brief preview of this tension during the debate surrounding the institu-tion of the EORTC data center in Brussels. Promoters saw the center as a step toward the establishment of a centralized organization capable of implementing a coherent research program. Detractors sought to limit the EORTC to a coordinating role and to maintain control of research in the disease-oriented cooperative groups, which were often organized on a national basis. As we will see, this debate continued in a different form during the 1980s, fueled by the observation that the organization's cooperative groups now operated independently and without strict quality control, rarely exchanged information, and generally ignored EORTC council directives.[1] A similar debate in the United States concerning the relations between the cooperative groups and the NCI or, more precisely, the Cancer Therapy Evaluation Program (CTEP) of the Division of Cancer Treatment (DCT) raised analogous issues. We begin with the European case.

Centralization or Coordination? The European Debate

In 1982 the EORTC asked its former president, Dirk van Bekkum, to prepare a short statement of the advantages of multicenter clinical trials ("the main activity of EORTC") over single-center trials.[2] As we saw in chapter 6, the Italian oncologist and member of the EORTC council, Gianni Bonadonna, had already embarked on a campaign to promote single-center trials. He was apparently not alone: other members of the EORTC council had also begun to challenge the rationale of multicenter and by extension multinational trials, a somewhat surprising development insofar as it ran counter to the very raison d'être of the EORTC. And yet it was only the opening shot of a wider debate that would lead to a profound transformation of the EORTC during the second half of the 1980s.

In his report, van Bekkum listed the drawbacks of single-center trials. To begin with, since any single institute had, by definition, a limited staff, it had equally limited opportunities for researcher interaction and the critical evaluation of clinical trial protocols. Limited input entailed a correspondingly narrow epistemic viewpoint and a paucity of alternative

---

1. "Proposal Veronesi: Revision of EORTC policy—Re-organization of EORTC," 12 May 1978, EORTC Archives (Brussels).

2. Letter from D. W. van Bekkum to M. Staquet, enclosing "Rationale for multicentre (international) clinical trials," 23 August 1982, EORTC Archives (Brussels).

ideas. Furthermore, in single-institution studies, there was no outside verification to counter breaches of discipline in patient selection, outcome reports, and the like. The diminished patient pool of a single center also meant that patient accrual was slower than in multicenter trials, leading to delays in completing the trial. In turn, the results of such trials generally failed to garner widespread acceptance without confirmation by other institutions. Finally, the pressure to publish that single institutions brought to bear on their members often resulted in the production of inconclusive preliminary reports that provoked concurrent me-too trials, wasting time and patients.

EORTC statistician Marc Buyse completed van Bekkum's litany of shortcomings with the example of a recent adjuvant trial in which the regimen under study showed benefits in one institution in one country and no difference when results were compiled from other countries and institutions. Single-institution results clearly did not always hold true in other countries and institutions. Multicenter, international trials were thus more likely to produce *convincing* results (notice here the implicit assumption that results should be *both* methodologically sound *and* persuasive).[3] And yet the debate did not end there. Other members of the EORTC council countered that single-center trials enforced uniform standards[4] and faulted multicenter trials—whose results also had to be confirmed by additional trials in order to gain general acceptance—for watering-down novel protocols by seeking multi-institutional common denominators.

The division between national and international trials raised yet another issue. Although defined as European and even though members and patients came from all European countries, over 60% of the patients registered by the EORTC between 1974 and 1992 came from France, Belgium, and the Netherlands. An obvious preoccupation for the EORTC leadership, board and council members complained about the asymmetry on a number of occasions and discussed possible causes. The United Kingdom, for example, had its own well-established clinical trials network and a tradition of work in this domain. Clinicians from countries without this tradition found other reasons to opt out of the EORTC.

3. Marc Buyse, "An illustrated comment of multinational versus national trials (July 1982)," EORTC Archives (Brussels).

4. "Summary of the papers submitted by members of the Board and Council (25 October 1982)," EORTC Archives (Brussels).

Working within national networks gave researchers access to sources of funding that were more generous than those available via the EORTC. This was certainly the case in Germany, a rich country with a large patient population.[5] More generally, the "subsidiarity" principle governing European policies meant that the European Community supported only those activities that could not be performed or funded at the national level. Finally, as in the original debate over cooperative research itself, researchers feared the inevitable dilution of academic credit: unlike EORTC trials, national trials restricted the number of authors in clinical trial publications and thus made the participants more visible as coauthors.[6] The existence of multiple structures for carrying out trials led some to advocate a reform of the European system that would unify them, while others, cherishing this plurality, argued that the EORTC should settle for a coordination role.[7]

In the wake of these criticisms, in 1986 the EORTC reorganized and established four branches: clinical trials, education (training and certification), research (to bridge the gap between fundamental and clinical research), and epidemiology and prevention. To highlight this transformation, the EORTC changed its name (but not its acronym), replacing the "on" in "research on treatment of cancer" with "and." The structural reorganization was less radical than one might have expected, because treatment, and thus the performance of clinical trials, remained the focal point of the EORTC activities. From our point of view, the most important transformation took place *within* the treatment branch. A clinical trial committee henceforth oversaw more specialized committees such as the protocol review committee, the data center committee, and the quality control committee, but in turn reported to the cooperative groups' chairmen's conference. The new structure thus embodied a tension between centrifugal and centripetal forces: quality control and protocol review gave the central office a handle on group activities, but these activities remained under the formal supervision of the groups' chairmen. Moreover, while poor patient accrual rates or a bad performance might lead the Clinical Trial Committee to withdraw financial or data center support for a given study, no action could be taken without

5. Letter from Theodor M. Fliedner to L. G. Lajtha, 7 January 1980, EORTC Archives (Brussels).

6. Interview with Jean-Claude Horiot, Dijon, 16 May 2005.

7. Letter from Theodor M. Fliedner to L. G. Lajtha, 7 January 1980, EORTC Archives (Brussels).

"detailed consultation with the individual group concerned."[8] The board of the EORTC, which oversaw all four branches of the organization, also had its say, as did the Scientific Advisory Committee that assisted the board in its deliberations.

This was the formal compromise. In practice, the Brussels data center continued to supervise studies that used the EORTC label, although it faced opposition or at least passive resistance from a number of cooperative groups. Consequently, the issue of quality control moved increasingly to the forefront of the EORTC agenda. The initial impulse to create a quality control program came, unsurprisingly, from radiotherapists who, as we saw in the United States (chapter 5) had pursued quality control in lieu of clinical trials. In the early 1980s, a group of young clinicians toured European centers, and armed with a single protocol launched a pioneering program originally confined to local equipment measurements (Horiot et al., 1986). The initial reactions bordered on surprise and incredulity: it would never work, colleagues argued, because a group of relatively junior radiotherapists could hardly pretend to instruct their older, more eminent colleagues. Despite the skepticism, the initiative succeeded, largely because the group eschewed the search for absolute standards and sought instead to establish relative standards based on local variation. After measuring how each center had implemented the test protocol, the team classified the results (and thus local institutions) into three categories: small, incompressible deviations due to instrumental differences; acceptable deviations that should nonetheless be monitored; and major, unacceptable deviations that undermined the validity of cooperative work between centers and had to be corrected if the local institution wanted to participate in a multicenter trial protocol.[9]

Because of its technical dimension, quality control acted as one of the ways the central office held cooperative groups to account for their activities, in the dual sense of providing an account and being held accountable. Indeed, as they lack central management, metaorganizations tend to use quality control and similar relational standards as a way to establish order among member organizations and make their activities observable and reportable. Within the EORTC, quality assurance

8. EORTC Institutional Committees, final draft (approved by Board on 17 August 1990), EORTC Archives (Brussels).

9. Interview with Jean-Claude Horiot, Dijon, 16 May 2005.

in radiotherapy went through three phases in the 1980s, moving from the early preoccupation with equipment (instrument calibration and dose measurement) to a focus on the delivery of treatments to patients, first by assessing the implementation of protocols within single institutions, and finally, at the end of the 1980s, by reviewing individual cases to determine the reliability of treatment procedures (Bernier, Horiot & Poortmans, 2002). The success of this initiative convinced the EORTC to extend the program to medical oncology and, unsuccessfully, to surgical oncology (Therasse, 2002). Like radiotherapy, quality control in medical oncology moved from a narrow focus on technical issues such as data collection and management (Vantongelen, Rotmensz & van der Schueren, 1989) to a more comprehensive interest in treatment delivery (Marinus, 2002; Vantongelen et al., 1991). In the case of chemotherapy trials, the evolution of quality control followed developments in computer software, in particular the introduction of the System for Management and Analysis of Randomized Trials (SMART) by the Brussels data center IT unit in the 1980s.

Quality control and its enforcement generated friction between the central office and individual cooperative groups. For instance, a 1992 data center report noted:

> These quality control measures have NEVER [their emphasis] been adequately enforced. First, the EORTC Data Center has not had the manpower or the means to carry them out in an efficient manner. Secondly, *the Cooperative Groups have resisted their enforcement* [our emphasis], especially with respect to the blocking of randomizations and the closing of poorly recruiting trials.[10]

There was more at stake, however, than the enforcement of rules. To act as the leader of an epistemic organization, the EORTC board had to supply a view of the future, and in this regard, the same report argued, there was little on offer:

> There is no coordination on the level of the EORTC of the scientific activities of the various cooperative groups and hence no well-defined, long-term strategy exists. This is true, for example, with respect to the development of common research strategies within the various cooperative groups and also with

10. "EORTC Data Center," March 1992, 25, EORTC Archives (Brussels).

respect to the evaluation of new drugs in a phase III setting after successful testing in phase II trials.[11]

The transformation of EORTC's cooperative groups into semiautonomous institutions that held the central office at arm's length was partly rooted in the commercial development of the successful regimens described in chapter 8. Those regimens, it will be recalled, targeted relatively common cancers. Until then, with treatment confined to rare cancers like leukemia, the pharmaceutical industry had largely ignored the cancer market. All this changed in the 1980s when it became clear on both sides of the Atlantic that chemotherapeutic regimens such as CMF and new substances such as cisplatin, while they did not "cure" at least prolonged life in common cancers such as breast cancer (33% of all cancers in the United States by the mid-1980s). The emergence of a younger generation of more aggressive surgeons also increased demand for drugs, as did the fact that systemic chemotherapy that was initially available only in specialized centers spread to general hospitals and office-based oncologists. By the end of the 1980s, Bristol-Myers' oncology division—which had managed to recruit prominent US and European trialists such as Stephen Carter and Marcel Rozencweig—had become the "undisputed leader" in the anticancer drug market (Laker, 1989). Given the limited financial resources of the EORTC, its groups welcomed the pharmaceutical companies' emerging interest in cancer as a new source of funds.

The cooperative groups were not necessarily eager to share these funds with the central organization. Some went so far as to seek independent legal status and approached funding agencies directly, bypassing the EORTC board.[12] Beyond the more immediate issues of money and influence, at stake, as in the case of quality control, was the ability of the EORTC to concentrate resources on promising new avenues of research. In this regard the EORTC confronted two problems. First, the trials that interested industry in the 1980s tended to be relatively mundane and thus produced little in terms of cutting-edge knowledge. Groups that pursued such trials ran the risk of drifting away from the

11. Ibid., 31.

12. Reconstitution based on the accounts of Jean-Claude Horiot (interview, Dijon, 16 May 2005) and Louis Denis (interview, Antwerp, 21 March 2005); "Minutes of the EORTC Board Meeting," 19 December 1995, EORTC Archives (Brussels).

research-oriented goals of the EORTC. Second, even nonindustry trials carried out semiautonomously by the groups posed obstacles to the implementation of EORTC research plans. A strong central office could in principle hand over a promising treatment developed through Phase I and II trials by the New Drug Development Committee to cooperative groups for Phase III trials. In practice, however, the existence of semiautonomous groups pursuing their own line of work preempted such plans insofar as the groups tended to cling to substances developed "in house." The ability to establish priorities and strategic initiatives implied, in other words, that the central office be able to intervene in the selection and planning of group trials.[13] As we will see in part 3, the EORTC took this direction at the turn of the century, when, in the throes of the molecular biology revolution, the four-branch structure was discarded and replaced by a focus on translational research.

In the meantime the scientific advisory committee continued to evaluate the performance of the cooperative groups. The committee's reports to the board on individual cooperative groups offer numerous examples of the problems faced by the EORTC in its drive to streamline its activities and, more importantly, to act as an epistemic institution.[14] Competing national and multinational groups, for instance, claimed the status of EORTC cooperative group for the same form of cancer. The multinational status of one group was questioned when it became clear that its membership came from only two countries. When national groups competed for patients, intergroup collaboration (for instance with the United Kingdom or West Germany) was a possible solution. In a multimodal group, radiotherapists and surgeons complained of the dominance of medical oncologists. Finally, the spread of oncology to community hospitals deprived academic groups of a source of patients, a problem shared with the United States, as we will see below.

Clinical Trials as Clinical Research: The US Debate

By the mid-1970s, the US cooperative oncology groups had evolved into robust operations that were relatively independent of their sponsor, the NCI. Multimodal groups often tested similar regimens in what could

13. Interview with Jean-Claude Horiot, Dijon, 16 May 2005.
14. Ibid.

be viewed as either a form of independent reality check, the development of an autonomous research program, or a wasteful redundancy amounting to nothing more than a series of me-too clinical trials. All three points of view circulated. Despite their independence, as members of a formal organization the groups were not impervious to censure or free from oversight. They were the subjects of almost constant review by various committees within the NCI. In what follows, we highlight some of the principal shortcomings and strengths of the groups as expressed through the major program revisions during the period under consideration. While these revisions may at first sight seem to have been simple administrative reshuffles, they also touched upon issues that partook of the epistemic core of the groups' activities. Notwithstanding initial resistance and skepticism, the kind of work carried out through the cooperative group program came to be widely accepted as an autonomous practice within the NCI; the distinction that we drew in part I between testing and research had become fully institutionalized by the mid-1970s.

The emergence of an autonomous form of research had begun, it will be recalled, in the mid-1960s when the cooperative group program was implicitly recognized as an independent research enterprise. Explicit confirmation of that first transition appeared in 1968 when the Clinical Trials Review Committee of the NCI observed: "Justification for the cooperative chemotherapy groups, formerly based on their role as a testing facility for the NCI's drug development program, now resides solely in the quality of the research they propose and conduct."[15] No longer restricted to testing screened products, the groups saw their charge expanded to include "exploring the application of the cooperative, multi-institute research techniques to the clinical investigation of cancer epidemiology, etiology, diagnosis and prevention as well as cancer therapy."[16]

Given this research mandate, the committee would henceforth review the groups' grant applications on "the inherent merit of the proposals and not on their relationships with the National Cancer Institute Drug Development Program."[17] The notion of a clinical trial was furthermore

15. "Report of the Cancer Clinical Investigation Review Committee on the 'Cooperative Studies Workshop' Held in Williamsburg Virginia, October 20–23, 1968," 1, NCI Archives.

16. Ibid., 2.

17. Memorandum from J. Palmer Saunders, Associate Director for Extramural Activities, National Cancer Institute, to Membership, Cancer Chemotherapy Collaborative

expanded insofar as it now included therapy without being reduced to therapy.[18] In keeping with this expanded mandate, the NCI further charged the committee with organizing seminars and symposia as it saw fit. And as befitting its research orientation, the committee's name was changed to the more wide-ranging Clinical Cancer Investigation Review Committee (CCIRC).

The new system lasted seven years. From 1968 until 1975, the CCIRC applied the above criteria in their review of cooperative group grant applications.[19] As a new style of practice, however, clinical cancer trials were also saddled with the fact that they were a collective undertaking and thus required coordination and administration that did not fit easily into the research framework created for the independent investigator, the dominant figure of the postwar research system. More than simply disbursing the funds, the CCIRC was also the de facto administrator of a program of research insofar as the committee also held the responsibility for reviewing the protocols submitted by the groups. The seven-year interlude thus constituted a sort of "research moment" for clinical cancer trials. This moment was doomed from the beginning; not only was there an inherent conflict of interest in the grantor being the administrator, but just as clinical cancer research could not be reduced to testing, it could not be easily characterized as pure research.

The system consequently underwent major restructuring in 1975 in the wake of the "War on Cancer" when administrative responsibility for the cooperative groups was transferred to the Division of Cancer Treatment (DCT) and its subdivision, the Cancer Therapy Evaluation Program (CTEP). To ensure that granting functions and administrative functions remained separate, the CCIRC retained granting power while explicit policymaking for clinical cancer research shifted from the CCIRC to the DCT. The renewed management of the groups resulted in a complex system of consultation and review that expressed both the hopes and misgivings of programmed research, the underlying tension between research and therapy, and the need for forums of collective reflection on a collective undertaking. Beginning in 1976, CTEP put to-

---

Clinical Trials Review Committee, "Charge to the Committee," 23 February 1968, 1, NCI Archives.

18. Ibid., 2.

19. Grant and Contract Review Committees, "Reports Relating to Phase I of NIH Advisory Committees," Box 17, RG 443, p. 5, Archives II, National Archives (College Park, MD).

gether cooperative discussion groups that included members of DCT and the cooperative groups. Described as "primarily responsible for monitoring and coordinating the grant-supported clinical trials performed by the various cooperative oncology study groups,"[20] the discussion groups were intended to give further direction to the cooperative groups and contractors.[21]

Organized around specific diseases, these groups supplemented a series of working groups organized along the same lines. Yet another prong of this disease-oriented strategy, as we saw in chapter 7, consisted in the appointment of personnel from CTEP as disease coordinators.[22] These coordinators interacted with the newly formed committee of cooperative group chairmen that met regularly to coordinate activities among themselves and to regulate their relations with the DCT and the CTEP. Management of the cooperative groups thus consisted of a "troika" (Davis, Durant & Holland, 1980: 383) of an executive committee of cooperative group chairmen, an extramural review committee (the CCIRC), and policy-formation groups populated by in-house NCI experts and administrators.

The 1979 NCI Review

In 1979 the NCI undertook a review of all clinical research conducted by the institute. Given that the groups had seen an increase in funds over that of other NCI programs, NCI planners singled them out for special attention and assigned the DCT's Board of Scientific Counselors the task of reviewing the groups' performance.[23] According to DeVita, the very existence of the groups was in play: "We are at a historical turning point in therapeutic research. We're struggling to put it all together. We're faced with the problem that if we were to start up a clinical trials program today, would we set it up as we did 20 years ago" (Group chairmen unhappy over H & N . . . , 1978: 8).

A number of overlapping issues occupied the review, the first of which

20. "The 1976 Report of the Division of Cancer Treatment, NCI, Vol. 3, Operational Report," 7.

21. Vincent T. DeVita, "Report of the Director—The 1977 Report of the Division of Cancer Treatment, NCI, Vol. 1, Summary Reports," September 1977, 40.

22. Ibid., 42.

23. The percent increase over other programs within individual groups ranged from 0% to 300% ("Minutes," Board of Scientific Counselors, March 26–27, 3, NCI Archives).

was how should clinical cancer research be evaluated? As a new form of research, the usual criteria did not necessarily apply. In preparation for the review, the chairman of the CCIRC compiled a list of thirty-six factors entering into the review of a cooperative group clinical trial grant application and submitted the list to CCIRC members asking them to rank the factors. As the following top ten factors show, while the clinical trial clearly remained a research instrument, the CCIRC viewed numerous aspects of clinical research other than originality as equally important:

1. The study design is good (realistic); the question(s) being asked will be answered.
2. The protocol is important (if only we had this information . . .).
3. Investigators participate in Group committees or Group administration, or both.
4. Investigators participate in study design.
5. Institution has high proportion of evaluable patients (i.e., quality data).
6. The study design accounts for important prognostic variables.
7. The study is unique (original).
8. Case evaluability is high.
9. Data is returned promptly.
10. Results are being analyzed to learn everything possible (ancillary information). (CCIRC members list . . . , 1978: 2)

The review mobilized the head of the committee of cooperative group chairmen, Barth Hoogstraten, who conducted an initial defense of the groups during a meeting held with DeVita and the CCIRC. In accordance with the promotion of the groups as research organizations, he began by listing group publications. Considering only the protocol studies, between 1973 and 1977, publications by the groups had risen fourfold, from fifty-three a year to 207 a year (CCIRC starts preparing . . . , 1978). But Hoogstraten went on to defend the groups not only for their productivity but also for their innovative structure. He pointed out how the cooperative groups had constructed an unprecedented network of associations instancing links with 493 affiliated hospitals, including forty-one foreign institutions, practically every medical school in the United States, and all but two of the recently created cancer centers (Walter & King, 1977).

Toward the end of October of 1978, DeVita let it be known that he would propose a major reorganization of the cooperative groups by transforming them into regional entities during the review that had been slated for March of 1979. Cooperative group chairmen already had reason to suspect that they were in for a rough ride. Henry Kaplan, a Stanford radiologist and radiotherapist and member of the board, had openly criticized the groups as useless, saying that they had conducted "what I call me-too trials, that have added trivial data and accomplished essentially nothing." In other words, even though the groups organized almost 75% of the clinical cancer trials in the United States, "there has been a paucity of ideas from Cooperative Groups and contractors. Most ideas have come from individual institutions" (Cooperative group reorganization . . . , 1978: 2). It was a criticism that would be repeated throughout the 1970s and 1980s. When, for example, the former head of NCI's cancer centers program, John Yarbro, gave his presidential address to the Association of Community Cancer Centers five years later, he asked rhetorically: "Of our curative cancer treatments, how many were developed by cooperative groups and how many by independent clinical investigators? Even neglecting surgical and radiation cures, ask yourself whether any of the following tumors were first shown to be curable by group research: uterine choriocarcinoma, Burkitt's lymphoma, histiocytic lymphoma, small cell cancer, germ cell cancer, ovarian cancer, osteogenic cancer" (Feds short on money . . . , 1983: 2).

In the months that followed his initial threat, DeVita scaled back his proposed intervention. The Board of Scientific Counselor's review came and went with little heat, and the structure of the cooperative groups remained relatively unchanged. Contrary to Kaplan's summary judgment, DeVita himself concluded the review saying: "It is my personal view that clinical trials often test fundamental therapeutic hypotheses. At the very least, they represent the effector arm of a wide variety of pre-clinical research programs and they have clearly yielded important results" (New funding mechanisms . . . , 1979: 4). According to the chair of the cooperative group chairmen's committee, DeVita admitted at a subsequent meeting, "that the results of the scientific presentations [by the groups] were sufficiently convincing to most everyone on the Board that a written review of the achievements of the groups seemed not to be appropriate" (Group chairmen balk at . . . , 1980: 5).

Although less than anticipated, significant changes did emerge from the process. To begin with the board managed to define the kind of research carried out by the cooperative groups not as research but as a new form of research that they denoted "clinical treatment research" and that included "any or all aspects of treatment involving individuals or groups of cancer patients" (New funding mechanisms . . . , 1979: 3).[24] Next, following the review, the board converted funding for the groups' trials from grants to cooperative agreements, a move that generated some controversy. While peer review governed both types of funding, some commentators saw the cooperative agreements as inferior to research grants. Commenting on the distinction, the chairman of ECOG offered the following special pleading: "I hope that in the long term strategy, the major cooperative groups can be looked upon in the same way as the ROIs [research grants]. Clinical research is being looked upon as something different than lab research. It is, but you can learn biology in clinical research" (Group chairmen unhappy over exclusion . . . , 1983: 5). Emil Frei of CALGB seconded: "The word is research. You can conduct a well-planned clinical trial that will tell you a lot about biology. Over the last 20 years, a lot of leads to major biological advances have come from the clinic. It goes both ways" (ibid.). In addition to the change in funding strategy, the board consolidated the review of clinical trial protocols and instituted a series of quality control exercises that included, in addition to on-site monitoring of clinical trials (Mauer et al., 1985; Itri et al., 1985), the surveillance of a number of scientific parameters such as pathology review and clinical laboratory data (DCT board approves . . . , 1980: 5).

The Reform of an Autonomous System

No sooner had the board's 1979 review and reform been completed than voices emerged to declare the reform incomplete. In particular, CTEP complained that the board had "made no attempt to evaluate the methodology used in the conduct, analysis and reporting of trials. Nor had it attempted to evaluate the quality of the ideas submitted to clinical testing or the timeliness of their implementation" (Wittes, Friedman & Simon, 1986: 241). Consequently, in 1984 CTEP undertook an exhaustive review of all clinical trials carried out by the cooperative groups between 1979 and 1984 to determine whether or not they had addressed questions

24. Vincent DeVita, "Opening Remarks—Minutes, Board of Scientific Counselors, Clinical Trials Review, March 26–27, 1979," NCI Archives.

of pressing importance in clinical research, the extent of duplication of work among the groups, and the quality of trials conducted.[25]

Between December of 1985 and November of 1986 panels of five to ten experts evaluated five disease groups that were considered a representative sample of cooperative groups' work. The panels' conclusions were less than flattering. Deficiencies in hypothesis generation, study design, and sample size varied from 25% to 80% according to disease group. Similarly, problems in patient accrual were noted in 35% to 80% of protocols.[26] While many had suspected that the groups had offered a less than stellar performance in the previous decade (Wittes, Friedman & Simon, 1986; see also Pater et al., 1986), rapid developments in molecular biology and immunology added further fuel to the fires of reform. CTEP noted: "Preclinical developmental therapeutics is moving very fast and basic biology is moving even faster. Thus the number of worthy new therapeutic hypotheses will increase dramatically in the next 5–10 years" (CTEP presents options . . . , 1986: 4). In other words, so much had been produced within the biological laboratories in the previous years that getting them into clinical trials would require some form of prioritizing (Translation of "new biology" into clinical use . . . , 1985). The question was how quickly could the clinical trial system adapt to the new biology. Similar issues confronted the EORTC, leading to the formation of a committee "in initiation of fundamental/clinical cooperation" in the early 1980s that following the 1986 reform morphed into the EORTC research branch.[27]

The ability of the groups to adapt depended in part on their ability to absorb novelty. According to CTEP, their capacity to do so was hampered on a number of fronts. First, many of the trials were too small, and the suboptimal accrual rates prolonged the trials unduly. This in turn led to a "long gestation time for new ideas." Administratively, the funding of the groups had created yet another bottleneck. The fixed relationship between the groups and the participating institutions meant that the

25. Associate Director for the Cancer Therapy Evaluation Program, "Summary Report—Annual Report, Division of Cancer Treatment, Vol. 1," October 1984–September 1985, 379.

26. "Clinical Investigations Branch, Annual Report, Division of Cancer Treatment, Vol. 1," October 1, 1985–September 30, 1986, 384.

27. "Report on the meeting of the committee in initiation of fundamental/clinical cooperation, Zürich," 26 April 1980, EORTC Archives (Brussels).

groups had difficulty reallocating funds according to emerging priorities. Group funds, in other words, were tied up in such a way that it was becoming increasingly difficult for clinical researchers to "get a reliable answer before you don't care what the answer is" (Wittes cited in CTEP presents options . . . , 1986: 4).

In April of 1986 the director of CTEP laid out a series of proposals designed to make the production of clinical trials more responsive to the new biology and less a mechanism for testing combinations of chemotherapy and other modalities (Changes in clinical trials . . . , 1986). The first proposal sought to develop the means to prioritize clinical trials by appointing strategy committees to advise CTEP on Phase III trials. To appreciate the importance of prioritizing, it is helpful to recall that at the time the cooperative groups attracted less than 3% of available cancer patients for common malignancies. At that rate, definitive Phase III studies of one thousand to three thousand patients required up to a decade to complete. Setting priorities would increase accrual rates within the system, and in the early years, it did; prioritized trials accrued patients three times as quickly as unprioritized trials.[28]

Two further proposals sought to make the groups less of a club and more of a free-floating association built primarily around individual protocols. The first of these proposals, described as an "administrative nightmare" by some present at the meeting, sought to restructure the financial ties between cooperative groups and clinical investigators. Heretofore, study investigators had been drawn solely from hospitals and cancer centers that were members in good standing of the groups. CTEP felt it should be possible to associate clinical investigators from any clinical setting with a given cooperative group depending on the needs of the study. In this way a group could be expanded or contracted according to the disease under study and the required science. It would also open the door to investigators becoming members of more than one group.

The most controversial proposal sought to further dissolve the institutional ties of the groups by paying for patients treated in clinical trials on a per capita basis. At the time, institutions such as hospitals received lump sum grants to defray costs and to compensate them for their participation in scientific research. By replacing the institutional grants with per capita reimbursement, CTEP would allow the groups to seek

28. "Director's Report, Annual Report, Division of Cancer Treatment, Vol. 1," 1 October 1989–30 September 1990, 19.

out participants from any source (HMOs, academia, etc.) and pay for them on an individual basis. Such a change put the groups in an awkward ethical position. As the president of the American Society of Clinical Oncology pointed out: "The work of the Cooperative Groups, being peer reviewed and with separate grants, is seen as science. If the system is changed so that the payment is just by patient accrual, Groups would lose that extremely sensitive position" (CTEP clinical trial . . . , 1986: 4; see also Muss, 2005). Such a change would also put the groups in direct competition with a pharmaceutical industry that already paid physicians for patient participation in clinical trials. In addition, given the existing rate structure, group rates would then have to be established according to such factors as the disease type, modality of treatment, anticipated length of follow-up, and so on, a major disadvantage noted by CTEP itself (Changes in clinical trials . . . , 1986).

Following further reaction by the cooperative groups and the cancer centers, CTEP replaced the "drastic changes" that had been announced with a "modification": the committees would "be based in the groups" (DCT board to study . . . , 1986: 4), the institutional grants would remain intact, and per capita reimbursement would be temporarily abandoned, as well as the attempt to undo the relations between the groups and their institutions. For the moment it was business as usual. Or so it seemed. Within two years all of the changes proposed by CTEP had been enacted, albeit in indirect ways. First, the DCT cut the number of groups from twenty-five to eighteen, with the number of studies falling correspondingly from around six hundred a year to under five hundred by 1988.[29] Strictly regional groups were terminated in recognition of the fact that many of the older groups such as ECOG had quasi-national coverage.[30] Second, the strategy committees moved quickly to designate a series of "high priority clinical trials."[31] Third, the cooperative chairmen and CTEP met and drew up a set of guidelines for the conduct of intergroup studies in order to overcome between-group variability in quality control and thus facilitate the conduct of studies involving more than

29. Associate Director for Cancer Therapy Evaluation (Michael Friedman), "Summary Report—Annual Report, Division of Cancer Treatment, Vol. 2," October 1988–September 1989, 614.

30. Clinical Investigations Branch, "Summary Report—Annual Report, Division of Cancer Treatment, Vol. 1," October 1987–September 1988, 295.

31. B. A. Chabner, "Director's Report—Annual Report, Division of Cancer Treatment, Vol. 1," October 1987–September 1988, 9.

one group.[32] Fourth, the DCT modified the terms of awards for the co-operative grants to allow "capitation forms of reimbursement" to "stimulate patient enrollment" in selected trials.[33] Finally, the larger, multidisease, adult cooperative groups such as ECOG, CALGB, and SWOG went on a membership drive among the four thousand members of the American Society of Clinical Oncology and garnered the adherence of a total of 157 new institutions and 4,500 new patients per year for priority trials.[34]

Oncology in the Community

We have so far focused on intra- and interorganizational oncopolitics. In the late 1970s the control of protocols and patients extended beyond the research centers and the cooperative groups. A number of European cooperative groups reported increasing difficulty in recruiting patients since the patients were often seen in community hospitals and no longer in the academic centers to which the trialists had access. In the United States, the cooperative groups faced a similar problem, but in addition, in the late 1970s they came under increasing pressure to move the latest results of clinical trials into the community at large and to translate their results from a relatively small network of academic centers to an extended network of community clinics and private practice. As a solution to this twofold problem, the NCI launched a program in 1982 to fund clinical trials in community hospitals. Known as the Community Cooperative Oncology Program (CCOP), community-based oncologists would manage the trials locally as affiliate members of the cooperative oncology groups and cancer centers.

The crisis had been in part created by the trialists themselves and had become apparent toward the end of the 1970s when complaints about the effects of community hospitals on patient accrual first surfaced. The chairman of the chairmen's committee noted that "we've all had trouble getting to the people who have control of patients . . . patients aren't coming to university hospitals anymore. One question that might

32. Associate Directory for Cancer Therapy Evaluation (Robert Wittes), "Summary Report—Annual Report, Division of Cancer Treatment, Vol. 1," October 1987–September 1988, 290.

33. Ibid., 297.

34. Ibid., 298.

be asked is, how good is cancer care in the community hospital? That's the kind of information we need" (New surgery grant applications . . . , 1979: 2). Upping the ante, he went on to suggest that without access to community cancer patients, the United States "will have trouble holding [its] leadership in clinical trials." Present at the same meeting, Bernard Fisher of the NSABP suggested that the loss of leadership was, in fact, well underway:

> We don't have the monolithic control of clinical research we once had. I was tremendously impressed by what I saw in Australia. They are just getting started with clinical trials and can do them as well as we can. That goes for Europe, too. (New surgery grant applications . . . , 1979: 2)

Thus while the community clinical oncology program targeted community needs, it also sought to fulfill cooperative group and cancer center research objectives. It was, in other words, a quid pro quo. The groups and centers had the technology and the clinical protocols. The community hospitals had the patients. In return for access to the patients, the central research organizations would give community oncologists some say in the development of protocols and the conduct of clinical trials. As a result of this exchange, community patients were deterritorialized, in the sense that they no longer necessarily belonged to the region in which the community was installed. As DeVita pointed out: "If you had a great source of breast cancer patients and there was no need for them in your area, there would be no restriction against affiliating with a national cooperative group" (NCAB . . . , 1982: 4).

Released in June of 1982, the request for applications announcing the CCOP generated 191 applications, fifty-nine of which were ultimately funded.[35] Had all the applications been funded, the participating hospitals could have potentially supplied 333,000 of the 800,000 new cancer patients per year in the United States (CCOP review completed . . . , 1983). While CCOP did succeed in attracting new patients to the groups' and centers' protocols, many were, unsurprisingly from an epidemiological point of view, lung and colon cancer patients for whom there were very few protocols (CCOPs at six . . . , 1984: 3). Nevertheless, the NCI announced plans to expand the program in 1984 with the hopes

---

35. Of the fifty-nine funded, nineteen were associated with ECOG and sixteen with SWOG (All 59 CCOP . . . , 1983: 2–3).

of doubling the number of CCOPs to 124 by 1990. As the cooperative groups had projected a similar doubling in their patient accrual by 1990, DeVita proposed significant increases in the groups' budgets, based on his belief that "the reduction in mortality we are seeing is largely due to the numbers of patients being seen by the cooperative groups" (Bypass budget . . . , 1984: 2).

The success of the program was further underscored the following year when following review only two of the CCOP groups were cut. Even when favorably reviewed, not all CCOPs chose to remain with the program. The Memphis group voluntarily withdrew after the first year saying that they saw no clinical advantage for their patients in the program. While they recognized the importance of clinical research, they did not believe that many of the ancillary tests developed to gain knowledge of the prognosis, staging, and biology of the disease offered much immediate benefit to their patients. The principal investigator for the program, for instance, described how a melanoma protocol required four CT scans a year for as long as the patients lived. Since, however, "some of them are cured . . . that would mean four CT scans a year indefinitely. Our surgeons wouldn't bite on that. They felt it was extravagant" (Memphis CCOP . . . . 1984: 3).

Although DeVita had hoped that as many as two hundred CCOPs would eventually see the light of day, funding for the program stagnated throughout the 1980s, and the number of program groups funded never again rose above sixty-three. Nevertheless, the chairmen of the cooperative groups that served as research bases for the program were delighted with the results. SWOG's chairman, whose organization served as a base for eighteen different community hospitals, enthusiastically declared: "The CCOP research data are unequalled in quality. It [the program] has been a major contribution to science. CCOPs have become an integral part of SWOG" (CCOP success . . . , 1986: 2). The program further demonstrated that not only could community physicians do research, but also that their application of clinical protocols often produced results superior to those obtained by the cooperative groups. The SWOG CCOPs, for example, produced better response rates and overall survival rates following protocols for acute leukemia and multiple myeloma than SWOG as a whole. Analysis of the data showed that it was not because the CCOPs received patients with a better prognosis. Pressed to explain the difference, SWOG's chairman could only speculate: "My guess is that CCOP physicians are directly involved in caring for patients. At the universities, under the direction of physicians, fellows take care of the patients. It's sad, but fellows just do not do as well" (CCOP success . . . , 1986: 3).

Even though the CCOPs became one of the pillars of patient recruitment for the cooperative groups, the penury of patients remained. ECOG's chairman reported that CCOP institutions generally entered only 10% of eligible patients into research protocols. In a list of obstacles to patient recruitment drawn up by the director of CTEP, physicians' attitudes headed the list: "They don't want to be bothered." Physician resistance was further reinforced by the fact that "many physicians are impressed with the need to individualize therapy. They don't want to have their hands tied" (Wittes, group chairmen agree . . . , 1987: 3). Not all groups suffered. Some were more successful in "squeezing the system." A member of the Children's Cancer Study Group (CCSG) reported: "If the percentage of pediatric patients at a hospital drops under 90, you can be darn sure [the CCSG chairman] will be on the phone wanting to know why" (Wittes, group chairmen agree . . . , 1987: 7).

DeVita drew the most pessimistic conclusion from this mix of responses to the patient recruitment crisis when he announced at a November 1987 meeting of the National Cancer Advisory Board that "it may well be that the cooperative group system is broken. We've been working on it for years and it is still not accruing patients as fast as it should to provide timely answers." While CTEP and the cooperative chairmen agreed that patient accrual was a major problem, they did not agree that the system was broken. The CTEP director instanced a CALGB breast cancer protocol launched in January of 1985 that would complete patient accrual in three years instead of the projected four. Deeply unimpressed, DeVita responded: "Why is that a success? With 100,000 patients a year, why shouldn't accrual be completed in one year? I see that as a failure. The fact is, in breast cancer, if we could accrue all the patients needed in one year, within 18 months you would usually know all you would at 10 years" (DeVita dissatisfied . . . , 1987: 2).

In the late 1980s the NCI used CCOP to solve another problem of American oncopolitics: the black/white differential in cancer survival rates and the relatively low participation of black oncologists and patients in the research process. Cancer researchers in the United States had long been aware of differential cancer survival rates that favored American whites over blacks. Historically, however, higher rates of cancer had prevailed among whites. The lower incidence of cancer in members of the black population, who, prior to the 1950s, often died at a precancerous age because of high rates of tuberculosis, had led early commentators to conclude that "Negroes" were immune to cancer (Wailoo, 2005; Clayton & Byrd, 1993; Byrd & Clayton, 2002). By the 1950s the epidemiological tableau had changed. As publicized in a 1973 *Cancer* article (Henschke et al., 1973), Afro-Americans consistently bore higher cancer incidence and lower survival rates. In 1972, the year before the Howard University researchers sounded the alarm, the NCI took the first steps toward tackling the unequal care issue through the creation of the Minority Biomedical Research Support Program. Designed to increase minority participation in cancer research, the program provided funds to train members of minorities in the field of cancer research. Originally restricted to "Blacks," the category of minority was expanded in the early 1980s to include "Hispanics," "Native Americans," and "Pacific Islanders."[36]

At the same time the NCI sought means to include "special populations" in clinical cancer research. In 1984 DeVita suggested that given the black/white cancer survival differential, the time had come to develop cancer centers to serve minority communities. Since creating clinical research programs for historically black colleges out of the blue seemed an unlikely undertaking, the proposal died on the table. The question reemerged in a different format several years later in 1988, however, when, following a five-year evaluation of the CCOP program, reviewers noted that whereas 20% of the US population had minority status, only 7% of CCOP patients were minorities (CCOP recompetition approved . . . , 1989). In 1989 the NCI created a separate CCOP program for minorities that generated considerable interest (Minority based . . . , 1989). Twenty years later, the minority based–CCOP program was widely considered a success. More than 70% of the patients enrolled in clinical trials through these institutions were minority patients even though they accounted for only 10% of all minority patients in clinical cancer trials (*Decades of Progress*, 2003).

Since the 1990s the community programs have increasingly specialized in cancer prevention and control programs, leaving recruitment to

36. NIH Guide for Grants and Contracts, vol. 13, no. 7, 25 May 1984, 11, at http://grants .nih.gov/grants/guide/historical/1984_05_25_Vol_13_No_07.pdf. In his recent analysis of the biopolitics of difference in the United States, Steve Epstein (2007) has examined the events surrounding the adoption of the 1993 NIH Revitalization Act that mandated the inclusion of underrepresented sectors of the population in clinical trials. As just mentioned, however, strategies of inclusion within the cancer domain date back several decades.

treatment trials in the hands of relatively few community practitioners (Cohen, 2003). Part of the problem lies with the new biology that began with innovations in immunology and gained speed throughout the 1980s as discoveries in molecular biology laboratories moved into the clinic. The increasing complexity of clinical trials that deploy the new biology has consequently put new burdens on CCOP practitioners: the new, targeted trials are not only more expensive, they are also more time consuming. One practitioner noted in this regard: "I give double bookings for my average patients on a complex trial like one with two targeted molecules for colon cancer chemotherapy. It takes me that much time just to sort through all the toxicities and do the dose adjustments for each drug" (Patlak, Nass & Micheel, 2009: 55). In addition, the growing number of ancillary studies that had already begun to generate complaints in the late 1980s forced local pathologists and radiologists to undertake numerous unpaid tests and procedures. It is thus to the rise of molecular biology and the oncogene theory of cancer and their effects on clinical cancer research that we now turn.

PART 3

Targeted Therapy, Targeted Trials (1990–2006)

Molecular Biology and Oncogenes

Introduction

T he third part of this book opens (chapter 10) with a series of overlapping clinical trials conducted between 1998 and 2001 to test the efficacy of a compound known as Gleevec. Designed to block an enzyme
produced by an oncogene, Gleevec was one of the first targeted cancer
therapies to gain FDA approval and thus signaled both the final conse-
cration of the oncogene theory of carcinogenesis and the beginning of a
new era of therapeutic agents produced using the tools of molecular biology. Readers unfamiliar with some of the terms used in the previous sentences will find in the present interlude a quick overview of the historical
and technical meaning of these notions.

As the title of this interlude suggests, we are concerned here with two
different yet interconnected topics: the introduction of molecular biology into clinical cancer research and the identification of cancer-causing
genes known as oncogenes (Fujimura, 1996; Morange, 1997). While both
events date back to the early 1980s, the relation between the two is relatively complex. In fact, the molecular biological turn in cancer research
emerged from the intersection of a number of distinct lines of work entrenched in different disciplines and specialties. These in turn gave rise
to research networks and consortia that cut across preexisting divisions.
We describe below a complex process, irreducible to a sequence of laboratory discoveries successfully applied to the clinic, whereby initially
unrelated domains and techniques—physicochemical, biological, and
clinical—interacted in often-unexpected ways to produce a new *biomedical* domain. Viewed from afar, the situation is quite straightforward:
the twenty years running from the isolation of the first human oncogene

in 1984 to the decoding of the human genome at the beginning of the twenty-first century changed the entire face of cancer research by carrying the level of investigation from the cellular to the subcellular level. From the point of view of cancer therapy, clinical research moved from the investigation of ways to kill cancer cells to the fabrication of the means to interfere with biochemical processes taking place inside, on the surface of, and between cells. The importance of oncogenes and of the new molecular approaches therefore cannot be underestimated: the suppression of oncogenes and their protein products has become one of the central concerns of clinical cancer research.

The Rise of Molecular Biology

As many readers will recall from their high school biology course, the original version of the central dogma of molecular biology (Strasser, 2006) was quite straightforward. Chromosomes contained in the nucleus of cells of higher organisms, including humans, are composed of two strands of DNA, a nucleic acid, twisted about each other in the form of a double helix. This iconic structure (Chadarevian, 2003) can be decomposed into chemical subunits called nucleotides, of which there are four kinds. Molecular biologists designate each kind with the letters A, T, C, G (for adenosine, thymine, cytosine, and guanine respectively). The individual nucleotides form the basis of the DNA code that is composed of sixty-four combinations of any three of the four nucleotides. Known as codons, these nucleotide triplets map onto the twenty amino acids that make up all human proteins. The process of protein production requires a number of intermediaries, the most important of which is RNA, yet another nucleic acid. In the process of producing proteins, a single strand of the DNA in the nucleus of the cell is transcribed into a complementary strand of RNA that then leaves the nucleus to migrate into the cytoplasm of the cell. Here the transcribed code is read and the amino acids aligned to create proteins.

The DNA code specifies the kind and order of the amino acids that when joined together create a protein. In one of the simplest human cases, fifty-one codons bear the code for the fifty-one amino acids that underlie insulin. The proteins of the body can be divided into structural proteins (e.g., hair and fingernails, proteins on the surface of the cell, etc.) and active proteins such as the enzymes and hormones that carry out the chemical reactions within the cell and allow cells to communicate with each other. These are the components of a system that flows in one direction: genes make proteins that make, organize, or power everything else that happens in the body. The body does not make genes: they are inherited from the parents of the organism in question. This barebones description summarizes the core knowledge of molecular biology circa 1965 (Watson et al., 2008).

Knowledge of the structure of DNA grew out of the analysis of the constitution of the key molecules of living organisms—DNA and proteins. These studies made great use of the technologies of fractionation and visualization—ultracentrifuges, x-ray crystallography, chromatography, radioactive tracers, and electron microscopy—most of which had been developed in the 1930s and 1940s and drew on related techniques in physics and chemistry (Olby, 1994; Chadarevian, 2002; Creager, 2002, 2009). In parallel, research into the function of DNA relied less on the physical and chemical technologies and more on techniques that used biological processes and accidents thereof (most often pathologies) as tools for the study of disease. Based on experiments with individual cells, bacteria, and viruses, the notion that DNA played a determinant role in cellular activities had been known in medical circles since the mid-1940s. Medical interest in DNA and disease extended to cancer via the study of cancer-causing viruses, a research tradition that reached back into the nineteenth century when researchers isolated the first cancer-causing viruses in plants, in particular the tobacco mosaic virus (Fenner & Gibbs, 1988; Creager, 2002).

The existence of cancer-causing viruses in animals remained hypothetical until 1911, when a researcher, Peyton Rous, at the Rockefeller Institute in New York isolated one such virus in chickens (Vogt, 2009; Javier & Butel, 2008). Eponymously known as Rous sarcoma virus (RSV), it would play an important role in the isolation of oncogenes several decades later (Rous was awarded the Nobel Prize for the isolation of RSV in 1966, i.e., almost sixty years after the fact). As for mammals, a similar cancer-causing virus was isolated only in the 1930s in rabbits. The field remained relatively barren until after World War II when researchers at the NCI described a number of mouse leukemia viruses. Taking a cue from the programs that had developed vaccines for the poliovirus and the cancer chemotherapy program, in 1959 the NCI established a laboratory of viral oncology to gather together researchers working on both the RSV and mouse leukemia viruses. Then in 1964 the NCI initiated a Special Virus Leukemia Program that used the same operations research approach—the *convergence technique*—that had been adopted for the chemotherapy program, as described in chapter 5. By the early 1970s researchers working at the NCI and those sponsored by the programs had "isolated and organized the production of viruses inducing tumors in mice, rats, cats, dogs and monkeys" but not humans (Gaudillière, 1998: 163). We will return to this work below.

Cancer and Molecular Biology: 1960–80

The Regulatory View

Despite the decipherment of the genetic code and of its role in the fabrication of proteins, molecular biology exercised little influence over cancer clinical research prior to the 1980s. It did, however, hold considerable sway over the theories of the causes of cancer. Most of these theories, as Michel Morange (1997) has argued, shared a common regulatory vision. This outlook amounted to a variety of competing hypotheses—including the ultimately successful oncogene hypothesis—held together by the notion that cancer could be described as a failure of regulation, or in other words, "as a disease of development" (1997: 6). In the 1960s, for example, molecular biologists at the Pasteur Institute, having shown that the transformation of DNA into protein was regulated by yet other strands of DNA that turned genes on and off, concluded that cancer was no doubt "a disease of genetic expression." Similarly, the fact that the fusion of cancer cells with normal cells repressed the cancer was interpreted as meaning that the cancer cells were missing some fragment of DNA or a protein that would normally block or repress cancer-causing genes, and that the normal cells supplied the missing elements (1997: 3). First isolated in the late 1960s in mouse-human hybrids, following the identification of human oncogenes these factors would become known as antioncogenes (Harris, 1995).

Pervasive and supple, the regulatory view often found residence in preexisting theories of cancer. Such was the case of chemical carcinogenesis. Chemists had been implicated in the study of the causes of cancer since the beginning of the twentieth century when researchers began to use a number of environmental irritants to produce cancer in mammals. The first convincing laboratory demonstration of chemical carcinogenesis took place in 1914 and involved painting coal tar into the ears of rabbits. Extended to mice, plants, and human organ systems, the chemical production of tumors became a mainstay in cancer research laboratories worldwide (Miller, 1994; Weinberg, 1996). By the 1970s a common framework for the understanding of chemical carcinogenesis known as the two-step hypothesis had become firmly embedded in the cancer studies literature. In the first step of the process, called *initiation*, the carcinogen, a category that now included radiation and viruses, produced an irreversible change, most commonly assumed to be a mutation, in the target cells (Yuspa, Hennings & Saffiotti, 1976). A possibly reversible stage termed *promotion* followed. This stage of the development of a cancerous tumor no longer required the original carcinogen for its progression; it could be maintained by other components of the cell's metabolism. Research into carcinogenesis could thus focus on one or both of the two steps of the process, and both stages lent themselves easily to the regulatory vision (Maskens, 1981).

Virology and Oncogenes

In addition to the aforementioned technologies of representation and fractionation developed in the 1930s and 1940s, cancer researchers developed a number of technologies for the manipulation of living organisms. Initially restricted to cell culture techniques (Landecker, 2007), by the end of the 1960s these technologies had grown to include subcellular components of viruses, bacteria, and human cells, and researchers used these components of life to analyze and manipulate the genes and proteins involved in cancer at a molecular level. In particular the new toolbox allowed researchers to take apart and reassemble the molecules that had been isolated and visualized in the previous decades. It also enabled researchers to segregate individual molecules, genes, and proteins acting in specific biochemical and carcinogenic pathways in living cells. So by the mid-1970s researchers had sequenced individual genes and proteins, reproduced them, and inserted them into other organisms. A decade later, by the end of the 1980s, researchers had automated many of the more routine aspects of this work, such as analyzing and synthesizing genes and proteins or producing many copies of a single gene.

In the midst of this work, researchers at the NCI advanced the viral oncogene theory according to which viruses caused cancer in both animals and humans in an indirect manner and in two stages. First, covert or undetected viral infections smuggled cancer-causing genes into humans. Once inside human cells, the genes remained dormant and were transmitted from parent cell to offspring along with the other cellular genes. On occasion a number of factors including "radiation, chemical carcinogens and the natural aging process" activated the genes (Huebner & Todaro, 1969: 1091). Cancer ensued. This hypothesis suffered two deficiencies: no cancer-causing genes had yet been found in viruses and no human cancers had been shown to be the result of viral infection (Weinberg, 1996).

These deficiencies were soon corrected by a consortium of California researchers housed at a number of elite institutions, all of whom conducted research on cancer viruses. Known as the West Coast Tumor Virus Cooperative, the group, which included the future 1989 Nobel Prize winners Michael Bishop and Harold Varmus of the University of San Francisco Medical School, met once every six weeks to compare notes (Varmus, 1989; 2009). Largely credited with having isolated the first viral oncogene, Bishop and Varmus had to overcome a major theoretical and technical obstacle. According to the original version of the central dogma of molecular biology, DNA made RNA and RNA made proteins in the course of the unidirectional process. The problem for researchers in the field of viruses was that many viruses appeared to contain only RNA and no DNA. How could those viruses incorporate themselves into the DNA of the cells they infected? The answer was provided by the discovery of an enzyme—reverse transcriptase—that transcribed the viral RNA into DNA, thus allowing the viral genome to integrate the infected cell's DNA genome and hijack the cellular machinery (Baltimore, 1970; Temin & Mizutani, 1970). The discovery, rewarded with the 1975 Nobel Prize, caused quite a stir. An editorial in *Nature*, cleverly titled "Central dogma reversed" (1970), made the significance of the discovery plain to all readers. This development solved a conceptual *and* a technical problem: since Varmus and Bishop used the aforementioned RSV (an RNA virus) for their experiments, in order to decode the viral gene they would have to translate the viral RNA into DNA.

Using reverse transcriptase and radioactive DNA probes complementary to the RSV oncogene, now known as src (an abbreviation of sarcoma), Bishop and Varmus showed that this cancer-causing gene was present not only in poultry but also in a number of other vertebrates. The ubiquity of src partly confirmed the viral oncogene theory (Stehlin et al., 1976). But Bishop and Varmus went further and showed, *contra* the theory, that the oncogenes had not been smuggled into the animal cells by viruses; they had been in the animal cells all along. In actuality, the traffic had moved in the opposite direction: the viruses had simply picked up the animal oncogenes in the course of infection and added them to their genomes. To distinguish them from the viral oncogene, the animal oncogene received the prefix c (for cellular), whence the abbreviation c-src (Varmus, 1985, 2009; Bishop, 1995, 2003). So where did the cellular oncogenes come from? They turned out to be mutated variants of normal genes: the products of radiation, carcinogens like tobacco, and accidents such as translocations (the shuffling of DNA from one chromosome to another) that occur during normal biological processes such as cell division.

The next question was: what biological function did the protein serve in a normal chicken cell? By 1978 the future CEO of the biotech start-up Genentech working in Michael Bishop's laboratory had shown that the protein acted as an enzyme called a kinase (Levinson et al., 1978). As such it modified the amino acids that composed other proteins by adding phosphate groups to them (a process termed phosphorylation). At the time only two amino acids were known to be targets of protein kinases (Uy & Wold, 1977), but soon a third one, tyrosine, was added to the list. At this point research on viral proteins hooked up with a previously unrelated field of research in biochemistry focusing on a substance known as epidermal growth factor (EGF). First observed in mice at the beginning of the 1960s by Stanley Cohen, EGF began its career as a glandular extract that provoked sharp growth in newborn mice (Carpenter & Cohen, 1979). Over the next twenty years, biochemists like Cohen, who received a Nobel Prize for his efforts in 1986, isolated an identical protein in humans and discerned that its function, like the c-src gene product, involved participation in the phosphorylation of proteins.

To sum up: by 1980 cancer researchers knew that mutated genes in a number of species gave rise to proteins that in normal cells acted as enzymes that phosphorylated proteins (Goustin et al., 1986). They also knew quite a bit about equivalent proteins in normal humans; namely, that they regulated cell growth. They also knew that these kinases sometimes acted in conjunction with proteins on the surfaces of cells—growth receptors—and sometimes with proteins inside the cells to carry out the phosphorylation. They knew, finally, that specific amino acids such as tyrosine could be the targets of the phosphorylation. Missing from this mass of information was a mutated gene in humans. In other words, as breathtaking as the discovery of cellular oncogenes had been from the biological point of view, from the standpoint of medicine it remained to be shown that oncogenes existed in humans. Researchers did not report the first human oncogenes until 1982, eight years after their isolation in lower mammals (Morange, 1997). Clinical researchers working in the field of cytogenetics supplied the necessary elements for this work. Indeed, the establishment of the oncogene paradigm in medicine depended crucially upon information available only to clinical researchers or researchers with access to clinical information. It was thus less the case that a biological insight had been applied to medicine than a question of combining biological and medical information into a biomedical form.

The Clinical Isolation of Human Oncogenes

Human oncogenes emerged from studies of the correlations between abnormalities in human chromosomes and cancer. While researchers had implicated chromosome abnormalities in the production of cancer since the beginning of the twentieth century (Boveri, 1914), their study changed significantly at the beginning of the 1960s. First of all, after many years of controversy cytogeneticists finally agreed on the exact number of human chromosomes. They also managed to craft a common nomenclature for their description, known as the Denver convention (Lejeune et al., 1960; Harper, 2006). Henceforth, observations of chromosome anomalies could be carried out using a common language. Second, clinicians on both sides of the Atlantic established a consistent association between an anomalous chromosome 22 and chronic myelogenous leukemia (CML). Termed the Philadelphia (Ph) chromosome after the Philadelphia team of clinicians who first noticed the connection in 1960 (Nowell & Hungerford, 1960; Nowell et al., 1988), scooping by a few months an Edinburgh team (Balkie et al., 1960), it became the first chromosome abnormality associated with a specific cancer. By the end of the 1960s, following further studies by the Philadelphia team and by researchers at the clinical center of the NCI, including George Canellos and Paul Carbone, the Ph chromosome acquired a prognostic value: against intuition, CML patients without the anomalous Ph chromosome had the shortest median survival (Whang-Peng et al., 1968). Despite this early progress, the anomaly remained a singularity: by the end of the 1960s, no other chromosome anomalies emerged as consistently associated with a single type of cancer.

In the course of cell division, chromosomes often exchange fragments, a process called crossing-over that has the generally useful effect of shuffling genes from corresponding (homologous) chromosomes in males and females and of stirring up the genetic pot from generation to generation. Sometimes, however, exchanges occur between different chromosomes. When this occurs, biologists speak of a translocation rather than crossing-over. In 1973 a Chicago pathologist, Janet Rowley, pinpointed the locations of the breaks on the two chromosomes involved in the translocation that produces the Ph chromosome (Rowley, 1973) and by the same token transformed a common chromosome anomaly, of which there were now many (Mitelman, 1983), into a well-defined pathological event. Nonetheless, how exactly the deranged chromosomes caused cancer remained obscure.

The institution, in 1977, of a series of six International Workshops on Chromosomes in Leukemia substantially sharpened the focus in the field (Meeting report . . . , 1978). While the first workshop brought together only seven laboratories, the second workshop, held two years later, attracted twenty-one laboratories. More than a convenient organizational arrangement, the workshops provided an important opportunity to collate data that could not otherwise have been collected together in the same place. For even though consistent new correlations between chromosome abnormalities and species of leukemia had been reported since Rowley's 1973 findings, those correlations had drawn on the relatively small numbers that a single laboratory could generate. Claims for significant changes required statistically significant numbers. Finally, at the beginning of the 1980s, cytogeneticists managed to demonstrate a relationship between chromosomal anomalies and oncogenes. In the process they isolated the first truly human oncogenes. Thus after years of collecting chromosome anomalies of unknown import, in 1982 investigators showed that the genes affected by the chromosomal translocations in two different cancers were identical to well-known viral oncogenes present in animals (Sheer et al., 1983; Nowell et al., 1983). According to Rowley (1983), things moved quickly after the conference. In other words, once cytogeneticists knew what to look for, they knew where to look. In a similar fashion, once virologists and molecular biologists found the human analogues of their cancer-causing genes, they knew which cancers were involved.

The isolation of the genes implicated in the Ph chromosome translocation is typical in this regard. In 1970 researchers funded by the Spe-

cial Virus-Cancer Program at the NCI's Viral Biology Laboratory isolated yet another cancer-causing virus that produced lymphoma in mice
(Abelson & Rabstein, 1970). The exact function of the cancer-causing
gene remained unknown until the late 1970s when further research identified the protein produced by the gene and determined that it was a tyrosine kinase. The gene was subsequently isolated and named *Abl* (after
the eponymously named Abelson murine leukemia virus, first described
in 1970 by Herbert T. Abelson) (Witte et al., 1978; Witte, Dasgupta &
Baltimore, 1980). Researchers then raced down the road from mice to
humans. In the fall of 1982, shortly after the announcement of the location of the first human oncogenes, an international research team that
included members of the NCI's Laboratory of Viral Carcinogenesis and
Britain's Imperial Cancer Research Fund announced the location of the
ABL gene on chromosome 9 in humans. It remained to be determined
whether or not the gene lay near the translocation point of the Ph chromosome found in CML (Heisterkamp et al., 1982). To access the human cells from CML patients they needed to pursue their task, the team
joined a Dutch group of cytogeneticists who had already characterized
the Ph chromosome using the latest chromosome mapping techniques
(Geurts van Kessel et al., 1981). The answer to the question was immediately forthcoming: the oncogene on chromosome 9 turned up in chromosome 22 in patients with CML or, in other words, the *ABL* oncogene was
translocated to the Ph chromosome (de Klein et al., 1982).

To recap: where did the oncogene come from? As we have just seen,
in the case of the Ph chromosome, two human chromosomes, chromosomes number 9 and 22, exchange fragments. The unhappy result is that
the genes from the two chromosomes actually fuse, creating a third, lethal gene that ends up on chromosome 22 as a result of the exchange.
The gene on chromosome 9, *ABL*, normally produces a tyrosine kinase.
The gene on chromosome 22, *BCR*, is a promoter, which means that it
activates the gene that comes after it on a chromosome. So when *ABL*
from chromosome 9 is translocated beside *BCR* on 22, the resulting fusion gene—*BCR-ABL*—is permanently activated and produces an un-
regulated tyrosine kinase. When this happens in human white blood
cells, they proliferate in the form of CML. As we will see in chapter 10,
the protein product of the *BCR-ABL* oncogene became the target of an
international research program that resulted in the production of the
first targeted therapy, Gleevec, which stops the process by blocking the
turned-on kinase.

Oncogenes and Oncoproteins since 1982

By the time the first human oncogene had been isolated in 1982, laboratory researchers had uncovered almost twenty other oncogenes in nonhumans. Enough was known about their function to generate considerable excitement and to prompt the formation of a special meeting devoted entirely to oncogenes. Researchers from the Salk Institute and the Frederick Cancer Research Facility in Maryland organized the first Oncogene Meeting in 1985 (Hunter & Simon, 2007), and a journal titled *Oncogene* began publication in 1987. The poster of the meeting read, in part: "In the beginning . . . was the word. And the word was . . . oncogene. On the seventh day he said 'Let there be . . . the first annual meeting on oncogenes'" (Hunter & Simon, 2007: 1261). While this kind of humor is not uncommon in science (Gilbert & Mulkay, 1984), it also suggests that meeting organizers were quite aware of the fact that they stood on the threshold of a number of important breakthroughs in cancer biology.

The Oncogene meetings quickly became the principal venue for the dissemination of results in the field worldwide. The first meeting drew 264 presentations, a number that grew to 457 by 1988. The Oncogene meeting remained the main forum for the presentation of findings in the field of oncogene research until the mid-1990s. A steady decline in the number of presentations began in 1995, reaching a low of 140 presentations in 2004, a trend that meeting organizers attributed to the concurrence of competing meetings specializing in subtopics and a concurrent tightening of research budgets (Hunter & Simon, 2007: online supplementary material). By then the oncogene theory of carcinogenesis had become firmly entrenched within the field of cancer biology and had, in effect, replaced the regulatory vision as the primary lens through which to view the generation of a cancerous cell (Croce, 2008). The hypothesis that mutated genes were directly related to biochemical growth factors went a long way toward explaining the intimate molecular connections between mutagenesis provoked by radiation, chemicals, or simple cellular accidents, on the one hand, and the out-of-control cellular growth displayed by cancers on the other.

While oncogenes remained an important focal point in the biology of cancer, the twenty years following the discovery of the first human oncogene revealed a picture considerably more complex than that first unveiled in the chicken. First, the simple "one gene—one cancer" tableau presented by RSV fell quickly to the wayside. By 2004 over one hundred different human oncogenes had been isolated (Soto & Sonnenschein, 2004) and work in the mid-1980s on colorectal cancer had already suggested that the development of a cancer tumor would require at least six to seven steps and at least as many cancer–causing genes with many other genes joining in the process as the cancer grew and developed (Vogelstein et al., 1988). Second, again by the mid-1980s, a second class of cancer-causing genes had emerged. Originally termed antioncogenes, these *tumor suppressor genes*, of which there are now more than thirty, perform normal biological activities when activated. When turned off, deleted, or corrupted, they allow cancerous cells to proliferate. One of the most common suppressor genes, p53, was originally thought to be an oncogene but turned out to work in the opposite manner as a tumor suppressor gene (Baker et al., 1990). Third, although oncogenes produce proteins that promote growth, the cellular reaction to these oncoproteins can be very complex. Cancer biology today situates these growth factor receptors and many other growth-signaling processes within a cascade of pathways linked through multiple feedback loops that again evolve as the tumor progresses. Finally, in therapeutic terms, while patients may start out with similar cancers, the evolution of the tumor may produce considerable heterogeneity in the cell lines that make up the tumor, which in turn may determine a patient's reaction to therapy. A common genetic cause, in other words, need not necessarily lead to a common therapeutic outcome. With these provisos in mind, let us now turn to the therapeutic research generated by the introduction of molecular biology and oncogenes into the cancer clinic.

CHAPTER TEN

Magic Bullets? The Gleevec Trials

Molecular Biology and Targeted Substances

The May 28, 2001, cover of *Time* ran the following headline: "THERE IS NEW AMMUNITION IN THE WAR AGAINST CANCER. THESE ARE THE BULLETS." Color coded in orange, the words AMMUNITION, CANCER, and THESE pointed to a small heap of orange pills identified in a smaller headline: "Revolutionary new pills like GLEEVEC combat cancer by targeting only the diseased cells. Is this the breakthrough we've been waiting for?" Two years later, Daniel Vasella, chairman and CEO of Novartis, the maker of Gleevec, reprised both the color theme and the war metaphor for the title of his book *Magic Cancer Bullet: How a Tiny Orange Pill Is Rewriting Medical History*.

Gleevec (imatinib), a small molecule, has revolutionized the treatment of chronic myelogenous leukemia. Along with Herceptin (trastuzumab), a monoclonal antibody used in the treatment of breast cancer patients, it is one of two much-vaunted examples of *targeted therapies*. The last part of this book will focus primarily on the search for such therapies, a search made possible by massive investments in molecular biology. But this will not be our exclusive concern, for although oncogenes and molecular biology colored the entire period, they were not the only cause of all that happened, nor did they capture all aspects of this most recent articulation of the new style of practice. In addition to novel targets and the new kinds of clinical trials designed to evaluate targeted interventions, new actors and institutions moved to center stage. To begin with, cancer drug development has undergone a sea change. Even though it maintains a significant noncommercial sector, pharmaceutical and biotech companies like the makers of Gleevec (Novartis) and Herceptin

(Genentech/Roche) must now be reckoned with. Patients have become bona fide participants in the research process, and a new professional figure, the oncologist, has redefined the specialty map in clinical cancer research and treatment. Finally, the data themselves no longer circulate on the tried-and-true routes of the past, as data centers and protocol statisticians have been forced to surrender control of interim results to data monitoring committees. These committees are a central component of a chain of custody of clinical trial data that did not exist in previous periods and that has shifted the surveillance of data beyond the cooperative groups stricto sensu to a larger collective of stakeholders directly or indirectly affected by the conduct of clinical trials.

The Trials of a Targeted Therapy

In 2001, the *New England Journal of Medicine* published the results of a Phase I cancer clinical trial that used a novel compound labeled STI571 (Druker et al., 2001). Observers hailed the substance—later named imatinib and sold under the trade names Gleevec (in North America) or Glivec (in Europe and Australia)—as the first of a new kind of anticancer drug: a targeted compound. The authors of the *NEJM* article included members of the Department of Oncology Clinical Research of the drug's developer and manufacturer (the Swiss multinational firm Novartis); Dr. Brian Druker, an oncologist at Oregon Health Sciences University; and his colleagues at the M. D. Anderson Cancer Center and UCLA. In their report the authors recounted how, between 1998 and 2000, Novartis and the academic clinicians had treated eighty-three patients suffering from chronic myelogenous leukemia (CML) who fulfilled two further criteria: they had failed interferon therapy, the standard therapy for patients in the chronic phase of CML, and they carried a cancer-causing chromosome known as the Ph chromosome (see Second Interlude). Unusually, and a clear sign of the drug's spectacular success, by the time the team had published the Phase I results, Novartis had already activated Phase II and Phase III trials for the drug.

The STI571 Phase I trial sought to determine the maximum dose of the drug that patients could tolerate. Normally such trials enrolled fewer than ten patients (Edler & Burkholder, 2006). The comparatively large size of the trial resulted partly from the fact that nobody knew what the maximum dose might be since the compound's low toxicity in prior laboratory studies had not suggested any upper limit. Tests on dogs, for example, had shown that liver problems set in only at very high doses (Druker & Lydon, 2000). As a result of these nebulous early indications, the protocol divided the patients into fourteen groups, each assigned a different daily dose size of STI571 running from 25 mg to 1,000 mg. A second finding also prompted the large number of patients: laboratory studies had suggested that the new drug would likely produce a remarkable outcome (Druker et al., 1996). Had this been a normal Phase I trial, Druker and his associates would have chosen patients in the terminal phase of disease who, while useful for determining toxicity, would most likely not have benefited from the treatment. Instead, hoping to show some curative effect, Druker and his colleagues selected patients who had failed standard therapy but were nonetheless healthy enough so that any positive effects would not be clouded over by the severe complications that affect patients at the end stage of CML (Sawyers, 2009).

The trial did produce unprecedented results: 98% of the patients who received doses of STI571 greater than 300 mg underwent complete hematologic remission, meaning that their white blood cell count fell to normal values. Despite this frank success, the authors of the *NEJM* article reported that they had failed to answer the original study question and identify the maximum tolerated dose (Druker at al., 2001). The inability to determine a clear toxic end point was, of course, good news. It heralded the arrival of an anticancer agent with such extreme specificity that the maximum-tolerable dose size no longer stood as a limit to the therapeutic use of the drug.

Another unusual feature of the STI571 treatment was its duration. When the Phase I results were published in 2001, patients had been receiving daily doses of STI571 for up to two years, and for most trial participants there was no end in sight. Like insulin and unlike standard chemotherapies, STI571 therapy was a lifetime prospect. Unlike insulin, imatinib did not supply a missing substance essential to human physiology. Instead, it blocked the production of the enzyme produced by the gene on the Ph chromosome responsible for the production of leukemic cells. In other words, rather than acting as a *cytotoxic* agent that kills cells, imatinib operates as a *cytostatic* agent, preventing their proliferation as cancer cells.

Because of its success, the Phase I trial folded easily into a Phase II trial enrolling patients who once again had failed interferon therapy and who bore the Ph chromosome. Implicating 532 human subjects, the

Phase II study was about five to ten times larger than a normal Phase II trial (Green, 2006). In order to conduct the trial, Druker and his colleagues formed the International STI571 CML Study Group funded by Novartis and composed of approximately thirty academic researchers in the United States and Europe. Novartis also provided the data management and statistical services. In addition, Novartis employees audited and reviewed the ancillary cytogenetic studies carried out by individual academic researchers (Kantarjian et al., 2002). In this regard, both the Phase I and Phase II trials deployed a novel cancer clinical trial organization: a private/public—or commercial/noncommercial—collective. At the same time, although both trials had eschewed the traditional cooperative oncology groups, they did borrow much of the organizational and statistical infrastructure developed by the groups.

Initiated in June of 1999 only a year after the beginning of the Phase I trial and just prior to the publication of the Phase I results, the international study group trial quickly recruited the requisite number of patients. Although enrollment did not begin until January of 2000, news of the results of the Phase I study had spread throughout the Internet in the fall of 1999, and an Internet petition initiated by a Montreal CML patient garnered over three thousand signatures in just three weeks. Forwarded to Novartis, the petition joined what the company's CEO, Daniel Vasella, recalled as a steady flow of "letters and emails . . . streaming into Novartis from patients and friends of patients, and relatives of patients—all wanting access to the drug" (2003: 117). The tipping point in the demand for STI571 came in December of 1999 when Druker and his colleagues presented early results of the Phase I trial at the annual American Society of Hematology meeting, attended by over 15,000 participants (Druker et al., 1999). STI571 soon sat on the lips of anybody who counted in the world of leukemia. Phone calls to Novartis inquiring about the drug that had been running at about fifteen a month jumped to two thousand a day during the meeting and six hundred a month afterward (Vasella, 2003: 121). Novartis submitted a new drug application to the FDA in February 2001, and it was approved in record time in May 2001, only three months later. It took less than four years to go from the first human dose to FDA approval, again a record time attributed by observers to the fact that the drug targeted a biologically specific population, thus decreasing the number of necessary trials and speeding up review cycles (Trusheim, Berndt & Douglas, 2007).

The swift turn of events caught Novartis by surprise. STI571 was an experimental compound, and the company had originally resisted its development reasoning that given the scarcity of CML patients, the limited potential market did not warrant the investment (Druker, 2007). The decision to support the STI571 project was risky and involved more than biological innovation: it also necessitated a change in management outlook and business model. The provision of 300 mg daily doses to over five hundred patients at such an early development stage required scaling up the twelve-step production process to an industrial level. To fulfill this demand the company rededicated a facility in Ireland. Once Gleevec was ready to be marketed, Novartis confronted yet another problem, a financial one that was tackled by developing an original price structure. This price structure was designed to recoup the investment in R&D and make a healthy profit, while taking into account both the small patient pool—about 3,000 new CML patients per year in America and about 1.3 per 100,000 worldwide as compared to markets for prostate and breast cancer that were about thirty times larger—and the patients' differential ability to pay (Vasella, 2003). Instead of fixing different prices for developing and industrial countries, the company established a single, universal price (on the high end: around $2,500 per month) and a Patient Assistance Program (PAP/GIPAP) that provided the drug for free or at a pro rata of the full cost depending on income and related conditions.

The novel approach seems to have paid off since Gleevec has become a multi-billion-dollar drug (Workman & Collins, 2008). In fact, observers speak of "a new economic model for the biopharmaceutical industry" corresponding to a *stratified medicine* approach. In contrast to traditional pharmaceutical markets based on blockbusters such as the anticholesterol drugs aimed at large patient populations, a "niche buster" such as Gleevec targets smaller populations in which it can demonstrate superiority. Higher prices and high adoption rates lead to higher revenues than those generated by a drug such as tamoxifen used for breast cancer patients (Trusheim et al., 2007: 289).

This success story has been partly clouded by the thriving generic pharmaceutical industry in India. Novartis's GIPAP program in India was designed, in part, to forestall the production of generics, and included the stipulation that the program would halt should a generic be made available in that country (Strom & Fleischer-Black, 2003: C7). Shortly after the initiation of the program, a Hyderabad firm released a generic version of imatinib at less than one-sixth the price of Gleevec. Attempts by Novartis to patent Gleevec and thus stifle the generic suffered a series of setbacks in the Indian courts and led to recriminations by Western NGOs such as Oxfam and Médecins sans frontiers (Mueller, 2007; Bate, 2007; Harrison, 2009). While NGOs maintained that Novartis's repeated attempts to enforce its position, if successful, would deprive poor populations of essential medication, the firm instanced the GIPAP initiative in its defense, arguing that India was simply trying to protect its thriving generic sector. Novartis is not alone in its predicament: another Swiss multinational firm, Roche, faces similar problems with its oncology drug Tarceva. The emergence of disputes such as these signals the fact that anticancer drugs, once a domain of little commercial interest, have now entered the maelstrom of commercial drug production.

Indeed, based on data from the US online registry ClinicalTrials .gov, a detailed analysis of industry-sponsored clinical cancer trials for the 2005–9 period concluded that "oncology has become the main focus of the industry in developing new medicinal products" (Karlberg, Yao & Yau, 2009: 415). Growing at approximately twice the rate of all other therapeutic areas combined, in 2007–9 oncology accounted for almost 21% of all industry-sponsored trials. Industry, however, faces stiff competition from public institutions—investigator-initiated, academic oncology trials—that account for 40% of the trials and almost 76% of the sites. Since data on nonindustry trials in ClinicalTrials.gov are mainly indicative of US activities, whereas data on industry trials cover industry's global activities, it is likely that noncommercial trials do exceed industry-sponsored trials. Most of these trials are Phase I–II—that is, more exploratory in nature—whereas industry trials focus more on Phase II and III trials. Because noncommercial trials recruit patient populations from the public and the academic sponsors' countries of origin, industry trials seek additional patients outside the Western world and thus fall to a large extent in the "global" category (Petryna, 2009).

Novartis launched its Phase III study in June of 2000, nine months before the publication of the Phase I results and a mere six months after the start of the Phase II trial. This time, STI571 went head-to-head with interferon in a study named IRIS, the International Randomized Study of Interferon versus STI571. A public/private collective once again designed and conducted the study that took place in 177 hospitals in sixteen countries with 1,106 patients. The trial sponsor, Novartis, supplied the drug, the statisticians, the data management and statistical support systems, while IRIS's academic investigators provided the patients and the data. Both members of the collective collaborated on design and interpretation. Unlike the Phase II trial, however, a third party, a data monitoring committee, oversaw the collection and interim interpretation of the data. Composed of two clinicians and the well-known statistician Marc Buyse, the former president of the International Society for Clinical Biostatistics, the committee had planned to review the accumulating data at six-month intervals. As events transpired, the committee met but twice. Recruitment for the trial ended in January of 2001; six months later, the committee concluded that the results of the trial were overwhelmingly positive and should be made public. As a result, the trial stopped, and when the dust settled, final results confirmed STI157 to be a significant improvement over the interferon regimen (O'Brien et al., 2003). There had not, in fact, been much of a trial left to pursue. Designed as a cross-over study, patients who fared poorly on one regimen crossed over to the competing regimen in the course of the trial. By the time the trial closed, 90% of the interferon patients had switched to STI571. While largely due to the superiority of STI571, movement had also been stimulated by interferon's toxic side effects and by the fact that following the FDA's accelerated approval of STI571, a significant number of patients had simply withdrawn consent to remain on interferon and dropped out of the trial in order to take Gleevec as nonprotocol patients.

Initially devised for a relatively small population base—CML patients—Gleevec has since moved into "a variety of cancers that share common molecular abnormalities" (figure 10.1) (National Cancer Institute, 2003: 31). Notice how a drug can redefine the contours of the disease(s) it targets—in this particular case, "molecular abnormalities" have replaced more traditional pathological criteria—and is in turn redefined by those reconstituted entities. In 2008, Gleevec earned Novartis $3.7 billion (US) and its success has reordered corporate thinking about where the future lies in drug development. As previously mentioned, a stratified medicine, niche-buster approach may produce results similar to the more traditional blockbuster approach, leading industry to conflate the two (Ray, 2009b). Novartis's CEO, however, observed in an interview that regardless of the size of the market, drugs that will treat only extremely rare diseases are now pursued if there is a clear mechanism or target involved (Whalen, 2009; IL-1β-targeted antibody . . . , 2009). This is in part due to *orphan drug* legislation in both the United States and Europe that provides incentives for the development of drugs against diseases affecting very small populations—for instance, by stipulating that regulatory agencies will not grant approval for other versions

Clinical Trials for STI571 have Mushroomed Since Early 1999

After success with a small Phase I clinical trial to test the safety of STI571 (Gleevec™) for treating chronic myelogenous leukemia, clinical investigators began testing the drug in a variety of cancers that share common molecular abnormalities. A rapid and broad expansion of clinical trials followed.

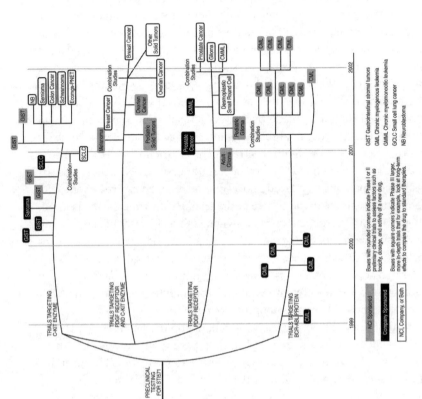

FIGURE 10.1. The "mushrooming" of Gleevec clinical trials, as represented in a NCI 2003 report. Reprinted from National Cancer Institute, *The Nation's Investment in Cancer Research: A Plan and Budget Proposal for Fiscal Year 2003*, NIH Publication No. 02–4373, Bethesda, MD: National Cancer Institute, p. 31 (public domain source).

of the same drug for the same indication. But Novartis's decision is also based on the hopes that what works initially in a restricted, yet targeted population can be easily extended to other targets, mechanisms, and populations (Capell, 2009).

Where Did Gleevec Come From?

In what follows, we offer a brief history of STI571, emphasizing its corporate origins and the shifting contours of the subsequent public/private collaboration. This description contrasts slightly with the narrative on offer in the scientific literature where the discovery of the compound and its use in cases of CML is usually described as the triumph of the *rational* over the *empirical* in the field of drug discovery. Although greeted as a paradigm or model for the rational development of molecular medicine and cancer therapy, STI571 did not initially flow from research on CML or its biopathological mechanisms (Capdeville et al., 2002; Sharifi & Steinman, 2002; Druker, 2002b; Sawyers, 2001). In fact, it first emerged somewhat serendipitously from a class of compounds that had originally been developed to treat heart disease. Those targets, in turn, had arisen in the course of a corporate strategy of cooperation and competition that began in the early 1980s at the Swiss company Ciba-Geigy, which merged with its rival Sandoz in the mid-1990s to form Novartis. Initial efforts to develop anticancer compounds had not been auspicious. At the beginning of the decade, Ciba-Geigy shut down its cancer research unit: returns on its investment to that point had been "paltry" (Vasella, 2003: 41). In 1983, as the first wave of biotechnology partnerships rolled over the pharmaceutical industry (Galambos & Sturchio, 1998), Ciba-Geigy reopened its cancer research division. By the mid-1980s it had developed a modest program searching for compounds to block tyrosine kinases. Although the project had marginal status within the company, tyrosine kinases were (and remain) major players in both the cellular and pharmaceutical economy, and numerous researchers had implicated the overproduction and underproduction of kinases in cancer, diabetes, and atherosclerosis.

In the early 1980s attention turned to a member of the receptor tyrosine kinases family known as the epidermal growth factor receptor (EGFR). The protein was categorized as both a cell-surface receptor for signals from other cells and as an enzyme that triggers a cascade of reactions from inside the cell wall once it receives a signal (Gschwind, Fischer & Ullrich, 2004), but the exact mechanism of these biochemical reactions was not known. Notice that the original growth factor, a protein isolated in a cell, has become just one step along a pathway. It would be an understatement to say that the pathway following from the initial signal to the final reaction by the cell is complex. Figure 10.2, which we will make no attempt to explain, should be viewed as a striking illustration of the intricacy of the EGFR pathway circa 2005 (Oda et al., 2005). The map describes 219 different biochemical reactions implicating 300 "species" or entities (proteins, oncogenes, etc.) that participate in the process initiated by the receptor.

Despite the ultimate complexity of the system, blocking a tyrosine kinase was, at the time, not an inherently difficult enterprise. Almost as soon as the biochemical function of the enzyme became known in the early 1980s, researchers identified naturally occurring compounds that interrupted its action. The natural compounds also interfered with a variety of related enzymes, rendering the compounds highly toxic (Graziani, Erikson & Erikson, 1983). Although researchers continued to identify new tyrosine kinases and their corresponding blockers throughout the 1980s, the number of kinases quickly outstripped the number of compounds capable of blocking them (Akiyama et al., 1987). For those who sought to block them, the problem lay in the fact that the part of the molecule that would have to be blocked to prevent it from functioning appeared to be almost identical in different kinases. Thus blocking one kinase would most likely lead to blocking many of the others and would consequently create considerable disruption in normal cellular processes (Hanks, Quinn & Hunter, 1988).

Ciba-Geigy was not alone in its pursuit of kinase blockers. Its rival in this area, Genentech, was already in hot pursuit of the genetic map and of the structure and function of EGFR, as well as a means of blocking it. Genentech's work was carried out in collaboration with a future Nobel Prize winner, Robert Weinberg, one of the discoverers of oncogenes at the beginning of the 1980s and a founding member of MIT's Whitehead Institute that had just opened in 1982. So when Ciba-Geigy began its search for a tyrosine kinase blocker in the mid-1980s, they chose to block one known as platelet derived growth factor (PDGF) that had yet to attract Genentech's firepower. PDGF also opened a possible breach into the much more lucrative market of heart disease: blocking specific

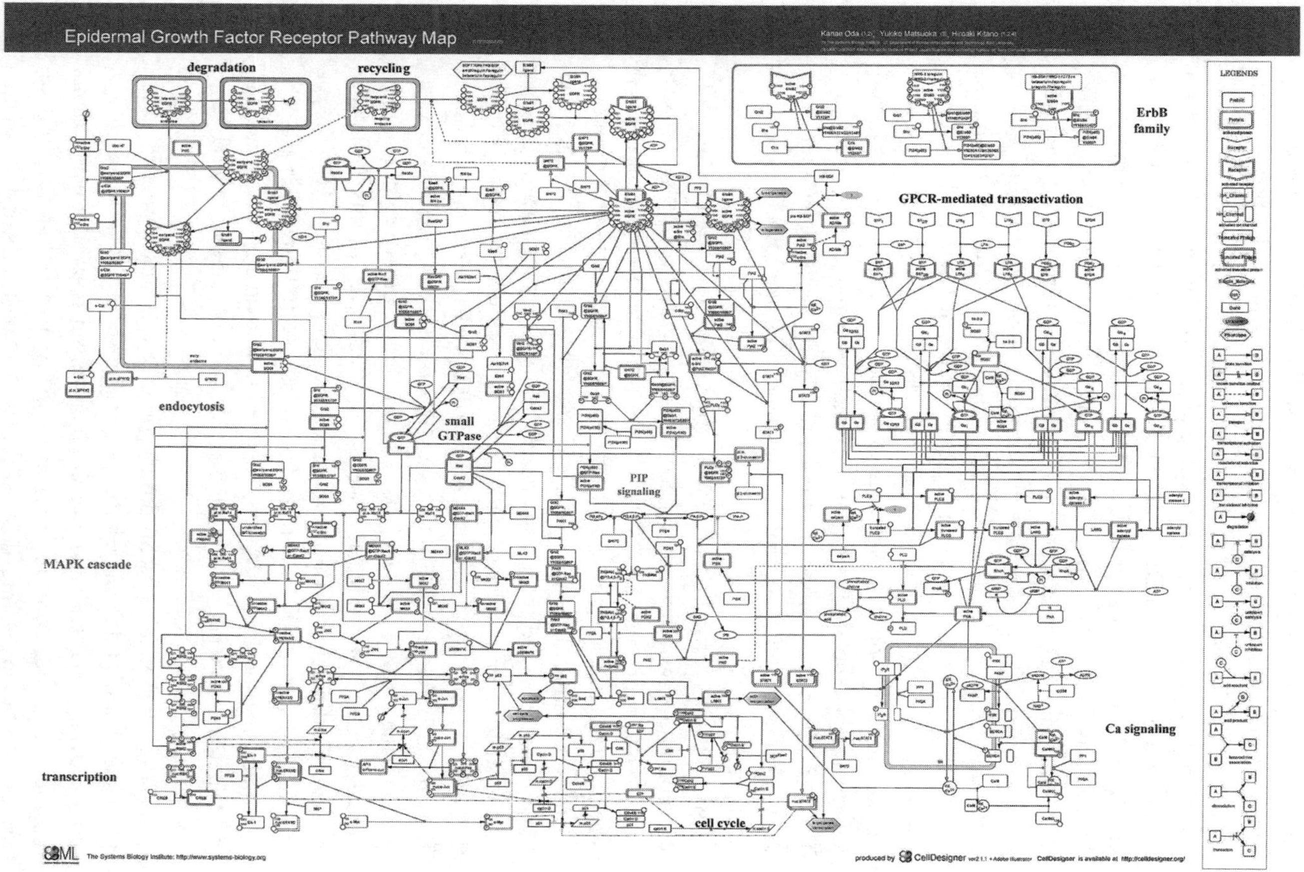

FIGURE 10.2. A computer-generated map of the EGFR pathway. Reprinted with permission from Macmillan Publishers Ltd., K. Oda et al., "A Comprehensive Pathway Map of Epidermal Growth Factor Receptor Signaling," *Molecular Systems Biology* 1 (2005): 2. © 2005.

receptors on the blood platelets would, so the theory went, prevent the reblocking of an artery following coronary angioplasty (Druker, 2002a).

Although the organizers of the tyrosine kinase program at Ciba-Geigy had originally sought something useful for the heart disease market, by the mid-1980s PDGF had also been implicated in human cancer. In 1988, therefore, the head of Ciba-Geigy's kinase program, Nick Lydon, traveled to the Dana-Farber Cancer Institute in Boston to create connections with experts in cancer-causing tyrosine kinases. His team had had difficulties producing the kinases using recombinant DNA techniques, and methods developed at Harvard provided a solution to their technical difficulties (Lydon, 2009). While at Harvard, Lydon also met Brian Druker, a fellow in medical oncology who worked on CML. Druker initially doubted the value of a tyrosine kinase as an anticancer treatment. Shortly thereafter an Israeli research group produced a kinase inhibitor that blocked an epidermal growth factor without damaging normal metabolism and thus convinced Druker otherwise (Yaish et al., 1988).

By a curious turn of events, by the time Ciba-Geigy had isolated the compound that would go on to become STI571, their collaboration with Druker had come to a temporary halt. In 1991, Dana-Farber had signed a celebrated exclusive agreement with Ciba-Geigy's competitor Sandoz. The $100 million, ten-year contract included a clause stipulating that Sandoz and Harvard share patent and royalty rights to "any drug found to be effective in interfering with the division of cancer cells" (Private dollars . . . , 1991: 1526). As a result, Druker was forced to cut off contact with Ciba-Geigy until 1993 when he went to the University of Oregon. During Druker's absence, work continued at Ciba-Geigy. The optimization of STI571 produced, once again, unexpected results. Following a minor modification of the compound, researchers observed that it specifically blocked a tyrosine kinase known as the BCR-ABL kinase. The reason for the specific inhibition remained unknown until after the clinical trials in 2001 when new protein structure data came to light (Lydon, 2009: 1153).

At the time of this crucial observation, Druker had returned to the Ciba-Geigy fold and the firm began to send him compounds to test. Luckily for Druker, the University of Oregon waived possible royalty rights on the substances delivered by the company: had the university done otherwise, "it would have delayed progress at best; at worst, this compound would never have made it into my lab or into the clinic" (Druker,

2009: 1151). Unlike industrial researchers, Druker had access to patients' cells (Lydon, 2009), and he soon found out that STI571 suppressed the CML cells themselves (Buchdunger et al., 1996). Following these in vitro tests, he conducted tests in animal models and toxicity studies in dogs (Druker et al., 1996). Human clinical trials soon followed, as previously described.

The story did not end here. In spite of the spectacular results obtained by Gleevec, approximately 2% to 4% of patients develop resistance to the drug annually and must turn to other measures. In approximately one-third of these cases, augmenting the dose reverses the resistance (Kantarjian et al., 2010). In the remaining cases the most promising avenue continues to be tyrosine kinase inhibitors that target other mutations to the *ABL* gene than the one blocked by Gleevec (Shah & Sawyers, 2003). Thus Bristol-Myers Squibb has developed dasatinib (marketed as Sprycel), while Novartis has marketed a second compound, nilotinib (marketed as Tasigna), developed the year after the launch of Gleevec. Events such as these suggest that the proliferation of targeted agents aimed at similar pathways or receptors may transform the oncology drug market. Rather than adding new products to standard-of-care combinations in a cumulative manner (as discussed in chapter 8 and implied by figure 8.1), new products displace older ones in a form of direct competition (Santoni & Foster, 2007). Does this also mean that the era of combination chemotherapy regimens is over and that we have entered the promised land of magic bullets, as implied by the *Time* article mentioned at the beginning of this chapter? Not quite.

Gleevec is an exception in the field of targeted therapy, for few, if any, compounds have produced positive results on their own. Given the multiple targets and multiple biochemical pathways implicated in any given cancer, it has become clear in the last decade that drug combinations and not single agents offer the most promising avenues of research. This is not the first time that the NCI has confronted the problem of how to test many possible combinations of anticancer agents. Unlike the 1970s and 1980s, the new era of combinations faces a number of unprecedented challenges. First is the sheer quantity of combinations possible; rather than dozens, thousands lie on the horizon. Second, given that most of the compounds have been produced relatively recently, many investigators seek to test combinations at a comparatively early stage in drug development. Thus rather than testing combinations of agents whose effects have been clearly evaluated, many trials test combinations of agents

that remain largely experimental. In addition to the scientific and administrative challenges posed by such an undertaking, the NCI initially
faced considerable intellectual property challenges. With agents flowing
in from both private and academic sources, the NCI's CTEP brokered
a template agreement that would allow academic researchers, industry researchers, and NCI researchers to undertake cooperative investigations of combinations and share both data and intellectual property.[1]
In 2008–9, for example, CTEP held cooperative agreements concerning
more than eighty different agents that were being tested in more than
five hundred Phase I/II trials.[2] The road since Gleevec now resembles a
labyrinth.

Gleevec and Herceptin (trastuzumab) are examples of the two main
categories of targeted agents: small molecules that inhibit molecular pathways within the cell or interfere with cell surface receptors, and monoclonal antibodies (biologic drugs) that block the latter. Herceptin reacts
with the HER2 receptor that is overexpressed in 20% to 25% of breast
cancers. It was developed by the American biotech company Genentech
in collaboration with a physician from UCLA's Jonsson Comprehensive
Cancer Center: again, a public/private configuration (Bazell, 1998). Co-
founded in 1976 by one of the academic pioneers of genetic engineering
and by a venture capitalist, Genentech is a prime example of the start-
ups that ushered in the biotech era. Since 2009, Roche—the other Swiss
multinational pharmaceutical company that has become a major player
in the field of anticancer drugs—has acquired full control of Genentech
and markets Herceptin internationally. The clinical career of Herceptin resembles, in part, the story of the CMF regimen discussed in chapter 6. Like the CMF combination, the biologic drug was first deployed
in clinical trials of metastatic (advanced) breast cancer—the first indication for which the FDA approved it in 1998. It was subsequently tested
as an adjuvant therapy, in particular in a trial called HERA (HERceptin Adjuvant), one of the largest breast cancer trials ever conducted
(5,081 patients at 478 sites in thirty-nine countries). Like Bonadonna's
CMF adjuvant trial, preliminary outcomes of the HERA trial indicated
a possible 50% decrease of the risk of relapse one year after surgery and

    1. At   http://ctep.cancer.gov/industryCollaborations2/default.htm#guidelines_for_
collaborations.
    2. At http://ctep.cancer.gov/MajorInitiatives/Phase_1–2_Early_Drug_Development
.htm.

prompted the trialists to rush to publication in 2005 (Piccart-Gebhart et al., 2005). An editorial accompanying publication of the report hailed the results as "not evolutionary but revolutionary" (Hortobagyi, 2005: 1736), but again as in the Bonadonna case, the initial enthusiasm quickly waned. Faced with incomplete results, critics maintained that it was too early to declare victory (Littlejohns, 2006; Picard, 2005). The absolute numbers, moreover, were far less impressive, and side effects such as cardiac toxicity made Herceptin a less than ideal drug. As in the case of CMF, the HERA trial had been preceded by a number of other trials on metastatic breast cancer, while other groups ran concurrent adjuvant trials (Tan & Swain, 2002). In particular, two US cooperative groups published the combined results of two adjuvant trials with trastuzumab in the same issue of the *NEJM* (Romond et al., 2005).

But the similarities stop here. The HERA trial was carried out under the auspices of a novel organization, the Breast International Group (BIG). Founded in 1996 during an EORTC breast cancer conference as an international nonprofit organization, BIG defines itself as a consortium of consortia—a metaorganization according to the terminology we used in chapter 9—connecting forty-five independent collaborative groups (from Europe, Canada, Latin America, Asia, and Australasia), each with their own data centers and networks of affiliated investigators, hospitals, and laboratories. In addition to the seventeen BIG groups that sponsored the trial, the HERA trial also mobilized nine other cooperative groups and ninety-one independent centers. Interestingly enough, a pan-European patient advocacy group, Europa Donna, was also represented on the trial's committee. A second noteworthy characteristic of the trial (at least for oncology) was its sponsorship by the pharmaceutical multinational Roche. In keeping with the BIG principle that clinical trials be conducted "with 'scientific independence' from the pharmaceutical industry" (Piccart, 2001: 1–2), the coauthors of the 2005 *NEJM* report stated that Roche had no access to the database or the interim analyses, and that independent statisticians presented analyses to an independent data monitoring committee without disclosure to the data center, the investigators, or the sponsor. The BIG principle, in other words, forced industry to delegate "significant control and management of trials to investigators." Here, however, BIG hit a temporary snag. Fearing that this arrangement might not lead to "data of regulatory quality within the anticipated time frame and within a reasonable budget," Roche decided in 2001 to terminate its partnership with BIG and run the trial "in house."

Later that year, a face-to-face meeting between six trialists and Roche upper management, and additional negotiations with the BIG board of directors, led to resumption of the collaboration (Piccart, 2001: 1).

What are the lessons for us here? First of all, notice the changes in the organizational architecture of Phase III trials. Not only has the number of local institutions, trialists, and patients grown to unprecedented levels creating a complex operation that takes years to plan and design, but also the trial collectives are now constituted on a trial-by-trial basis (see chapter 11). Organizations such as the EORTC or its constituent cooperative groups increasingly conduct their activities as parts of intergroup, international endeavors that include patient advocacy groups. Second, both the Gleevec and the HERA trials bear witness to the increasing importance of industry in a field long dominated by public and academic institutions. While joint commercial and not-for-profit initiatives may result in conflicts, more remarkable is the management of those conflicts via the development of novel endogenous regulatory institutions such as the data monitoring committees.

Monitoring Data, Managing Risks

As we just saw, one such Data Monitoring Committee (DMC) oversaw the trial comparing Gleevec and interferon and took the decision to end the trial before all the results were in. A novel institution in the field of clinical cancer trials, most cooperative groups adopted DMCs as oversight mechanisms for clinical trials in the late 1980s with the express purpose of restricting access to clinical trial data to a committee of experts not directly involved in running the trial. While the reasons for this will become clear below, the constitution and the continuing debate over the structure and deployment of DMCs offer a telling illustration of the variety of institutions that embed the practice of oncology.

According to the trialists themselves DMCs originated in the late 1960s at the National Heart Institute. The founding document is the 1967 Greenberg Report ([1967] 1988), an internal NIH report produced by a committee headed by Bernard Greenberg, then a professor of biostatistics at the University of North Carolina and a former NIH statistician. The report has been cited ever since as the first official document to recognize "the need for a mechanism for terminating a trial early if it became evident that it could not meet its objectives or new information rendered it superfluous" (Ellenberg, Fleming & DeMets, 2002: 5). A close reading of the report, however, shows the preceding statement to be somewhat anachronistic. Briefly, the report defined members of the policy or advisory board as "senior scientists, experts in the field of study but not data-contributing participants in it," and they were to "review the overall plan, make recommendations on any possible changes (including changes in protocol and operating procedures), adjudicate controversies that may develop, and advise the National Heart Institute on such matters as the addition of new participants or the dropping of nonproductive units [i.e., those that fail to enroll enough patients]" (Greenberg Report, 1988: 146). No mention here of interim review. Although the report recommended periodic review of the data, that activity fell within the purview of the chairman of the clinical study and his Executive Committee and included "the responsibility . . . to review the data analysis coming from the Coordinating Center and to prepare *frequent reports* to the participants, as well as annual reports to the National Heart Institute" (142).

In other words, the purpose of periodic reviews in 1967 was not to restrict access to data to a designated committee but to put as much data into circulation as possible. While the question of early termination of a trial was discussed, the issue was viewed separately from the continual review of data and was seen as something that might occur only under "unusual circumstances"; that is, "if the accumulated data answer the original question sooner than anticipated, if it is apparent that the study will not or cannot achieve its stated aims, or if scientific advances since initiation render continuation superfluous." As to who should make this decision or how it should be made, the Greenberg report gave no answer. It merely suggested that "a mechanism must be developed for early termination" (1988: 146). It is thus doubtful that Greenberg had present-day DMCs in mind as a possible mechanism.

In order to analyze interim data, DMCs use statistical methods devoted to that purpose. Readers will recall that methods for analyzing clinical trial data as they accumulate with a view toward stopping trials once a yes or no answer had been reached had ostensibly been available since the early 1960s in the form of sequential methods (chapter 4). In their initial formulation, these methods required the analysis of patient data as each new pair of patients was treated in a single institution. As clinical trials became multicentered and as the time lag between the start of treatment and the measurement of patient responses grew, the early sequential methods fell into disuse. In the mid-1970s, however, Stuart Pocock, an ECOG statistician, generated the first group sequential designs (not to be confused with the previously mentioned sequential methods) for clinical trials (Pocock, 1977). The new designs allowed trialists to analyze groups of patients enrolled in different institutions at predefined intervals, say every six months, to correspond with cooperative group meetings. While these statistical tools would later became the mainstay of DMCs, they did not at the time prompt their formation.

Even though the institutional and technical conditions necessary for the creation of DMCs existed in the late 1970s, DMCs did not become widely adopted in the field of cancer until the end of the 1980s. The reason for this decade-long delay lies in the absence of a clear motive for the creation of DMCs. This was duly supplied by headwinds that buffeted the clinical trial enterprise in the 1980s. Remember: by the 1970s, cooperative oncology groups ran hundreds of Phase II and Phase III multi-institutional trials. The subsequent spread of clinical cancer trials into the community oncology groups gave these trials a national audience. The problem DMCs ultimately came to solve was this: during the 1980s, a steadily increasing number of trials failed to reach completion because of the widespread knowledge of interim results presented at cooperative oncology group meetings. Favorable or unfavorable, knowledge of either result led the growing number of clinical oncologists to the same conclusion. If interim results of a new treatment showed that it was not an improvement over standard treatment, then there was clearly no advantage for their patients to enter the trial. If the treatment did show an improvement, then there equally was no point for a patient to enter the trial and risk being assigned to the old treatment.

The cases of SWOG and ECOG are instructive in this regard. Until the mid-1980s, SWOG had conducted all trials without monitoring committees. Aware of the problem of widespread knowledge of early results, SWOG compared a series of fourteen SWOG trials with those sponsored by the North Central Cancer Treatment Group (NCCTG). The NCCTG had instituted a DMC composed solely of the study investigators and statisticians from the group's data center in order to restrict circulation of interim data rather than sharing with all members of the group, as was the usual practice (Ellenberg, Fleming & DeMets, 2002). As suspected, unlike the NCCTG many of the SWOG trials had suffered declining accrual rates that had followed what appeared to be convincing positive or negative results. Thus to save future data from similar corrupting influences, SWOG adopted a policy of restricted circulation of interim results, confining them to members of the DMC (Green, Fleming & O'Fallon, 1987). ECOG followed suit in 1990 (Rosner & Tsiatis, 1989: 513–14).

In 1992, after the cooperative groups had begun to use DMCs, the NCI made them mandatory for all groups. The guideline mandating the DMCs required them to take steps to eliminate possible conflicts of interest and to establish DMCs "independent of trial investigators," reasoning that "because clinical trials are under increasing public scrutiny, we must use procedures that *protect our research* against the *appearance* of impropriety" (Simon and Ungerleider as cited in Green, Benedetti & Crowley, 1997: 90). The NCI guideline sparked debate among clinical trial administrators over the proper form of a DMC (Meinert, 1998; Ellenberg & George, 2004). Should it be an outgrowth of the organizational machinery of the trial or should it consist of independent, external experts as outlined in the NCI guideline? Was the primary purpose of the DMC to restrict go/no-go decisions to trial experts or was it to prevent interested parties, now the trialists themselves, from pursuing a trial for reasons such as research that did not directly benefit the patients? Proponents of the external committees suggested that an "internal" DMC composed of researchers directly involved in the trial might be willing to pursue a trial in spite of serious problems.

Supporters of internal committees countered that an "external" independent committee might lack both "clinical sensitivity" and a sufficiently intimate knowledge of the trial's intricate design and machinery to make a reasoned decision (Simon, 1993: 525). In the ensuing, sometimes acrimonious, debate, a number of issues emerged, including the relation between statisticians and clinicians and the question of whether concerns about possible risks to patients could be convincingly separated from scientific issues. The 1997 version of the NCI guidelines brokered a compromise between the two extremes by mandating that the DMCs be composed of a majority of external members with a few insiders, as well as a patient representative. Partly as a result of these changes and partly because of experience, some of the original opponents of independent committees have come to see them as a virtue. According to the former chief statistician at ECOG, the change in viewpoint owed much the fact that the external monitors have learned to do what the internal monitors had done previously: acquire the expertise needed to understand the trial and not mechanically apply stopping rules.[3]

In chapter 1 we noted that a style of practice engenders its own criteria for producing and assessing evidence and thus leads to new institutions. Although DMCs are a late addition to the clinical trial machinery, they are yet a further component of the arrangements ensuring that the clinical trial system will generate *evidence*—clinical and regulatory— about new therapies and, more recently, the biological mechanisms underlying them. And while DMCs guarantee, in principle, the integrity of the clinical trial process through the objective assessment of interim data, the search for objectivity requires constant reflection on the structure and function of DMCs. Such a search raises questions concerning the evolving definitions of the qualifications of DMC members, their attributes and prerogatives, and the extent to which a DMC should be able to redefine the parameters of a trial and to restrict the circulation of data. This ongoing concern is reflected, for instance, in discussions prompted by the issuance in 2001 of an FDA draft guidance on DMCs (FDA, 2001; Keating & Cambrosio, 2009).

A Medical Oncologist and His Patients

We turn now to two final aspects of the Gleevec story: the role of the medical oncologist and the role of patients. When he assumed responsibility for the Gleevec trials in the late 1990s, Dr. Brian Druker was a seasoned medical oncologist. He had graduated in medicine at the University of California, San Diego almost twenty years earlier in 1981, and following his internship and residency he had won a three-year fellowship in medical oncology at the Dana-Farber Cancer Institute at Harvard University. To become a specialist in blood cancers Druker trained in the mixed specialty of hematology/oncology. In 1987 he found a position both as an instructor at Harvard Medical School and as a laboratory researcher. In the lab he boarded the molecular biology juggernaut and immediately began work on oncogenes and tyrosine kinases (Roberts et al., 1988; Druker, Mamon & Roberts, 1989). By his own estimation he was quite successful, but the Dana-Farber leadership did not share this self-assessment and in 1993 told him in no uncertain terms

---

3. Interview with David Harrington, Boston, 25 April 2005.

that he "didn't have a future at Dana-Farber." He met with a similar fate at Boston's Beth Israel Hospital, and obviously devastated opted for a position at the less prestigious Oregon Health & Sciences University, a move he retrospectively saw as a turning point (Druker, 2007) insofar as the transfer to Oregon allowed him to resume his collaboration with Ciba-Geigy. Druker claims that during the difficult period when he "felt abandoned by some people [he] once respected," he found solace in the unquestioning support he received from his patients (Druker, 2007). Almost a decade after the initial Gleevec trial, Druker continues to end his talks on Gleevec with thanks to and a photograph of the patients who enrolled in that trial.[4] The link between Druker and his patients reveals a reality that goes beyond the anecdotal. Druker, as a medical oncologist, acted both as chemotherapist and as the long-term physician for the patients. Medical oncologists perform this dual function even in multimodal protocols:

> The medical oncologist usually serves the traditional role of internist in the multi-disciplinary management of cancer. Whereas the surgical procedure or even the radiotherapeutic treatment course is of short duration, the medical oncologist has continuing responsibility that may stretch over months or years of therapy, and decades of follow-up, depending on the neoplasm. (Holland et al., 2000: 988–89)

The status and function of the medical oncologist is a relatively recent development. The idea of medical oncology as a separate subspecialty first surfaced in the United States on the margins of the NCI's National Cancer Chemotherapy Service Program in the early 1960s when an Ohio physician and a professor at the University of Pittsburgh Medical School first proposed the creation of a national association of chemotherapists (Krueger et al., 2006). Fueled in part by their enthusiasm for the anticancer drug 5-FU, they convened a small number of like-minded practitioners on the sidelines of the 1963 AMA meeting to draw up a list of potential members of a new society they named the American Society of Clinical Oncology (ASCO) (Krueger, 2004; Armitage, 1998). Things moved quite quickly after the formation of ASCO. In 1966 the NCI be-

4. Zach Goldsmith, "Meeting the Minds behind Gleevec," *Oncology Updates: Documenting the Evolution of Cancer Medicine*, 15 September 2006, http://oncologyupdates.blogspot.com.

gan funding fellowships for oncology training, and in 1971, following representations by ASCO, the American Board of Internal Medicine recognized oncology as a new subspecialty (Kennedy, 1999).

While ASCO, and by definition clinical oncology, included chemotherapists, surgeons, and radiotherapists, medical oncology defined itself as something smaller than clinical oncology and somewhat larger than chemotherapy. On the eve of certification in 1971, it was not obvious to all participants in the field of cancer what that something might be. In an interview with the editor of a prominent cancer journal for clinicians, a Minnesota oncology professor defined medical oncology as "a subspecialty of internal medicine" that "involves the study of the prevention of cancer, its detection, definitive and palliative treatments and care of the patient dying from cancer" (Kennedy, 1970: 368). Two characteristics set the medical oncologist apart from other cancer specialists. First, the "myriad medical complications" of cancer required the supervision of a specialist with training in internal medicine over and above the technical intervention of a surgeon or radiotherapist. Second, given the medical oncologist's research orientation, the interest in chemotherapy was but part of a larger interest in "molecular biology, mechanisms of cell kinetics, biochemistry of tissue growth, epidemiology, detection and natural history of cancer" (1970: 370). When the interviewer pointed out that many other specialties such as hematology treated cancer and that surgeons and radiotherapists also used chemotherapy, the nonplussed interviewee replied that while all those practitioners had a "special interest" in chemotherapy, it was not their "specialty."

That view did not prevail. When ASCO began promoting the formation of a distinct specialty in the late 1960s, the prior implication of hematologists in chemotherapy led many to question their possible exclusion from oncology (Kennedy, 1970; Shaw, 1977). As a result, the Board of Internal Medicine created two separate oncology specialties: one for hematology (hematology/oncology) and one for the newcomer, medical oncology. Medical oncology and hematology would henceforth require two years of training following certification in internal medicine; hematology/oncology would require three years (Krueger, 2004). The last option gained considerable traction during Druker's training in the mid-1980s, when hematology/oncology pulled ahead of oncology as the specialty of choice. In 1987, for example, many more first-year fellows enrolled in the combined specialty (271) than in oncology alone (98) (Lyttle et al., 1991; American Board of Internal Medicine, 1998). Despite this popularity or perhaps because of it, less than a decade later the relative proportions of oncologists and hematologists/oncologists had been inverted. In 1994, when ASCO polled its members with regard to their specialties, roughly 70% of their respondents considered themselves medical oncologists, 25% hematologist/oncologists, and the remaining 5% hematologists (Ad Hoc Committee . . . , 1996).

The emergence of medical oncology was more problematic in Europe, beset with differing national regulatory frameworks and traditions. Several nations had centralized cancer treatment in radium and radiotherapy centers, and radiotherapists from those countries had come to see chemotherapy as one of their tools. Noting this difference, a 1971 editorial in *The Lancet* suggested that the only reason for the emergence of medical oncology in the United States was the relatively backward state of radiotherapy as a specialty (Medical oncology, 1971: 419). One could argue that, *au contraire*, the Europeans paid for earlier leadership in radiotherapy by trailing the Americans in the official recognition of medical oncology. In any event, the origins of the European equivalent of ASCO, the European Society for Medical Oncology (ESMO), go back to 1975, when it was instituted as the Société de Médicine Interne Cancérologique, an association of primarily French-speaking internists. Following the initiative of young clinicians involved in the EORTC and who felt the need to establish a European counterpart of ASCO,[5] in 1980 the Société expanded to become ESMO. By 1989 ESMO had begun to offer European-wide certification in medical oncology (Wagener et al., 1998) and in 1990 launched its own official journal, the *Annals of Oncology*, the counterpart of ASCO's *Journal of Clinical Oncology*. The European and American societies, ESMO and ASCO, have recently undertaken a joint initiative toward international harmonization by proposing a "global core curriculum" (Hansen, 2004: 177; ESMO/ASCO Task Force . . . , 2004).

As suggested above, neither ASCO nor medical oncologists saw themselves as mere chemotherapists. Both ASCO and medical oncology endorsed clinical research as a constitutive component of their activities. The election of eminent researchers such as Emil Frei and Joseph Burchenal as society officers was but one symptom. Henceforth, medical oncologists would be viewed as potential researchers in the sense

5. Interview with Franco Cavalli, Bellinzona, Switzerland, 22 December 2006; see Saul (2002).

that in addition to patient care, medical oncologists would maintain an "investigational orientation" (Kennedy et al., 1973: 130), a position continually endorsed by ASCO.[6] A similar position was entertained by the EORTC: medical oncologists' training ought to include the relevant basic sciences, and training in research was "highly desirable."[7] The rise of medical oncology and its orientation as a research/management specialty has tended to reduce the other modalities to the status of technical interveners. Surgeons themselves have observed that "multiple influential entities are forcing us into a background role" (Copeland, 1999: 425). As the president of the Society of Surgical Oncology pointed out in his 1999 presidential address, even the NCI portrayed the surgical oncologist almost as a technician who reported to the medical oncologist. According to a NCI pamphlet for breast cancer patients, the medical oncologist, not the surgeon, was the person who could "put together all of the information about your case and can discuss your treatment choices with you" (NCI, 1998: 21–22, cited in Copeland, 1999: 425).

Thus even though chemotherapy is often an adjuvant or accessory to surgery and radiotherapy, the medical oncologist usually maintains responsibility for the total management of the patient insofar as, in addition to the aforementioned side effects, the oncologist handles the pain produced either by the cancerous growth or by the therapeutic interventions themselves. The importance of pain management for the status of medical oncology cannot be underestimated. Some observers suggest that the advent of new, cytostatic drugs with relatively few side effects may ultimately reduce the importance of the medical oncologist in cancer therapy (de Vries, Mulder & Sleijfer, 1999: 312). Nonetheless, even low toxicity drugs such as Gleevec have side effects that may include a reduced white blood cell count (Sneed et al., 2003) and thus require constant monitoring by a medical oncologist.

As an oncologist, Druker is one of tens of thousands who meet regularly at the annual meetings of the American Society of Clinical Oncology (ASCO) that serve both as an archive of the previous year's accomplishments and as a worldwide announcement of what will happen in the year ahead. As noted in chapter 1, since their inception in 1964

6. E.g., Presidential Symposium: Report of the Clinical Trials Subcommittee (1999), available at http://www.asco.org and cited in de Vries et al. (1999: 315).

7. *EORTC Newsletter*, no. 92, March 1981, "Criteria and definition of Medical oncology as a subspecialty within clinical oncology."

with a group of fifty participants, ASCO meetings have attained gigantic proportions. The 2003 meeting attracted 26,000 attendees, about half of whom were foreign and about an equal number of whom were certified medical oncologists (surgeons, radiotherapists, nurses, patients, and a significant press corps also attend these meetings). Together they heard 3,641 papers. The meeting proceedings occupied an 11,061-page book, 1,000 pages of which consisted of disclosures of ties between oncologists and drug companies (27% of abstract authors had ties to commercial enterprises) (Hoogstraten, 2005). Five years later, growth continued unabated with attendance topping 34,000 (see figure 1.3, chapter 1). As a result, ASCO meetings have become so large that organizers now hold follow-up, "best of" sessions where the four hundred best papers of the original meeting are re-presented (Simone, 2004).

ASCO meetings bring together not only the global community of oncologists but also, through the media, oncologists and "the public." The press has had its own meeting room since 1985, attendance in the previous year having broken the 5,000 mark (ASCO, 2004). According to a former ASCO president, the media presence has consequences beyond mere publicity "because ideas get out into the world very quickly, and you get feedback very quickly too. . . . You need one major international event every year, together with all the hoopla and drama that goes with it" (Larry Norton . . . , 2000: 192). Beneath the excitement, oncologists exchange information, and in this respect the ASCO meetings participate in a network of conferences where oncologists meet on a frequent basis:

> I have a certain set of colleagues, numbering in the hundreds, but we see each other every few weeks, somewhere in the world. I remember when I was very young and first got into this area, it would shock me when the older doctors and scientists that I used to hang out with, would see their colleagues and they wouldn't even say hello—they would just pick up the conversation where they left off 3 weeks earlier in Oxford or Bangkok. (Larry Norton . . . , 2000: 192)

Patients and Activists

Just as patients hold a place of pride in Druker's slide show, so too do they have a place at ASCO and related meetings. ASCO has, in fact, solicited patient groups since the beginning of the 1990s as part of a larger move to make the society more politically active. According to the 1990 presidential address, "political, legal, economic and regulatory issues now loom as large in our focus as do educational and academic pursuits" (ASCO, 2004: 25). Shortly after the 1990 address, ASCO established a Government Relations Office in Washington and a Patient Advocacy Committee (2004: 26).

Although patients cannot join ASCO, the ASCO website maintains a special section for patients that contains over 120 patient guides for different cancers, information concerning clinical trials, and offers the possibility of downloading ASCO's patient newsletter, *People Living with Cancer*. The newsletter also contains verbatim reports of online chats between patients and ASCO's designated experts. The discussions are often quite technical and implicate patients as active participants in clinical cancer treatment. We saw previously how CML patients lobbied Novartis for greater access to Gleevec. Patients involved in other Gleevec trials have been equally as active, not only in demanding products and answers but also in creating their own knowledge base as part of an emerging epistemological activism. In 1999, for instance, patients of a relatively rare form of gastrointestinal tumors known as GIST—which EORTC studies found to be responsive to Gleevec (van Oosterom et al., 2001; Verweij et al., 2004)—created a patient group known as the Life Raft Group, devoted, inter alia, to sharing information on clinical trial results and side effects. While "it can take years for patient trial data to filter down into the general medical community and even longer to reach the general public," Life Raft's data-sharing techniques have enabled patients to access "that data in real-time." More importantly, Life Raft members collect their own data and edit their newsletter: "When we look at side effects . . . we believe that we are more sensitive and accurate in reporting on this from the perspective of the patient. . . . Our ability to collect timely side effects information cannot be replicated by the medical community. We are the front line. We raise red flags" (Vasella, 2003: 131). Life Raft prides itself in its independence and actively counsels its members on which clinical trials to enter by helping them to select the one that has the best chance of improving survival.

The Life Raft Group is one of over 850 American cancer advocacy organizations and associations—most founded after 1990—that sometime cooperate and sometime compete in the field of cancer (Mortenson, 2002). Not all of these organizations are solely devoted to the needs of patients, although all claim to have patient needs uppermost in mind. Like oncologists, cancer patient advocates are of relatively recent coinage. Present-day patients and their advocates find themselves implicated at all levels of the decision-making process in clinical cancer research. Historians and advocates usually trace the beginnings of the patient advocacy movement back to the activities of a breast surgery patient, Terese Lasser, who in 1952 began to offer her services (occasionally unsolicited) to women recovering from breast cancer surgery (Collyar, 2005; Timothy, 1980; Lerner, 2001). Despite initial resistance, Lasser developed her activities into a program called Reach to Recovery (R2R) that gained such widespread international popularity that in 1969 the American Cancer Society made it its own (Boehmer, 2000).

Instituted on a worldwide basis, the R2R program is not without its critics, particularly since the 1990s when a renewed activism among breast cancer patients became sharply critical of the normative nature of several components of the program. Some breast cancer activists are consequently skeptical of the R2R movement in its present form and have gone so far as to characterize it as an "object of ridicule and anger among women with breast cancer" (Batt, 1996: 237). Whereas an early activist like Lasser focused on the consequences of cancer for the personal life of individual patients, the new generation of activists have shifted their attention to treatments: the choice of specific drugs and regimens, the enrollment of patients in clinical trials, and more importantly, research topics previously deemed the exclusive domain of professionals. One of the characteristics of new patient advocacy groups is that they tend to blur the distinction between lay and experts through the establishment of *hybrid forums* and *hybrid research collectives* (Epstein, 1996; Callon, Lascoumes & Barthe, 2009; Callon & Rabeharisoa, 2008). Not only are patients represented in institutions such as DMCs or the scientific committees of clinical trial organizations, but they also co-organize research meetings such as the European Breast Cancer Conferences where all stakeholders (patient representatives, clinicians, health professionals, and scientists) mingle and debate.

Patient advocates entered the cooperative groups at the beginning of the 1990s in the form of patient representatives. At ECOG, for example, the group charter was modified in 1993 to attach a Patient's Representative Committee to the Group Chair. In addition, ECOG added patient representatives to the disease committees, thus allowing them to participate in protocol selection, review, and evaluation.[8] CALGB and other cooperative groups followed suit in the mid-1990s choosing, however, to name their patient representatives *patient advocates* (Schilsky, 1997). As for the NCI, it undertook parallel efforts in the early 1990s to draw consumers and advocates into the decision-making and evaluative process. In 1993, following a survey of a number of advocacy groups, the National Cancer Advisory Board suggested that the NCI establish channels of communication with advocacy groups and seek ways to incorporate group representatives into the NCI. Since then the NCI has carried out a number of programs to include members of cancer advocacy groups beginning with the establishment of an Office of Liaison Activities in 1996 and the NCI Director's Consumer Liaison Group, a group of sixteen consumer advocates, in 1997.[9]

The NCI has established a number of other structures to promote lay participation in its activities. The CARRA program, set up in 2000, has trained approximately two hundred members of advocacy groups to sit on progress review groups that assess research needs for specific cancers, to participate in peer-review grant committees, as well as to evaluate patient-oriented materials. In addition, NCI staff may request the participation of advocates in scientific meetings, site visits, and workshops organized by the NCI. Between the inception of the program in 2001 and 2008, NCI staff made approximately 750 such requests for advocate participation. By the beginning of the twenty-first century, patient advocates had gained positions at all levels of the clinical trials process: study design, steering committees, institutional review boards, and DMCs.[10] Patient advocates have also been integrated into the NCI's preclinical translational research program known as SPOREs (Specialized Programs of Research Excellence), launched in 1992.[11] In an attempt to bridge the fundamental/clinical divide, the SPOREs also conduct clinical trials to study new technologies and to explore significant genetic pathways. Patient advocates are more than passive onlookers: as members

8. Eastern Cooperative Oncology Group, "Chairman's Report," 8, and "Operations Office," 6, in *Progress Report, 1989–1993*, vol., 1, *Overview and Central Resources*.

9. At http://dclg.cancer.gov.

10. At http://carra.cancer.gov/about/whatiscarra/history.

11. Andrew C. von Eschenbach, "SPOREs: A Force in Translational Research," Director's Update: 13 July 2004, http://www.nci.nih.gov/directorscorner/directorsupdate.

of the Patient Advocate/Research Team (PART) program, they actively participate in a number of substantive ways.

In many instances, patients groups have become their own sources of information. The National Breast Cancer Coalition (NBCC) has taken the process of engaging knowledge production one step further with its Clinical Trials Initiative. Their three-pronged strategy begins with training seminars for advocates in basic biology, epidemiology, and clinical trials methodology. Graduates go on to advanced study in a number of topics, including protocol formulation in breast cancer trials, overviews of drug development, and the analysis of end points and interim results. The second prong—public policy—covers a number of legislative initiatives concerning clinical trials. In 2005, for example, the NBCC targeted the Fair Access to Clinical Trials Act (FACT) as one of its legislative priorities. Although it did not become law, the act would have required that all clinical trials for all life-threatening diseases be made publicly available through a public data bank, including those trials whose results were not published in a journal. Finally, on the scientific collaboration front, the NBCC Foundation (NBCCF) has its own criteria for participation in a clinical trial: "Once a trial meets NBCCF criteria, advocates work alongside scientific and medical professionals in every step of the trial process" (Visco, 2007: 152).

In Europe, organizations similar to the NBCC, such as the aforementioned breast cancer group Europa Donna, emerged in the early 1990s. ED rapidly evolved into a metaorganization consisting of forty-two autonomous national groups and established partnerships with clinical cancer organizations such as the EORTC (Buchanan et al., 2004a, 2004b; Steinle, 1996). In turn these organizations have become members of a number of higher-level umbrella organizations such as the European Cancer Patient Coalition. While oncologists and their organizations may have initially questioned the motives and agenda of patient advocates (Baum, 1997), clinical researchers have since learned to actively seek patient participation, and professional bodies have established multiple interfaces with patient groups on both sides of the Atlantic. In 2002 the European Society for Medical Oncology added a patient seminar to their annual meeting and created a cancer patient working group designed to lead from patient-physician communication to patient-physician partnership (Mellstedt, 2006). Partnership and active involvement are now common tropes when professionals discuss patient involvement in research, in Europe as well as in the United States (Collyar, 2005). In return, patient advocacy groups have actively promoted measures to increase the enrollment of patients in cancer clinical trials (Sinha, 2007; Cauhan & Eppard, 2004).

State agencies in Europe have also multiplied patient interfaces. The creation of the European Medicine Agency in 1995 was greeted not simply as a way of promoting innovation in the pharmaceutical sector but also as a means of streamlining discussions between regulatory authorities and patient representatives from the different European countries (Houÿez, 2004). The European Commission has since supported the establishment of the European Cancer Patient Coalition in 2003 and the European Patients' Forum in 2004 as a representative body to which the commission could turn for advice (Rice, 2004a; 2004b).

More controversial are the relations between pharmaceutical companies and patient groups, whereby the latter, by pressuring for the release of new drugs, act as innovators *and* as promoters of a company's product, as was the case with Gleevec (Kent, 2007; Mintzes, 2007). It would be naive, however, to describe patient groups at the beginning of the new century as victims of manipulation by medical practitioners and/or pharmaceutical companies. The 2005 and 2006 editions of the "Masterclass in Cancer Patient Advocacy" organized by the European Cancer Patient Coalition, for instance, included, in addition to sessions on the problems of discrimination faced by cancer patients, teaching periods covering access to cancer drugs and trials, and, in particular, the issue of patient group–pharmaceutical company relations.[12] In other words, cancer patient groups, as in the case with other disease groups (Rabeharisoa, 2006), are presently engaged in a form of collective mobilization that can be best characterized as one in which the group, far from simply representing the interests of patients and their families or yielding to commercial strategies, mediates between different "stakeholders" and institutions.

Conclusion

During the last two decades, the new style of practice analyzed in this book underwent major transformations both in terms of drugs and regi-

---

12. At http://www.cancerworld.org/CancerWorld/home.aspx?id_stato=1&id_sito=9.

mens following the translation of molecular biology into targeted treatment and in terms of the trial machinery as exemplified by the rise of DMCs. In addition, oncologists and patients were also transformed. The success of the new style of practice generated a large number of clinical and medical oncologists, who have in turn resorted to more traditional forms of professional organization, thus transforming the medical division of labor and increasing the institutional heterogeneity of the centers that provide cancer care and treatment. There is no cancer treatment without patients; patients, however, once acted largely as passive bodies on which trialists intervened. They have since become informed research subjects; education and information are now the sine qua non for their participation in clinical trials. Beyond that, through their organizations they have become research partners, represented on the scientific committee of clinical trial networks. In addition to these new actors, clinical trial protocols have been fundamentally transformed by the molecular biology turn. Chapter 11 thus examines how both cooperative groups and the NCI attempted to incorporate molecular biology into their research programs and created new forms of clinical trials to deal with the transformed nature of the therapeutic substances issued from molecular biology.

CHAPTER ELEVEN

Targeted Therapy and Clinical Cancer Research

Introduction: Molecular Biology and Translational Research

In the late 1990s, faced with tightening finances and a changing bio-medical landscape, the EORTC underwent yet another reform intended to refocus the organization on its core mission, "the testing of new therapeutic regimens . . . through the execution of large, prospective, randomized, multicenter, cancer clinical trials."[1] As shown in figure 11.1, the reform structured the group's activities around the notion of *translational research* and a "network of scientists and clinical investigators" that spanned laboratory and clinical research.

Translational research can be defined in a number of ways, not all of which are convergent. Some suggest that the practice entails the unidirectional transfer of tools and concepts from the laboratory to the clinic. Others present the activity as a two-way traffic between bench and bedside wherein clinical insights serve as starting points for laboratory investigations just as the results of scientific investigations serve as the immediate basis for clinical interventions. The second view has been enshrined in an NCI Translational Research Working Group definition established in 2005: "Translational research transforms scientific discoveries arising from laboratory, clinical, or population studies into clinical applications to reduce cancer incidence, morbidity, and mortality" (Brown, 2007: 11). Regardless of its exact meaning, translational

---

1. F. Meunier, "The challenges and opportunities of cancer clinical research: The EORTC perspectives (1995)," EORTC Archives (Brussels).

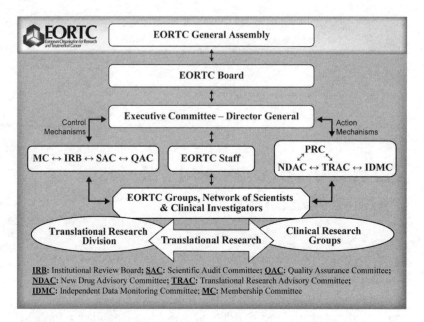

FIGURE 11.1. The EORTC structure in 2009–10. This updated version is similar to the slightly simpler diagram originally published in Meunier & van Oosterom (2002: S6). Courtesy of Dr. Françoise Meunier.

research inevitably requires clinical trials "to incorporate the to-ing and fro-ing between lab and patient" (12), a requirement that increases the intricacy of the rococo machinery of clinical studies.

The catchphrase *translational research* first appeared in the early 1990s at the NCI and then in the new century spread quickly from the field of cancer to other biomedical domains. Its early use described work conducted on cancer susceptibility genes such as the breast cancer *BRCA* genes and more generally on oncogenes and their possible clinical implications (Butler, 2008). More than a mere descriptor, the NCI subsequently embedded the notion of translational research in programs such as the Specialized Programs of Research Excellence (SPOREs) created in 1992 and the Translational Research Initiative established in 2001. Translational research gained further traction in the 1990s as molecular biologists and oncogene researchers came to occupy key positions in the biomedical establishment. Following his 1989 Nobel Prize, Harold Varmus, for example, went on to become head of the NIH in 1993. By then the NIH leadership had to face the fact that the research pendulum had swung away from clinical research to basic research in biomedicine, largely because of the massive funding of molecular biology. The success of molecular biology in capturing research funds raised two issues. First, the rationale for the public funding of the NIH—to find cures for diseases—had been undermined by the emphasis on basic research, and the NIH, according to its critics, was now clearly producing knowledge for the sake of knowledge. Second, the turn to basic research had shifted research personnel from the traditional physician-scientist to highly specialized researchers motivated by a reward system based on individual, investigator-initiated grants and publications. Numerous observers of the biomedical scene warned of the impending demise of clinical research and of the physician-scientist who, in the 1950s and 1960s, had conducted clinical research while treating patients (Nathan, 2002). Armed with the new "mantra 'bench to bedside and back,'"[2] the NIH and other organizations such as the Institute of Medicine of the National Academy of Science issued a number of reports (Kelley & Randolph, 1994) and offered a number of solutions, the most wide-ranging of which was the 2003 "NIH roadmap" (Zerhouni 2003; 2005), an ambitious project to reengineer the clinical research enterprise through translational research.

Since the turn of the century, and in keeping with its reorganization of the late 1990s, the EORTC has also sought to promote translational research through the creation of institutions such as a Translational Research Advisory Committee, a Translational Research Unit, and a Translational Research Fund aiming at increasing cooperative groups' integration.[3] In addition to these internal adjustments, other initiatives have enrolled outside organizations such as cancer research institutes. While the EORTC consists mainly of disease-oriented cooperative groups, it now supplements the groups with a translational research Network of Core Institutions (NOCI) that cuts across the groups in order to further cooperation between scientists who work in "large institutions with patient-accruing potential as well as sophisticated laboratory facilities."[4]

2. Eric Schaffer, "Translating translational research," *The NIH Catalyst* (July–August 2008), available at http://www.nih.gov/catalyst/2008/08.07.01/page01_translational.html.

3. Minutes of the EORTC Board Meeting, 27 March 2003; Minutes of the EORTC General Assembly Meeting, 28 March 2003, EORTC Archives (Brussels); see also Lehmann et al. (2003).

4. F. Meunier, "Translational research: Time for action," *Hospital Management* (2006), available at: http://www.hospitalmanagement.net/features/feature751/.

Translational research is but the most recent organizational expression of the ongoing molecular biology turn and the concomitant restructuring and retooling of clinical oncology with the products of molecular genetics and genomics. The process did not take place overnight; it began in the 1980s even though one of the earliest products of this revolution, Gleevec, did not go to clinical trial until 1998. Reforms at all levels of the clinical trial process, from the procurement and sourcing of drugs to the Phase III trial, produced new systems of clinical research on both sides of the Atlantic. This chapter retraces the major turning points in this ongoing process and examines the consequences of the oncogene revolution for the conduct of clinical cancer trials. We begin with the introduction of molecular biology into the American cooperative groups.

Molecular Biology and the US Cooperative Groups

In 1982, the year cytogeneticists reported the first human oncogenes at a conference in Chicago, ECOG held an investigators' retreat. In a section titled "New science as it interfaces with clinical trials," retreat organizers invited a number of scientists "to give presentations on potential areas of interest for ECOG."[5] Those areas, it turned out, did not include molecular biology, cytogenetics, or oncogenes. Instead, immunotherapy, immunology, and biological response modifiers dominated the agenda. Following the Chicago conference, molecular biology moved swiftly to center stage. The next year the ECOG education committee invited the codeveloper of the first version of the oncogene theory, George Todaro, to give a talk titled "The Oncogene Today."[6]

Although numerous oncogenes had been identified by 1984, their clinical significance remained unknown in part because studies carried out in animals or human cell lines did not tally easily with studies conducted on human tumor samples (Wong et al., 1986). A significant threshold in the clinical acceptance of oncogenes was breeched in 1987 when a public/private consortium of researchers at the biotech company Genentech and two universities (UCLA and the University of Texas) established a clinical correlation between a proto-oncogene named HER-2/neu

5. Eastern Cooperative Oncology Group, *Progress Report, 1980–1984*, 9–10, Frontier Science and Technology Research Foundation Archives (Boston).
6. Ibid., 178.

and breast cancer: the greater the number of copies of the HER gene in tumor samples, the worse the prognosis (Slamon et al., 1987). ECOG quickly adopted the technique of counting gene copies in one of their first clinical studies to look at the role of oncogenes in cancer.[7]

The members of the cooperative groups most immediately hit by the molecular biology revolution were the pathology committees whose primary function to this point had been quality control. In ECOG, for example, prior to 1988 a pathology committee reviewed pathology slides in approximately 60% of trials conducted by the cooperative group. Far removed from the front lines of research, it constituted "the bulk of the pathology activity in ECOG" (Gilchrist et al., 1988: 862). A major review in 1988 showed, however, that the quality control made little difference to the final statistical analysis. Most trials had 90% concordance rates between the local pathologists' initial diagnosis and the group pathologists' review. ECOG thus abandoned central review and dispersed the pathologists among the disease committees. In these committees the pathologists turned from quality control to scientific studies that included not only staying abreast of the continuing expansion of immunology and the techniques of immunohistochemistry but also adapting to the tests and hypotheses issued from molecular biology laboratories.[8]

The reorientation of the pathology committees formed part of the group's response to the criticism that there was a "lack of science in ECOG." Such criticisms were not new. The NCI cancer clinical investigations review committee that had assessed ECOG's performance during the first half of the 1980s had concluded that the group had innovated little in cancer research except in the conduct of standard trials. In its response ECOG instanced an increasing number of correlative studies that used immunological markers.[9] But the NCI continued to press ECOG to establish closer links with university scientists and biological research, suggesting that "there is ample (and largely unrealized) potential [within the groups] for performing correlative scientific studies."[10] So in 1991 ECOG established a laboratory science committee mandated with supporting laboratory research and establishing consortia

7. Ibid.

8. Ibid., 8.

9. Ibid., 7–9.

10. Associate Director for Cancer Therapy Evaluation (Michael Friedman), "Summary Report," Division of Cancer Treatment, *Annual Report*, 1 October 1988–30 September 1989, vol. 2, 615, NCI Archives.

implicating ECOG and "nationally recognized laboratories at ECOG institutions."[11]

The laboratory science committee had been part of a larger, scientific renewal of ECOG initiated in the fall of 1989 when the head of ECOG appointed a strategic planning committee to reflect upon the future of the group. The committee admitted that not only was there "a perception that ECOG has not been fast enough to incorporate laboratory/basic science correlative projects into its clinical trials,"[12] but also that "principal investigators of some major ECOG institutions have not fulfilled their leadership role of bringing new science and new scientists expeditiously into the Group."[13] These observations took place within the growing realization that not only had industry now taken the lead in the development of new anticancer compounds, but also that "the scientific revolution at the molecular level . . . presents the challenge of introducing basic science into our clinical trials."[14]

In order to take advantage of the molecular revolution, the committee proposed a number of new directions. First, ECOG made laboratory studies a priority by reducing clinical trials by 25% and establishing a scientific advisory committee to set scientific goals for the group. In addition, ECOG created the aforementioned laboratory science committee that set about funding a number of laboratory projects concentrating on "new studies which emphasize, but are not limited to, new opportunities presented by advances in molecular biology." The new studies differed from previous laboratory studies by shifting the emphasis from natural history to mechanisms. The change in emphasis was clear to organizers:

> historically, the scientific activities of the group have generally revolved around descriptive studies often including various immunohistochemical associations relating to disease stage or prognosis. It has been anticipated that the activities of the Laboratory Science committee and its members would lead to an increasing presence of more hypothesis-driven research. Such research would rely on the resource of the group to provide the numbers of

11. Eastern Cooperative Oncology Group, *Progress Report, 1989–1993*, vol. 1, Overview and Central Resources, Chairman's Report, 4, Frontier Science and Technology Research Foundation Archives (Boston).

12. Ibid.; Strategic Planning Report, October 1990, Attachment A, 8.

13. Strategic Planning Report, October 1990, Attachment A, 10.

14. Ibid., 1.

observations required to test clinically relevant hypotheses in a peer review framework.[15]

Other cooperative groups followed a similar trajectory. CALGB's correlative science committee, instituted in 1975, had begun as an immunotherapy committee that studied BCG immunotherapy. In 1979, with the advent of new immunological methods to detect markers at the surface of cells (Keating & Cambrosio, 2003) the committee became the immunology committee that expanded in 1982 to include a cytogenetics section charged with tracking chromosome anomalies, principally in leukemia. Studies conducted by the committee members remained, however, primarily descriptive throughout the 1980s.[16] At the beginning of the 1990s, following ECOG's lead, interest turned to molecular markers and the molecular basis of the mechanisms of leukemia (Bloomfield, Mrozek & Caligiuri, 2006).

In the course of reorienting their work toward molecular biology in the first half of the 1990s, the cooperative groups faced continual pressure to adapt to an environment that often expressed contradictory tendencies. In 1995, for example, the director of CTEP, Michael Freidman, described the opposing forces the groups would have to confront. On the one hand, in the new basic science environment clinical cancer trials would be subject to increased scientific scrutiny: "Things are going to be science-based. Are you asking clear [research] questions, are the answers useful?" On the other hand, in a context of reduced resources, the groups might be forced to package their trials as more concerned with the evaluation of therapy than science. Freidman suggested, for example, that the groups try to convince insurance companies that clinical cancer trials constituted a form of administrative evaluation known as "outcomes research" that would have the added advantage that "if trials are kept cheap and simple, you can show that trials don't cost much or maybe even less than standard care" (Cooperative groups advised . . . , 1995: 2).[17]

15. Ibid., vol. 3, Modality and Standing Committee Reports, Laboratory Science Committee, 4.

16. Immunology and Cytogenetics Committee, CALGB, *Scientific and Administrative Report, 1981–1989*, pp. F-1–F-90, James Holland personal papers.

17. From an industry point of view, cooperative group trials were and remain a bargain. Patient reimbursement rates have remained frozen at $2,000 per patient since 1999, whereas industry can pay as much as $15,000 per patient (Patlak, Nass & Micheel, 2009).

In addition to pressure from CTEP, the groups also confronted a new director of the NCI. The nomination of Richard Klausner, a molecular biologist, as director of the NCI in 1995 signaled a final and definitive turn toward molecular biology. As a bench scientist, Klausner was aware that his lack of clinical experience put him at a disadvantage in the eyes of some NCI workers. As he pointed out, it would have been impossible for an individual to represent all the interests invested in the NCI (White House to appoint . . . , 1995: 1). In keeping with his status as new director and giving voice to the growing chorus of complaints surrounding the cooperative groups, Klausner undertook a major review of the entire clinical trials program. Convened in 1996, the Clinical Trials Working Group called for a major restructuring of clinical cancer trials in the United States in its report tabled in 1997.[18]

Uppermost on the group's agenda was the accelerated development of the molecular biology of cancer that had provided a plethora of agents and targets and thus required a clinical trial system capable of producing a significantly higher throughput. Between 1991 and 2001, the yearly production of anticancer agents had increased fourfold, rising from one hundred to four hundred. The number of companies producing anticancer agents had similarly risen from forty-five to seventy (National Cancer Institute, 2003). By 2003, 88% of Phase III cancer trials involved studies of molecularly targeted therapies as opposed to conventional cytotoxic therapies (Roberts, Lynch & Chabner, 2003). A single agent could, moreover, become the source of many trials, as we saw in the case of Gleevec that had by 2002 generated forty-four different trials in a variety of cancers ranging from prostate cancer to breast cancer and gastrointestinal stromal tumors (GIST), all of which shared common molecular abnormalities.

In its fall 1998 report, the implementation committee charged with the realization of the recommended reforms[19] redistributed the control of clinical cancer trials from the "troika" mentioned in chapter 9 to a larger, yet even more centralized, research and clinical base. The com-

18. The group was named the Armitage group after its director. It consisted of twenty-eight members drawn from the NCI's Board of Scientific Advisors, the National Cancer Advisory Board, directors of cancer centers, and patient advocates. See Working group to review . . . , (1996).

19. J. H. Glick & M. C. Christian (Co-Chairs). (1998). *Report of the National Cancer Institute Clinical Trials Implementation Committee Presented to the NCI Board of Scientific Advisors* (September 23), NCI Archives.

mittee walked a fine line between therapy evaluation and research. In order to produce more relevant hypotheses, it proposed state-of-the-science meetings to generate ideas for Phase III trials. Whereas the groups themselves had previously generated trial concepts, the science meetings would now include scientists (clinical and basic) from university and industry as well as patient advocates. Furthermore, the committee proposed the formation of disease-specific concept review committees that would similarly include outside investigators and members of the advocacy community (NCI overhauls . . . , 1999). Finally, a third organizational innovation aimed at allowing the entire clinical trials process to unfold outside the cooperative groups. Termed a national Network of Physicians, it authorized all physicians—not just members of cooperative groups—to enroll a patient in any clinical cancer trial sponsored by the NCI. To bypass the administrative structure of the groups, a cancer trials support unit would register patients, credential institutions and individuals, and collect data on standardized forms. The unit itself would be contracted out to the private sector, and while ostensibly relieving the groups of administrative duties when unaffiliated physicians put patients on trials, it would also diminish the importance of the groups in their conduct.

By March of 2002, concept evaluation panels that included members from outside the NCI had been created to review Phase III protocols and to streamline concept review and protocol development.[20] The contract for the Cancer Trials Support Unit had been awarded to the Westat Corporation of Rockville, Maryland, in 1999 (NCI funds . . . , 1999) and the first patients enrolled in 2000. After two years of discussion, a Central Institutional Review Board that substituted for individual IRB's was finally up and running and had approved nineteen protocols.[21] Five different State of the Science meetings had been held that included patient advocates as participants. While all these measures initially appeared to some as a way of marginalizing the cooperative groups, the groups emerged from the exercise reinforced. The overall cooperative group budget increased by 60% ($58 million) in the years immediately following the review committee report,[22] and far from reducing clinical trials

20. Board of Scientific Advisors, Meeting Minutes, 25–26 March 2002. Available at http://deainfo.nci.nih.gov/advisory/bsa/bsa0302/0302min.pdf.

21. Ibid.

22. Ibid.

to a form of therapy testing, the NCI continued to flaunt the National Clinical Trials Program operated by the cooperative groups as a "laboratory without walls" (National Cancer Institute, 2003: 29). Over half of the treatment trials initiated by the groups in the year 2000 contained correlative laboratory studies, prompting a participant to claim that the "beauty of Cooperative Group trials is that they allow examination of epidemiology, etiology, and molecular [sic] at the same time they seek to determine which therapies are best."[23]

The Reorganization of the Drug Discovery Process and the Transformation of the NCI Screen

In the 1980s and 1990s, the NCI vigorously pursued research in the area of tyrosine kinase blockers. As we know, however, Novartis developed Gleevec. Had the NCI thus missed the boat in the development of this class of anticancer agents? Not quite. Like Novartis, the NCI saw the early promise of oncogenes and reacted promptly to the flood of tools and concepts produced by molecular biology laboratories around the world. Unlike Novartis, the NCI did not get lucky, for even in the age of rational therapy, empirical problems continue to bedevil researchers.

The NCI began acquiring biological, as opposed to its existing chemical, expertise just prior to the oncogene revolution at the end of the 1970s when it set up a program to screen and test biological response modifiers (BRMs); that is, substances—such as the interferon used to treat CML before Gleevec became available—produced in small amounts by the body and reputed to stimulate the body's response to disease. Although run by the division of cancer treatment, the Biological Response Modifiers Program (BRMP) had its own screening and experimental programs (Oldham & Smalley, 1983). The program also fostered new designs of early stage (Phase I) clinical trials tailored to the body's distinctive reactions to these agents.[24] In addition to academic and in-house sources of biological molecules, the BRMP quickly established relationships with a number of pharmaceutical companies, which supplied immuno-

23. J. M. Bennett & P. Valent, "Breakout Session A: Etiology/Molecular Biology/ Pathology," at http://www.webtie.org/sots/Meetings/Leukemia/04-29-2002/transcripts/02/ transcripts.htm.

24. Associate Director for the Biological Response Modifiers Program, "Summary Report," in Division of Cancer Treatment, *Annual Report, 1984–1985*, 847, NCI Archives.

logical substances for testing. The group, which initially consisted of a select number of companies, quickly grew as more and more pharmaceutical and biotech companies moved into the field of cancer. By 1993, the BRMP could claim that it had "established relationships with most of the [more than forty] biotechnology and pharmaceutical companies which produce BRMs."[25]

The BRMP had been the first to go molecular when in 1985, following the initial turn to immunology and shortly after the discovery of the first human oncogenes, it set out to put the oncogene theory to therapeutic use. While the NCI had the necessary experimental expertise, the translation of that knowledge into therapy required profoundly different resources. After issuing a request for research proposals on "the use of oncogene related products for cancer therapy," the program opened its own laboratory of biochemical physiology devoted to the study of oncogenes. To head the lab, the NCI lured a senior researcher away from the Department of Molecular Genetics and Molecular Oncology of the Swiss multinational Hoffmann–La Roche.[26] Despite these intentions and although the NCI laboratory studied the mechanisms of oncogene activation and the signaling pathways in which the oncogene products participated, it did not immediately seek to produce therapeutics in this area.[27] As a consequence the first tyrosine kinase blockers entered the NCI through the back door of a public/private collaboration in the early 1980s, when the NCI initiated funding for the National Cooperative Drug Discovery Groups (Hallock & Cragg, 2003). Intended to bring together researchers from academia, industry, and government, the original call for proposals asserted that "since it is clear that few single institutions possess a critical mass of all the varied talents for effective drug discovery, a new instrument that permits the combination of the available expertise from diverse institutions is required" (Participants sought . . . , 1982: 6–7).

Among the earliest groups funded was a team that brought together Michael Bishop, one of the pioneers of oncogene research, members of the Northern California Cancer Center, and industrial researchers at Bristol-Myers. The group planned to synthesize tyrosine kinase inhibitors, and the NCI involvement consisted of screening the candidate

25. Associate Director for the Biological Response Modifiers Program, "Summary Report," in Division of Cancer Treatment, Annual Report, 1993–1994, 109, NCI Archives.

26. At http://nstl1.nstl.gov.cn/pages/2008/191/95/7(9).pdf.

27. Laboratory of Biochemical Physiology, "Summary Report, 1993–1994," in Division of Cancer Treatment, *Annual Report*, 223–31, NCI Archives.

compounds. The group thus entered the field well before Ciba-Geigy, even though it ultimately failed to bring a tyrosine kinase blocker to market.[28] In the twenty years that followed, the NCI funded sixteen similar groups that directly produced four new substances and indirectly participated in the production of twenty-one others.[29] By 2000 the annual budget had risen sixfold to $120 million.[30] Considered extremely successful, the program served as a model of public/private partnerships for other institutes at the NIH. In addition to establishing interfaces between university and industry researchers and the NCI, the program created new kinds of collective resources such as libraries of natural compounds.

The EORTC approached the public/private divide in a somewhat different manner. Readers will recall that the EORTC had established a New Drug Development Office (NDDO) in the 1980s to handle the early testing of novel molecules. Although the NDDO conducted its activities under the official EORTC label, the new organization quickly became a "state within the state," with separate offices in Amsterdam—dubbed "the European capital of new drugs" (New optimism . . . , 1998: 789)— rather than at the EORTC headquarters in Brussels. The secessionist tendencies intensified following the sharp increase in the number of new molecules generated by targeted therapy approaches, and the EORTC and the NDDO ultimately parted ways in 1998 following the emergence of "different approaches to the philosophy of their respective activities" (NDDO splits . . . , 1998: 1145–46). The NDDO viewed the EORTC as just one of many cooperative groups and commercial institutions in the European cancer field and argued that there was thus no reason for an exclusive relationship. The EORTC, in contrast, framed the issue as one of independent versus commercial research.

Commercial research grew significantly in the 1990s as pharmaceutical firms conducted an increasing number of cancer clinical trials, often resorting to the services of contract research organizations (CROs); that is, commercial firms specialized in this domain (Mirowski & Van Horn, 2005). In 2005 a CRO called INC Research bought NDDO Oncology,

28. "Annual Report of the Drug Evaluation Branch," in DCT *Annual Report, 1984– 1985*, 101–7; 104, NCI Archives.

29. At http://dtp.nci.nih.gov/branches/gcob/gcob_web3.html.

30. Request For Applications (RFA), National Cooperative Drug Discovery Groups, Release Date: 14 April 1999. Available at http://grants.nih.gov/grants/guide/rfa-files/RFA-CA-99–010.html. This RFA is a reissue of CA-95–020, published on 22 September 1995 and available at http://grants.nih.gov/grants/guide/rfa-files/RFA-CA-95–020.html.

therefore completing its transition to the commercial domain. The European field has undergone further restructuring through, for instance, the creation of the Southern Europe New Drug Organization (SENDO) (A new drug office . . . , 1996). More recently, the EORTC has established a New Drug Advisory Committee (NDAC) whose role, in addition to providing an interface with industry, is to channel new drugs into the organization's clinical trial system, to establish priorities, and to make recommendations for all EORTC cooperative groups.[31] In so doing the EORTC has repositioned itself vis-à-vis the growing commercial involvement in cancer research by defining its role "as an academic research organization (ARO) as opposed to a contract research organization (CRO)": an ARO should privilege "pivotal studies" addressing "fundamental issues for public health and patient management . . . [whose] results can be published in high cited journals and [whose] outcome will influence oncological health care,"[32] and even when collaborating with industry, insist on the necessity to retain "control over design of the study and key parameters of the data bases, and the interim and final analyses as well as right to publish" (Saul, 2008: 320).

The search for new molecules had repercussions not only at the institutional level. It also prompted a thoroughgoing revision of the informational enrichment process to which candidate compounds were submitted. Specifically, the NCI redesigned its drug screening system by shifting away from whole animals to focus on the definition of the molecular nature of the compound tested and the specific kind of cancer. The movement culminated in 2000 when, after forty-five years of service, the NCI scrapped its linear array and the associated decision network committee. Some features of the old system remain nonetheless unchanged. The NCI continues to source substances from the pre-2000 network and move them through a series of predefined stages similar to the ones outlined in part 2 of this book. Likewise, a central committee continues to follow the substances through the various testing stages, to evaluate evidence, and to decide whether or not to invest further in individual compounds. Despite these continuities, the changes remain considerable and

31. EPOD: The Early Project Optimization Department, *EORTC Headquarters Newsletter*, 11 March 2009, available at http://www.eortc.eu/Newsletter/March%202009/ Default.html.

32. Reflections from Professor van Oosterom and Françoise Meunier about the core mission of the EORTC, 20 October 1995, EORTC Archives (Brussels).

the number of substances tested has dropped 90% from 40,000 a year to 4,000.

To make the program more responsive to the surrounding research environment, a committee known as the Drug Development Group now seeks extramural guidance and review. Substances are prioritized on the basis of the originator's proposal and scored according to "credentials, novelty, cost and benefits and need for NCI involvement" (Reynolds, 2000: 1555). Once in the system, candidate compounds are processed through an in vitro stage that uses three different human cancer cell lines, followed by a second in vitro stage using sixty different human cancer cell lines, and by an in vivo stage consisting of mice bearing transplanted human cancers (Alley, Hollingshead & Dykes, 2004; Hollingshead et al., 2004). As can be seen, the most important change in the screening system was the substitution of the test tubes for mice in the first stages of the process. Critics had pointed out repeatedly over the years that animal tumors did not grow like human tumors (Chabner, 1990). As cell culture methods improved, screeners agreed that sooner or later the human would replace the animal as a model, and that subsequently the in vitro would replace the in vivo. Work on the in vitro screen began in 1990 (Monks et al., 1991), and it took six years to complete.

More than just a cost-cutting exercise, the in vitro screen constituted a radical departure from previous practices. The NCI expected an in vitro screen to sort candidate compounds according to molecular variables such as growth-factor dependence, oncogenes, suppressor genes, drug resistance factors, and drug inactivating enzymes. The data would allow researchers to choose compounds on the basis of their ability to act on subcellular mechanisms and not simply to kill cells and reduce tumor size. The new approach also shifted the emphasis of screening from the substance screened to the disease. By testing the differential sensitivity of a compound against sixty different human cancers, the in vitro screen now did the work of determining which tumors were most sensitive. The new set-up simultaneously acted as an experimental system by allowing screeners to follow a single active compound across a pattern of sixty different reactions, possibly leading to experimental insights.

The overhauling of the NCI's drug discovery program transformed the drug *screening* into a program of drug *research*. Overall, through the incessant tightening of protocols, a quintessentially random activity has evolved into a quasi-experimental activity that can be usefully qualified as the functional equivalent of a human clinical trial. The screen remains a screen, and its ability to isolate useful compounds on the basis of molecular rather than cellular criteria still rests on an empirical footing. The ultimate proof of the pudding remains in the clinic.

Clinical Trials in a Targeted Age

In order to solve problems raised by the number and the nature of targeted agents, clinical cancer researchers have set about redesigning clinical trials. The problems are numerous and vary according to the type of trial conducted. In Phase I trials, for example, toxicity may not necessarily be an appropriate means of measuring dose size, as in the case of traditional cytotoxic agents. Absent any simple dose/response relationship, establishing an appropriate dose size for cytostatic agents may require a larger number of patients (Korn et al., 2001). Similarly, tumor regression may no longer offer a useful end point for Phase II trials testing cytostatic agents since these agents may stop growth without reducing the size of the tumor (Therasse et al., 2000). Finally, in Phase III trials, given the mechanism of action of the new cytostatic agents, there may simply not be enough time to evaluate all the possible combinations of patients and drugs according to the old system that required thousands of patients, hundreds of millions of dollars, and a decade of research per drug (Schilsky, 2002). The present complexity of this process cannot be underestimated. Experts in operations management have estimated that it now takes approximately 810 steps to open a Phase III trial, implicating as many as thirty-eight different individuals and groups in the decision-making process (Dilts et al., 2006; 2008). With these time delays and the sometimes-baroque additions to the original protocol, it is little wonder that 64% of all Phase III cancer trials sponsored by the NCI between 2000 and 2007 failed to meet their minimum accruals (Patlak, Nass & Micheel, 2009).

Problems such as these have prompted numerous initiatives in the search for solutions. In 2000, for example, the NCI developed a program to assess novel clinical trial end points, and in 2002 the members of EORTC's New Drug Development program called for new forms of clinical trials—new study templates with redesigned study statistics and study performance—in order to deal with the new types of compounds (Reynolds, 2000; Lacombe et al., 2002). Our purpose here is not to lay out the full menu of proposals and projects that have emerged over the last fifteen years. Rather, this section seeks to give the reader a sense of how targeted therapy is changing clinical cancer trials as a system, for, as just noted, all phases of clinical trials have been affected. In order to do so, we first consider the most striking reform: the creation of a novel phase of clinical research that lies between the laboratory and Phase I trials: Phase 0.

Phase 0 trials are also known as microdosing studies: subtherapeutic doses of an experimental drug are administered to gather data on the drug's behavior in the human body, in particular to assess whether the drug has had an impact on its intended molecular target (Bates, 2008). As described in an FDA (2006) guidance, the new category emerged as an attempt to pry open the bottleneck located between the isolation of a promising compound in a screening system and its first use in human beings in a Phase I trial. Given that the vast majority of compounds are expected to fail somewhere along the Phase I–III path—some estimates go as high as 90%—research administrators seek ways to eliminate drugs earlier rather than later in the process in order to save both time and money (Kola & Landis, 2004). The use of subclinical doses minimizes the danger of adverse reactions thus eliminating a number of toxicity studies in animals, and laboratory procedures can easily supply the amount of drug needed for the trial, thus bypassing scaling-up processes. The FDA did not set out precise rules on how to proceed with Phase 0 trials. Instead it offered three examples of what a Phase 0 trial might look like and invited investigators to experiment with the paradigm outlined by the three exemplars (Patlak & Nass, 2008). Industry has responded favorably to the FDA initiative, applauding the spirit of flexibility touted in the guidance and displayed on the ground in day-to-day dealings with the FDA (Robinson, 2008). Similar approaches have since been incorporated into international standards regulating clinical trials.[33]

Phase 0 trials raise interesting questions for patients. Since they test drugs at doses well below the level of clinical effects, the trials clearly have no therapeutic intent. What then would motivate patients to enroll in a Phase 0 trial? While altruism might seem to be the obvious answer, the era of patient activism has offered another. According to Deborah

33. See, e.g., "Guidance on Nonclinical Safety Studies for the Conduct of Human Clinical Trials and Marketing Authorization for Pharmaceuticals (Draft). International Conference on Harmonization Consensus Guideline," 15 July 2008, available at: http://www.fda.gov/CDER/guidance/8500dft.htm.

Collyar, the founder of Patient Advocates in Research (Collyar, 2008), Phase 0 trials allow patients "to act as participants instead of research subjects" (Gutierrez & Collyar, 2008: 3689). Other observers, however, maintain that Phase 0 trials are not entirely without danger for participants. In addition to the (relatively small) risk posed by repeated tumor biopsies, a much larger risk concerns the ability of patients to understand that "no therapeutic intent" really means what it says. Experience with Phase I trials shows that even though the latter also have no therapeutic intent, patients often enrolled in the trials believing that there was a possibility that their cancer may be alleviated if not cured (Miller, 2001). In order to avoid all confusion, the NCI Ethics Committee that reviewed one of the Phase 0 protocols insisted that the patient consent forms clearly specify that participation in the study "will be of no therapeutic benefit to [the patient] but may be of benefit to others" (Gutierrez & Collyar, 2008: 3690).

Phase 0 trials are but the most visible change in the clinical trial system. Changes in the other phases are sometimes more subtle. We saw in the previous chapter that the Phase I Gleevec trial failed to establish the dose at which the compound caused toxic effects. In that case researchers proceeded on to Phase II trials using the smallest effective dose, which did prove to be effective and nontoxic (Talpaz et al., 2001). Patients who later developed resistance to the drug sometimes overcame that resistance simply by switching to a larger dose (Kantarjian et al., 2003). Should they have taken a larger dose to begin with? Such a question would not have arisen in the cytotoxic era when Phase I trials often used toxicity as a surrogate for setting effective dose levels. Absent clear toxicity points, what other end points should be used to determine effective dose levels for cytostatic drugs (Sleijfer & Wierner 2008)? In the meantime, failure to establish an ideal dose in Phase I puts further pressure on Phase II trials that are now forced to test for efficacy and to seek the optimal dose.

Studies of alternative end points for Phase I trials (Eisenhauer, 1998; Korn, 2004) require considerably closer contact between the laboratory and the clinic than had the testing of cytotoxic drugs, adding considerable expense to the Phase I trials (Ratain & Glassman, 2007) but also opening up new perspectives for academic research organizations. The research infrastructure needed to conduct Phase I trials is now significantly greater than that needed for Phase II trials despite the larger number of subjects recruited for the latter (Craft et al., 2009). The use of molecular markers as surrogate end points restricts the study to those patients who express the marker, thus limiting patient pools already under stress. The same restriction does not, however, apply to Phase II/III trials where restricting the patient pool to those bearing the marker of interest will most likely enhance the difference in treatment effects and thus reduce the size of the population needed for the study. Clinicians are thus urged to select patients whose tumor cells overexpress the target in question and who are thus more likely to respond to the targeted agent (Sonpavde et al., 2009).

Finding a molecular marker that adequately reflects the anticancer properties of a targeted drug is not as straightforward as it might appear. A 2008 editorial noted: "Despite all major leaps in cellular and molecular biology, there is currently no agent for which a biomarker has been identified that is known to adequately reflect antitumor activity of that particular agent in humans" (Sleijfer & Wierner, 2008: 1577). Even if more biomarkers were available, as the director of the FDA's Center for Drug Evaluation has pointed out, as far as clinical cancer research is concerned: "We don't have a lot of experience linking drugs and diagnostics during development when they're both experimental" (Janet Woodcock, as cited in Hampton, 2006: 1951). Clearly, considerable time and resources will be invested in developing this new form of experiment. Meanwhile, the slow progress in the search for biomarkers for Phase I cancer trials has provided fertile ground for skeptics who have argued that simply confirming that a target has been blocked by using a biomarker is no predictor of clinical outcome (Goulart et al., 2008; Ratain & Glassman, 2007), Indeed, despite a mass of publications, the FDA has approved only a small number of biomarkers since the mid-1990s. Nonetheless, poor results and the accompanying skepticism have yet to dampen the enthusiasm of researchers and research administrators. In 2007 the NCI, the FDA, and the American Association for Cancer Research (AACR) launched the Cancer Biomarker Collaborative (Khleif, 2009; Yu & Veenstra, 2007) to speed movement through the biomarker pipeline by developing guidelines to integrate biomarkers into clinical trials.

Like Phase I trials, Phase II trials of targeted therapies raise issues with regard to the end points measured and to the type of trial conducted. While Phase II trials have traditionally used standard measures of tumor reduction in solid tumors or complete remission in the case of leukemia to gauge therapeutic efficacy (Simon, 1989), targeted therapies tend to produce growth delay rather than growth reduction. In addition, the most effective targeted drugs such as Herceptin, Erbitux, and Avastin did not work well as single agents and produced few clinical responses in Phase II studies (Chabner, 2007). In other words, it may well be the case that unlike Gleevec, most targeted molecules will find their true efficacy in a multidrug regimen and not as a single agent. As a result, Phase II clinicians have likewise been forced to search for alternative ways of assessing the efficacy and therapeutic benefits of targeted molecules. Generally categorized as time-to-event measurements, the new end points have sought to test more global responses to treatment such as disease stabilization, time to progression, and progression-free survival (El-Maraghi & Eisenhauer, 2008). Researchers have consequently called for "more complex Phase II designs with time to progression end points" in order to save drugs that show great promise in the laboratory and Phase I but fail to show up in traditional Phase II trials (Chabner, 2007: 2307). Such trials often require recourse to randomization, a controversial issue in the field of cancer (as we already saw in chapter 7), for unlike other specialties where randomized Phase II trials are the norm, in oncology they remain the exception (Michaelis & Ratain, 2007).

The testing of multidrug regimens in an age of molecular targets raised the problem of which combinations should be tested. While in theory the pinpoint accuracy of targeted drugs should have eliminated much of the guesswork of what a successful combination might be, in practice incomplete knowledge of the mechanisms and molecular pathways of the cancer process has left the door open to considerable empiricism. In 2003, following the failure of a number of single target therapies to show effectiveness in clinical trials, the NCI organized a critical molecular pathways retreat to discuss strategies for the development of successful combinations of targeted drugs. Reaching back to the strategies deployed in combination chemotherapy in previous decades, the NCI launched a screening program to sift through the five thousand plausible combinations of the hundreds of new compounds. Results to date have yielded some successes, many failures, and a few "beneficial combinations that were completely unexpected" (Kummar et al., 2010: 851).

The gold standard for a Phase III trial is survival. This can be measured directly in terms of overall survival, but measuring such an end point entails a long and costly process. So, as in the previous phases, trialists now look to surrogate measures to reduce costs and to speed up study time. Among these measures are disease control, progression-free survival, and time to progression. Which of the latter may be a useful surrogate depends in part upon the mechanism of action of a targeted compound. Some compounds, like Gleevec, do produce a reduction in gastrointestinal tumors; others reduce tumor size in only a subpopulation of patients, whereas still others simply stabilize the disease (Gutierrez, Kummar & Giaccone, 2009). Research is ongoing and varies with tumor type (Buyse et al., 2007; Burzykowski et al., 2008). Even when overall survival is used as an end point, cytostatic therapies do not always measure up. With the exception of Avastin, all of the large trials of cytostatics reported in 2002–3 failed to show an increase in overall survival. While this may be proof that the drugs did not work, there was at least some evidence at the Phase II stage that they did. The fact that they all failed has suggested to some commentators that the problem perhaps lay more with the measured end point than with the drug itself (Millar and Lynch, 2003). The FDA has noted that the lack of positive survival data should not be seen as an obstacle to getting drug approval, since more than 75% of all approvals given since 1990 have been based on alternate data. Gleevec, for example, was approved based on hematologic and cytogenetic evidence, not overall survival rates (Johnson, Williams & Pazdur, 2003).

Phase III trials of targeted therapies have also raised the issue of biomarkers. In cases where overall survival is used as an end point, biomarkers enable researchers to enrich the study population by selecting those patients known to carry the lesion targeted by the drug. European trialists went so far as to suggest that "it is no longer considered ethically acceptable to support studies that do not include a translational research component" (Fricker, 2007: 8). Given the skyrocketing costs of targeted drugs, trials should select only patients who carry the biomarkers that make them likely responders and should use "patient material" only in study designs likely to achieve "maximum output," options possible only within the framework of translational research (Fricker, 2007: 8). Recent studies have shown that targeted agents that failed to show any advantage over traditional cytotoxic therapies in an undifferentiated patient population were clearly superior when used in patients who carried a specific mutation, or vice versa, that patients who carried a mutation had worse outcomes when treated with a targeted agent (e.g., Mok et al., 2009; Gazdar, 2009; Amado et al., 2008). As a result, regulatory agencies such as the FDA may now approve a drug only when a diagnostic test has simultaneously been made available for testing the presence of the relevant marker in patients (Ray, 2009a, 2009c).

More Oncopolitics

As we have observed several times in this book, innovations in one domain have instant repercussions in others. The material discussed in this chapter is no exception. Institutional reverberations of the molecular biology turn can be found in the organizational arrangements surrounding anticancer drug regulation and the cooperative groups themselves.

In the wake of the targeted therapy revolution, the FDA decided in 2004 to establish a new Office of Oncology Drug Products (OODP). Cancer organizations such as the American Association for Cancer Research immediately applauded this decision, which they saw as a "significant step toward speeding the delivery of new drugs for the treatment and prevention of cancer."[34] Observers and protagonists accounted for the FDA's initiative with a variety of factors and motives. The FDA was undeniably under attack by a coalition of conservative critics of state regulation (including editorial writers at the *Wall Street Journal* [FDA to patients . . . , 2002]), the pharmaceutical industry, and some patient advocacy groups, who, taking advantage of the favorable climate created by the Bush presidency, promoted easier and faster drug approval procedures (Goldberg, 2005). Depicted as a bureaucratic body insensitive to patients' needs, the FDA reacted by stating its intention to increase the "transparency" of its decisions and to "provide consistency in the review of oncology products" via the establishment of a streamlined structure such as the OODP (Pazdur, 2005: 612).

Such a purely political explanation is, however, insufficient. The emergence of targeted drugs had clearly transformed the field. As we saw, the testing and approval of these drugs required, among other things, the definition of new end points to replace overall survival, the classical end point for cytotoxic drug testing. The search for alternative end points

---

34. "AACR applauds FDA decision to create new office of oncology drug products," *Science Blog*, July 2004, http://www.scienceblog.com/community/older/2004/1/2004006 .shtml. R. M. Lowe, "New government office seeks to speed cancer drug approvals," *CancerPage*, 16 July 2004, http://www.cancerpage.com/news/article.asp?id=7283.

consequently became a recurring theme of the articles and presentations authored by the OODP director (FDA examining . . . , 2003; Dagher & Pazdur, 2006). Establishing new end points, it was hoped, would expedite drug reviews and keep the regulatory body abreast of current oncology drug trends, such as the development of "companion biomarker tests" to select the subpopulation of patients for which a targeted drug should be expected to work (Pazdur, 2005). The development of targeted compounds that ranged from small molecules (e.g., Gleevec) to large biological compounds such as monoclonal antibodies (e.g., Herceptin) had the additional consequence of challenging the traditional division of labor between the regulators of chemical substances and biological products. The OODP thus houses the previously separate divisions of Drug Oncology Products, Biological Oncology Products, and Medical Imaging Drug Products.[35]

As for the cooperative groups, here is what the former NCI director and Bush nominee John Niederhuber had to say about them shortly before leaving office in 2010: "If you think about where our science is done, it's not in a hotel room in Chicago or San Francisco" (K. B. Goldberg, 2010c: 2). The comment referred to the fact that cooperative group members located in geographically diverse institutions meet regularly in hotels to discuss and plan their work and suggested that the clinical trials system should be based in cancer centers rather than "siloed cooperative groups." The former cooperative group's chairman's reply to Niederhuber was as flippant as the initial comments: "having just spent the weekend in a hotel room in Chicago with 700 of my closest colleagues from CALGB, representing 25 comprehensive cancer centers, I can tell you that is where team science is done in clinical research" (K. B. Goldberg, 2010a: 4). Niederhuber's statement followed the April 2010 release of an Institute of Medicine (IOM) report highly critical of the state of cooperative group operations (K. B. Goldberg 2010b). From a historical vantage point, the IOM document was but the latest in a series of expert panel reports that have recommended changes in the structure and operations of the cooperative groups. In 1997 the Armitage report worried

35. Pamela Holland-Moritz, "Moving biotech products from CBER to CDER: A work in progress," *Pharmaceutical Regulatory Guidance Book* (July 2006), 70–73. Available at http://www.advanstar.com/test/pharmascience/pha-sci_supp-promos/phasci_reg_guidance/articles/Manufacturing7_Moritz_rv.pdf. In the latest turn, the FDA has reorganized the oncology office by *disease type* thus focusing on specific malignancies (P. Goldberg, 2010b).

about a lack of responsiveness by the groups to advances in molecular biology, and in 2005 a working group of the National Cancer Advisory Board asked for important administrative changes. Underlying these events was a longstanding "tension over the control of the clinical trials structure" between cancer centers and cooperative groups (K. B. Goldberg, 2010c: 2) also expressed, in a different guise, in Europe.

Contrary to what one could surmise from Niederhuber's comments, however, the IOM report did not call for the dismantling of the groups but for a consolidation of their operations and for increased funding. Insisting on the necessity of a "strong publicly supported clinical trials systems," the IOM report recommended improvements in the speed and efficiency of trials; the incorporation of innovative science and trial design; improvements in the prioritization, selection, and completion of trials; and more incentives to enroll more patients and physicians in the trials (K. B. Goldberg 2010b). Faced with a decrease in the absolute and relative number of trials sponsored by the NCI, a concomitant, rapid increase in the number of commercially sponsored trials, and warnings that academic oncologists were "losing control" over Phase I trials (Scoggins and Ramsey, 2010; P. Goldberg, 2010a), the IOM defended the existence of a public system on the ground that questions that are important to patients are not necessarily a priority for industry, an argument shared by trialists on both sides of the Atlantic. This is not the place for a detailed discussion of the report's recommendations, but it is important to note that in addition to administrative streamlining, such as the sharing of statistical operations, on which the groups were quick to act (P. Goldberg, 2010c), a number of suggestions addressed issues directly related to the requirements of molecular translational research, chiefly among them new trial designs (made necessary both by the attributes and quantity of novel targeted substances) and repositories of patients' tumor samples (a rare commodity required by the molecular investigation of the targets of the new drugs and the search for biomarkers to guide therapy).

Conclusion

The incorporation of molecular biology, its tools, and the new oncogene theory of carcinogenesis affected the entire system of clinical research from the procurement and screening of substances to the clinical trials that investigated their potential and the regulatory arrangements that surrounded these investigations and substance approvals. We have seen that pressure for change came and continues to come from numerous directions: often the sheer number of targeted agents produced by the public/private sectors served to foster attempts to streamline the organization of clinical trials. In tandem with this onslaught of new agents and the renewed understanding of how cancer emerges and develops was the wholesale movement of private enterprise into the field of clinical cancer research. Promoted by numerous public incentive programs, clinical cancer research has become a hybrid activity of considerable import, and this in turn has had profound repercussions on patients, oncologists, and their organizations.

Conclusion

CHAPTER TWELVE

Looking Back and Looking Forward

We have described the constitution of a new style of practice. At the center of this new practice lies the clinical trial. Our view of the clinical trial and its role in clinical cancer research differs clearly from other analyses of the clinical trial process, some of which eschew epistemological issues and adopt a decidedly accusatory tone (Abadie, 2010; Fisher, 2009; Petryna, 2009). But even those analyses that show a clear understanding of the epistemic challenges (Greene, 2007; Marks, 1997) tend to portray clinical trials as an administrative tactic designed to regulate the safety and efficacy of new drugs or as the empirical icing on a more rational scientific-biological cake. We have shown instead how medical oncology and its core dynamic of clinical research elude such characterizations. Clinical trials that appear merely to register successes and failures have other facets of their design and performance that go beyond simple accounting. First, they draw on and feed back into biological and pathological models of disease and therapy. These models are not static, and we have tracked their evolution from the early kinetic models of cancer therapy to the current use of a plethora of genetically manipulated animals. Second, clinical trials evolve in relation to previous and concurrent trials that generate findings that are often independent of biology (hence the possibility of feedback).

As noted by Harry Marks, one of the reasons for the difference between our analysis and the analysis provided by other social scientists lies in the fact that, contrary to standard assumptions, "a culture of using clinical trials to promote innovation can develop independent of any substantial commercial interests" (Marks, 2009). The cancer cooperative groups, in Marks's terms, are an example of "non-commodified networks" that have "fostered trials as an instrument of drug discovery and biological research" and by the same token redefined disease entities even "when the opportunities for profit were weak or non-existent." This was certainly true in the 1970s when, as we saw, the pharmaceutical industry shunned the cancer field, but it also applies to the most recent period when, as mentioned in chapter 11, a complex division of labor has emerged between pharmaceutical companies, contract research organizations and academic research organizations. The current situation remains fluid, structured by the emergence of new, hybrid arrangements and networks; but this is the topic for another book, focusing on the configurations of cancer genomics.

In the course of this book, we have drawn attention to a number of clinical cancer research routines and innovations. Of course, the opposition innovation-routine belies the fact that the routines themselves were at some point innovations and that none of the routines resist innovation as they are constantly remade and refashioned. Many of these routines-innovations are overlooked in summary overviews of the field that tend to concentrate on breakthrough therapies. Viewed as a new style of practice, the range of innovations in clinical cancer research is as broad as the sources are varied. As a style of practice that draws freely from other practices, clinical cancer research has profited from advances in the realm of the clinical (e.g., the definition of categories of disease that capture their treatment history as opposed to their natural history), the statistical (e.g., the development of new techniques to deal with problems like censored data), the biological (e.g., the steady production of new models of diseases and disease therapy), the organizational (e.g., the creation of central data processing centers and data managers), the socioepistemic (e.g., the restructuring of cooperative groups as epistemic organizations and the redesign of clinical trials to incorporate molecular approaches), and political activism (e.g., the inclusion of patients and patient representatives at all levels of the trial process). This is what we meant when we said the new style of practice was and remains open-ended.

Open-ended does not mean amorphous. Although subject to continuous change, clinical cancer research has maintained its identity as a style of practice. Contrary to more traditional specialties that first underwent local and national phases of development, medical oncology and cancer clinical trials have been an international or, at least, a multicenter endeavor from the outset. This is not to say that medical oncology has no local, national, or "glocal" dimension; we maintain, however, that the management and alignment of clinical and research practices crosses many (inter)national sites and has been a constant preoccupation for the expanding array of practitioners involved in this field. The elimination of clinically meaningful differences between sites has long been a concern for oncologists and trialists who have dedicated considerable effort to the establishment and maintenance of standards within the cooperative groups. Similar efforts have been deployed at the international level since the mid-1960s, for instance, by transnational organizations such as the Union for International Cancer Control (UICC), but also via ad hoc, bilateral or trilateral agreements between the NCI and partner institutions in North America, Europe, and Japan. Some of these agreements concerned the procurement and testing of promising chemotherapy compounds, but the de facto harmonization of practices under the "one community" heading has always been high on the agenda (Wittes & Yoder, 1998). Major annual meetings such as ASCO and SABCS have ripple effects that pervade the entire oncology community on both sides of the Atlantic. Last but not least, transnational efforts have been long underway to homogenize training in medical oncology and thus to normalize practices by streamlining practitioners. In sum, the development of the new style of practice has resulted in the formation of an extended international network, characterized by a few major hubs and the circulation of oncologists, protocols, drugs, and so on within this network.

While for obvious reasons much of our discussion has focused on the United States, we have consistently referred to concurrent and subsequent developments in Europe. We have not done so in the search for a comprehensive narrative but rather to show how different paths to similar practices were possible and to highlight the contingent yet convergent nature of the developmental process. Skeptical readers may question the extent to which clinical oncology practices are truly similar in different countries. There are variations in the way clinical cancer research is conducted both between and within countries; we have on several occasions pointed them out. We also recognize that despite the circulation of protocols and the general adoption of multimodal therapies, patients in different countries or hospitals are subject to different treatments. One could therefore multiply differences and consequently multiply styles ad infinitum. In the final analysis, however, our thesis holds insofar as findings made in one cooperative group or country are understood and used in the others *as if they were their own*. And while criticisms are often made of others' findings (including one's own previous trials), this is part of the enterprise and not a negation of it. Indeed, cancer researchers have long been intrigued by the possible impact of practice and research variations on therapeutic outcomes.

To put the matter in a more practical light: Do the many differences in oncology research and practice among groups or countries make a difference? This question begs the question of what should count as a (meaningful) difference and according to which parameters: How should one measure outcomes? Should nonclinical elements such as cost be included? Answers to these and related questions are framed by the fact that once the new style of practice gained a foothold in the field of cancer research and treatment, it replaced alternative ways of producing evidence and became self-vindicatory. In turn, the evolution of the new style, for instance, its incorporation of molecular practices, redefined evaluative issues, more recently via comparative effectiveness research (CER). Below we consider an early alternative to clinical cancer trials and the current interest in CER.

The Observational Alternative to Clinical Cancer Trials

There are two kinds of answers to the question of whether local contingencies make a difference in clinical outcomes. The first is generated from within medical oncology and is part and parcel of international clinical trials. Significant differences between sites raise the question of whether the populations treated are truly different or whether the treatments offered are similar and/or delivered in the same manner. Such questions are, in other words, built into the new style of practice. A second way of looking at this question involves observational studies. By focusing on comparisons of the results of treatment as recorded in cancer registries, epidemiologists are in a position to uncover disparities and consequently differences in the ways patients with the same disease are treated in different locales and different countries. While large-scale comparisons of cancer outcomes are useful from a public health point of view, following the widespread acceptance of clinical cancer trials as the gold standard for the evaluation of therapies in the 1970s, they appear to have comparatively little to say to practicing oncologists.

One might be tempted to conclude that observational survival studies

and clinical trials are just different ways to get an answer to the same
question: What are the different treatments available and which treatment is the best treatment? Such a conclusion glosses over the fact that
the two practices instantiate two very different kinds of rules (Searle,
1995; Ricciardi, 1997). The rules stipulated by observational studies to
select and compare cases from a registry are *regulatory* rules: cancer
treatment will continue and the registries will continue to collect cases.
The practice and the collection of cases historically and epistemologi-
cally predate the practice of constructing and comparing survival rates.
The rules that are used in clinical trial protocols are different: they are
constitutive rules in the sense that the data collected would not exist
without those rules. Without the trial there are no clinical trial data. By
following the rules in a clinical trial protocol one produces the division
between the biological and the social that the protocol sets out. In a survival study or a pattern of practice study, one can only observe such a
difference and speculate about its causes.

Clinical trials and observational studies were initially deployed in
parallel. In 1956, following the inauguration of the Cancer Chemotherapy National Service Center and the associated cooperative groups, the
NCI planned to use cancer survival data compiled in three US central
state registries to draw a general picture of the efficacy of cancer treatment. Data inadequacy, however, precluded a detailed analysis of cancer
survival. In order to garner more uniform and hence comparable data, in
the late 1950s the NCI launched the Cancer End Results Evaluation Program (Cutler & Latourette, 1959), which in 1973, following the 1971 National Cancer Act (War on Cancer), transmogrified into the first national
registry, known as the Surveillance, Epidemiology, and End Results Program (SEER). These initiatives had been prompted by the fact that the
"rapidly growing chemotherapy research program" meant that the "evaluation of chemotherapy had to be related to the end results achieved by
other means of treatment" (Bailar, 1964: 18–19). In other words, the observational program set itself up as an alternative to clinical trials for the
evaluation of therapies at the level of modalities: Was chemotherapy better than surgery? Was surgery better than radiotherapy? Such a comparison presupposed the collection of comparable data, and thus the rules
established for the collection stipulated, for example, that cases certified
by pathologists had to be separated from nonconfirmed cases, and that a
simple staging dichotomy—localized/not localized—ought to be used to
classify cases within a diagnostic category.

At a 1963 meeting, participants exposed numerous international treatment differences (Cutler, 1964). Yet they also noted that it was impossible to know whether or not countries should adopt the more successful approaches since it was not "permissible to conclude that these successes are the result of treatment rather than the selection of cases for treatment" (Campbell, 1964: 205). To know that would have required a clinical trial. Assessment of the clinical outcomes of national differences in therapeutic approaches was rendered problematic by the nagging issue of whether data generated by the various country registries were in fact comparable (Lourie, 1964). Following the 1963 meeting, almost forty years would elapse before a similar intercontinental comparison was attempted for a second time. This lack of urgency can be accounted for by several factors, but a key element is to be sought in the alternative provided by clinical trials, the crucial component of oncology practices *and*, simultaneously, a device for assessing those same practices. By establishing the means to separate social from biological differences through the control of practice variables, clinical trials had emerged as the principal producer of differences in oncology practices and thus the primary forum for their comparison.

Registries have not abandoned attempts to act as alternative sources of therapy and practice evaluation. Initiated in 1990, EUROCARE is a research project funded by the European Union implicating members of a consortium of over forty cancer registries in seventeen European countries. Unlike clinical trials, the group insists, "only population-based studies, using data from cancer registries, are capable of quantifying the effectiveness of cancer treatments in the population as a whole, allowing survival to be compared between populations and measuring the extent to which survival has improved" (Berrino et al., 1995: ix). The first studies, published in 1995, prompted considerable discussion with regard to the quality of cancer care in the European community (Berrino et al., 1995). The study's conclusion that "quality of care, in the widest sense, is the major determinant for the variation in the figures" seemed to raise more questions than it answered. Absent the requisite clinical statistics, one could, for instance, argue that quality of care was merely a proxy variable for access to care. Once again the effect of therapy on survival remained nebulous given the uneven nature of the data collected and the absence of major clinical variables. Over the last fifty years the diagnostic procedures that allow oncologists to assign individual patients to a given pathological stage (an indication of how advanced their disease is) have continuously evolved. When they evolve, so do patient assignments, a phenomenon known as *stage migration* (Feinstein, Sosin & Wells, 1985). Improved diagnostic procedures generally lead to early diagnosis and treatment, and statistical records will necessarily show that patients treated earlier live longer with the disease than patients who have not been diagnosed and treated. As a result the increased survival time in these cases is not necessarily the result of improved therapy; it may just be the result of earlier detection.

In order to overcome obstacles such as these, a subgroup of EURO-CARE launched a project to compare European and North American data (Welch, Schwartz & Woloshin, 2000). The project required registries to return to the original data and to submit original slides to an international panel of pathologists in order to ensure uniformity of diagnosis at the histopathological level. The project has proceeded in fits and starts, with preliminary results published in 2008 (Coleman et al., 2008). Final results are expected to "facilitate joint assessment of international trends in incidence, survival, and mortality as indicators of cancer control" (Coleman et al., 2008: 730), but oncologists themselves may not find them directly relevant to their practice, both for the previously mentioned reasons and for the fact that in the meantime the rise of targeted therapies and molecular biomarkers has shed new light on the issue.

Comparative Effectiveness and Clinical Trials

Social scientists have no crystal ball, and we are no exception, but the aforementioned feud between observational and experimental approaches has taken an interesting turn with the emergence of comparative effectiveness research (Sox, 2010). CER is the latest example of a set of technologies that include health technology assessment, evidence-based medicine, and clinical guidelines, which have been designed to evaluate competing medical interventions and guide medical decision-making. There is still confusion surrounding the actual goals and methods of CER, and this is not the place to rehearse these issues, but CER's emergence is clearly linked to the development of scores of competing targeted agents and biomarkers, in short, to the emergence of the promise of a personalized medicine.

Now, as mentioned above, supporters of observational studies claim that only their approach is able to assess the effectiveness of cancer treatments in the population as a whole. Clinical trials, from their point of view, are experiments performed on selected subjects under controlled conditions; their results do not necessarily apply to the real world. This argument mobilizes the distinction between *efficacy* and *effectiveness*: while both terms denote the ability of a drug or procedure to produce a given effect, the former refers to an experimental setting while the latter to routine clinical practice (e.g., Albertsen, 2010). Yet CER did not emerge as an observational alternative to clinical trials, but rather as a part of clinical trial technology. One of the very first initiatives financed by the $1.1 billion allocated to CER in the United States (Sox, 2010) is a clinical trial carried out by a cooperative oncology group (SWOG) to assess a biomarker testing which patients will benefit from chemotherapy. The incorporation of CER in new trial designs, in turn, hypothesizes that whereas past designs were largely empirical, comparing treatment A versus treatment B and only subsequently investigating the modus operandi of the successful agent, the novel approach will start by identifying molecular targets and design smaller trials comparing targeted agents in the relevant population (Kaufmann, Pusztai & Biedenkopf Expert Panel Members, 2011). In other words, while in the past diagnosis led to therapy, the new design will involve a more complex sequence, including focused screening and early detection followed by individualized therapy and therapeutic monitoring. From this point of view, the "demographics" of the patient population will be redefined in molecular terms, and this is what will allow experimental methods to embody, at least in principle, a pragmatic dimension that was previously confined to observational studies.

This is the theory. As exemplified by two examples in a recent editorial on CER and targeted therapies (Pritchard, 2010), however, numerous problems arise when one moves from schematic designs to the real world performance of clinical trials. US centers, for example, "consistently" continue treatment with the targeted agent Herceptin (see chapter 10) once a patient's tumor reemerges, whereas many European countries and Canada do not. Initial attempts to use a clinical trial to settle the issue failed because clinicians were apparently "unwilling to randomly assign women to the option in which [Herceptin] was discontinued." A recent trial "strongly suggested" that the continuation option was superior and thus "reinforce[d] the pre-existing bias of most North American oncologists," (including the author) favoring the Herceptin-forever strategy (Pritchard, 2010: 1090). That trial, however, was both "underpowered and terminated earlier because of problems with accrual"; in other words, it did not pass statistical muster. Cognitive dissonance between clinicians and statisticians, as in the old-style trials, thus lingers (Marks, 1997: chap. 7). The second example drew on a trial in which Herceptin and a second targeted agent were to be given alone or in combination. Due to "investigator feedback" that such an option would not be helpful to patients, the Herceptin-alone option was deleted from the trial design, thus biasing the results in favor of the combination. As a result, the question of whether Herceptin alone was equally beneficial was left "unasked and unanswered" (Pritchard, 2010: 1090). The lesson we can draw from these examples resonates with the main theme of this book: as experiments, clinical trials require more than the mechanical application of routine methodologies; they require the definition of "appropriate" research questions.

Coda

Cancer clinical trials have obviously matured, but this is not the end of the story: as just argued, cancer genomics adds a new spin to the domain, and new entities and issues continuously inform the design of trials. Clinical trials, in the meantime, have become an irreplaceable means for the production of evidence about the nature and the treatment of cancer.

In attempting to characterize this new style of practice, we have outlined several of the components and their interconnections. One of the themes that surfaced consistently during the initial configuration of the new style was whether clinical cancer research should count as true research, or if it should be viewed as the routine testing of substances for therapeutic effects. While only a few decades ago clinicians and health administrators entertained both perspectives, and while critics will argue that routine testing still defines many of the commercial trials, much of the work carried out by academic networks qualifies as bona fide research in the eyes of most observers. Clinical cancer research, despite almost fifty years of criticism and reform, has emerged as an autonomous enterprise that typifies modern biomedicine. Viewed from a distance, and from the perspective of Phase III clinical trials that compare well-known procedures and combinations of known substances, clinical cancer research and its incremental improvements in cancer therapy may look somewhat mundane and trivial. But this perspective misses two important things: first, it severs the connection between the trials and all the research that has both led up to that point and that will, ultimately, point the way to the future; and second, it overlooks all the work—inventive and mundane, laboratory and institutional—that goes into the production of even negative results.

Readers might want to ask a final question that we have not dealt with directly: Has this new style of practice been successful? There is a succinct response to this interrogation. Since scientific practices are self-vindicating, by definition they set their own criteria for success, and so in this respect, the answer is yes. In this case, the existence and extent of the practice supply the answer.

References

A new drug office for Southern Europe. (1996). *Annals of Oncology* 7: 983–84.

Abadie, R. (2010). *The Professional Guinea Pig: Big Pharma and the Risky World of Human Subjects*. Durham, NC: Duke University Press.

Abelson, H. T. & Rabstein, L. S. (1970). Lymphosarcoma: Virus-induced thymic-independent disease in mice. *Cancer Research* 30: 2213–22.

Ad Hoc Committee on the Medical Oncology Workforce. (1996). Status of the medical oncology workforce. *Journal of Clinical Oncology* 14: 2612–21.

Ahrne, G. & Brunsson, N. (2006). Organizing the world. In M.-L. Djelic & K. Sahlin-Andersson, eds., *Transnational Governance: Institutional Dynamics of Regulation*. Cambridge: Cambridge University Press, 74–94.

Akiyama, T., Ishida, J., Nakagawa, S., Ogawara, H., Watanabe, S., Itoh, N., Shibuya, M. & Fukami, Y. (1987). Genistein, a specific inhibitor of tyrosine-specific protein kinases. *Journal of Biological Chemistry* 262: 5592–95.

Albertsen, P. C. (2010). Efficacy vs. effectiveness in prostate-specific antigen screening. *Journal of the National Cancer Institute* 102: 288–89.

All 59 CCOP awards announced. (1983). *The Cancer Letter* 9 (22): 1–3.

Alley, M. C., Hollingshead, M. G. & Dykes, D. J. (2004). Human tumor xenograft models in NCI drug development. In B. A. Teicher & P. A. Andrews, eds., *Anticancer Drug Development Guide: Preclinical Screening, Clinical Trials, and Approval*, 2nd ed. Totowa, NJ: Humana Press, 125–52.

Altman, D. G., Whitehead, J., Parmar, M. K., Stenning, S. P., Fayers, P. M. & Machin, D. (1995). Randomised consent designs in cancer clinical trials. *European Journal of Cancer* 31A: 1934–44.

AMA Therapeutic Trials Committee. (1960). Androgens and estrogens in the treatment of disseminated mammary carcinoma: Retrospective study of 944 patients. *Journal of the American Medical Association* (hereafter *JAMA*) 172: 1271–83.

Amado, R. G., Wolf, M., Peeters, M., Van Cutsem, E., Siena, S., Freeman, D. J., Juan, T., Sikorski, R., Suggs, S., Radinsky, R., Patterson, S. D. & Chang,

D. D. (2008). Wild-type KRAS is required for panitumumab efficacy in patients with metastatic colorectal cancer. *Journal of Clinical Oncology* 26: 1626–34.

American Board of Internal Medicine. (1998). Some questions about ABIM examinations. *Perspectives* (Spring–Summer): 6–7.

Amin, A. & Cohendet, P. (2004). *Architectures of Knowledge: Firms, Capabilities, and Communities.* Oxford: Oxford University Press.

Andrews, J. R. (1978). *The Radiobiology of Human Cancer Radiotherapy,* 2nd ed. Baltimore: University Park Press.

Anscombe, F. J. (1963). Sequential medical trials. *Journal of the American Statistical Association* 58: 365–83.

Ansfield, F. J., Ramirez, G., Mackman, S., Bryan, G. T. & Curreri, A. R. (1969). A ten-year study of 5-fluorouracil in disseminated breast cancer with clinical results and survival times. *Cancer Research* 29: 1062–66.

Armitage, J. O. (1998). American Society of Clinical Oncology: Where have we been? Where are we going? *Journal of Clinical Oncology* 16: 381–86.

Armitage, P. (1960). *Sequential Medical Trials.* Oxford: Blackwell.

——. (2003). Fisher, Bradford Hill, and randomization. *International Journal of Epidemiology* 32: 925–28.

Armitage, P. & Schneiderman, M. A. (1958). Statistical problems in a mass screening program. *Annals of the New York Academy of Sciences* 76: 896–908.

Arnold, H., Bourseaux, F. & Brock, N. (1958). Neuartige Krebs-Chemotherapeutika aus der Gruppe der zyklischen N-Lost-Phosphamidester. *Naturwissenschaften* 45: 64–66.

Aronowitz, R. A. (2001). Do not delay: Breast cancer and time, 1900–1970. *Milbank Quarterly* 79: 355–86.

ASCO. (2004). *40 Years of Quality Cancer Care.* Alexandria, VA: American Society of Clinical Oncology.

Aub, J. C. & Nathanson, I. T. (1951). Neoplastic diseases. *Annual Reviews of Medicine* 2: 343–66.

Bailar, J. C., 3rd. (1964). History of the Ad Hoc Group on International Cooperation in Evaluation of End Results. In S. J. Cutter, ed., *International Symposium on End Results of Cancer Therapy. National Cancer Institute Monograph* 15. Washington, DC: Superintendent of Documents, U.S. Government Printing Office, 5–20.

Baker, C. G. (1977). Cancer Research Program strategy and planning: The use of contracts for program implementation. *Journal of the National Cancer Institute* 59: 651–69.

Baker, S. J., Markowitz, S., Fearon, E. R., Willson, J. K. & Vogelstein, B. (1990). Suppression of human colorectal carcinoma cell growth by wild-type p53. *Science* 249: 912–15.

Balkie, A. G., Court-Brown, W. M., Buckton, K. E., Harnden, D. G., Jacobs, P. A. & Tough, I. M. (1960). Possible specific chromosome abnormality in human chronic myeloid leukaemia. *Nature* 188: 1165–66.

Baltimore, D. (1970). Viral RNA-dependent DNA polymerase. *Nature* 226: 1209–11.

Barnard, G. A. (1946). Sequential tests in industrial statistics. *Journal of the Royal Statistical Society, Series B, Statistical Methodology* 8: 1–26.

Barry, A. (2005). Pharmaceutical matters: The invention of informed material. *Theory, Culture & Society* 22: 51–69.

Barthes, R. (1987). *Mythologies.* New York: Hill and Wang.

Bate, R. (2007). India and the drug patent wars. *Health Policy Outlook* 3: 1–5.

Bates, S. E. (2008). From the editor. *Clinical Cancer Research* 14: 3657.

Batt, S. (1996). *Patient No More: The Politics of Breast Cancer.* Melbourne: Spinifex.

Baum, M. (1997). Who truly represents the needs of the consumer diagnosed with breast cancer? Who are these patients' advocates? How are they informed? What, if any, are their secret agendas? *European Journal of Cancer* 33: 807–8.

Bayart, D. (2005). Walter Andrew Shewhart, economic control of quality of manufactured product. In I. Grattan-Guinness, ed., *Landmark Writings in Western Mathematics, 1640–1940.* Amsterdam: Elsevier: 926–35.

Bazell, R. (1998). *Her-2: The Making of Herceptin, a Revolutionary Treatment for Breast Cancer.* New York: Random House.

Beatson, G. T. (1896). On the treatment of inoperable cases of the mamma: Suggestions for a new method of treatment, with illustrative cases. *The Lancet* 2: 104–7, 162–65.

Beauchamp, T. L. & Childress, J. F. (2001). *Principles of Biomedical Ethics*, 5th ed. Oxford: Oxford University Press.

Bennett, D. J. (1999). *Randomness.* Cambridge, MA: Harvard University Press.

Berg, M. (1997). *Rationalizing Medical Work: Decision Support Techniques and Medical Practices.* Cambridge, MA: MIT Press.

———. (1998). Order(s) and disorder(s): Of protocols and medical practices. In M. Berg and A. Mol, eds., *Differences in Medicine: Unraveling Practices, Techniques, and Bodies.* Durham, NC: Duke University Press, 226–49.

Berkson, J. & Gage, R. P. (1950). Calculation of survival rates for cancer. *Proceedings of Staff Meetings of the Mayo Clinic* 25: 270–86.

Berlivet, L. (2000). *Une santé à risques: L'action publique de lutte contre l'alcoolisme et le tabagisme en France (1954–1999).* Doctoral thesis (Science politique). Rennes: Institut d'Études Politiques de Rennes.

———. (2002). Déchiffrer la maladie. Epidémiologie et cultures de santé publique. In J. P. Dozon & D. Fassin, eds., *Les cultures de la santé publique.* Paris: Balland, 75–102.

———. (2005). Exigence scientifique et isolement institutionnel: L'essor contrarié de l'épidémiologie française dans la seconde moitié du XXe siècle. In G. Jorland, A. Opinel & G. Weisz, eds., *Body Counts: Medical Quantification in Historical and Sociological Perspective*. Montreal & Kingston: McGill-Queen's University Press, 335–58.

Bernard, J. (1967). Les nouveaux traitements des leucémies et des hémopathies malignes. *Assises de Médecine* 25: 279–83.

Bernier, J., Horiot, J. C. & Poortmans, P. (2002). Quality assurance in radiotherapy: From radiation physics to patient and trial-oriented control procedures. *European Journal of Cancer* 38: S155–58.

Berrino, F., Sant, M., Verdecchia, A., Capocaccia, R., Hakulinen, T. & Estève, J. (1995). *Survival of Cancer Patients in Europe*. The EUROCARE Study, No. 132. Lyon: IARC Scientific Publications.

Bierman, H. R., Cohen, P., McClelland, J. N. & Shimkin, M. B. (1950). The effect of transfusions and antibiotics upon the duration of life in children with lymphogenous leukemia. *Journal of Pediatrics* 37: 455–62.

Bisel, H. F. (1956). Criteria for the evaluation of response to treatment in acute leukemia. *Blood* 11: 676–81.

Bishop, J. M. (1995). Cancer: The rise of the genetic paradigm. *Genes and Development* 9: 1309–15.

———. (2003). *How to Win the Nobel Prize: An Unexpected Life in Science*. Cambridge, MA: Harvard University Press.

Bland, J. M. & Altman, D. G. (1998). Survival probabilities (the Kaplan-Meier method). *BMJ* 317: 1572.

Bloomfield, C. D., Mrozek, K. & Caligiuri, M. A. (2006). Cancer and leukemia group B leukemia correlative science committee: Major accomplishments and future directions. *Clinical Cancer Research* 12: 3564s–3571s.

Bodey, G. P., Watson, P., Cooper, C. & Freireich, E. J. (1971). Protected environment units for the cancer patient. *CA: A Cancer Journal for Clinicians* 21 (4): 215–19.

Boehmer, U. (2000). *The Personal and the Political: Women's Activism in Response to the Breast Cancer and AIDS Epidemics*. New York: State University of New York Press.

Bonadonna, G. (1985). This week's citation classic. *Current Contents* 5: 22.

Bonadonna, G., Moliterni, A., Zambetti, M., Daidone, M. G., Pilotti, S., Gianni, L. & Valagussa, P. (2005). 30 Years' follow up of randomised studies of adjuvant CMF in operable breast cancer: Cohort study. *BMJ* 330: 217–20.

Bonadonna, G., Rossi, A., Tancini, G. & Valagussa, P. (1983). Adjuvant chemotherapy in breast cancer (Letter). *The Lancet* 1: 1157.

Bonadonna, G., Rossi, A., Valagussa, P., Banfi, A., Veronesi, U. (1977). The CMF program for operable breast cancer with positive axillary nodes. Updated analysis on the disease-free interval, site of relapse, and drug tolerance. *Cancer* 39 (6 Supplement): 2904–15.

Bonadonna, G. & Valagussa, P. (1978). The logistics of clinical trials. *Biomedicine* 28 (Special issue): 43–48.

Bonadonna, G., Valagussa, P., Moliterni, A., Zambetti, M. & Brambilla, C. (1995). Adjuvant cyclophosphamide, methotrexate, and fluorouracil in node-positive breast cancer: The results of 20 years of follow-up. *New England Journal of Medicine* 332: 901–6.

Bourret, P. (2005). BRCA patients and clinical collectives: New configurations of action in cancer genetics practices. *Social Studies of Science* 35: 41–68.

Boven, E., Winograd, B., Fodstad, O., Lobezzo, M. W. & Pinedo, H. M. (1988). Preclinical Phase II studies in human tumor lines: A European multicenter study. *European Journal of Cancer* 24: 567–73.

Boveri, T. (1914). *Zur Frage der Entstehung maligner Tumoren*. Jena: Gustav Fischer.

Bradbury, W. (1966). The all-out assault on leukemia: Two views—the lab, the victim. *Life* 61 (November 18): 88–107.

Brady, L. W., Kramer, S., Levitt, S. H., Parker, R. G. & Powers, W. E. (2001). Radiation oncology: Contributions of the United States in the last years of the 20th century. *Radiology* 219: 1–5.

Breslow, N. (1978). Perspectives on the statistician's role in cooperative clinical research. *Cancer* 41: 326–32.

Brock, N. (1989). Oxazaphosphorine cytostatics: Past, present, future. *Cancer Research* 49: 1–7.

Bross, I. (1952). Sequential medical plans. *Biometrics* 8: 188–205.

———. (1954). Statistical analysis of clinical results from 6-mercaptopurine. *Annals of the New York Academy of Sciences* 60 (2): 369–73.

Brown, H. (2007). Translational cancer research in the USA. *Molecular Oncology* 1: 11–13.

Buchanan, M., Kyriakides, S., Fernandez-Marcos, A., Horvatin, J., Mosconi, P., O'Connell, D. & Zernik, N. (2004a). Breast cancer advocacy across Europe through the work and development of EUROPA DONNA, the European breast cancer coalition. *European Journal of Cancer* 40: 1111–16.

Buchanan, M., O'Connell, S. D. & Mosconi, P. (2004b). EUROPA DONNA, the European breast cancer coalition: Lobbying at European and local levels. *Journal of Ambulatory Care Management* 27 (2): 146–53.

Buchdunger, E., Zimmermann, J., Mett, H., Meyer, T., Muller, M., Druker, B. J. & Lydon, N. B. (1996). Inhibition of the ABL protein-tyrosine kinase in vitro and in vivo by the 2-phenyaminopyrimidine derivative. *Cancer Research* 56: 100–104.

Bud, R. (1978). Strategy in American cancer research after World War II: A case study. *Social Studies of Science* 8 (23): 425–59.

Burchenal, J. H. (1975). From wild fowl to stalking horses: Alchemy in chemotherapy. Fifth annual David A. Karnofsky Memorial Lecture. *Cancer* 35: 1121–35.

Burchenal, J. H., Heyn, R., Sutow, W. W., Freireich, E. J., Whittington, R. & Louis, J. (1960). Investigations in acute leukemia. *National Cancer Institute Monograph* 3: 149–68.

Burzykowski, T., Buyse, M., Piccart-Gebhart, M. J., Sledge, G., Carmichael, J., Lück, H.-J., Mackey, J. R., Nabholtz, J.-M., Paridaens, R., Biganzoli, L., Jassem, J., Bontenbal, M., Bonneterre, J., Chan, S., Basaran, G. A. & Therasse, P. (2008). Evaluation of tumor response, disease control, progression-free survival, and time to progression as potential surrogate end points in metastatic breast cancer. *Journal of Clinical Oncology* 26: 1987–92.

Butler, D. (2008). Translational research: Crossing the valley of death. *Nature* 453: 840–42.

Buyse, M., Burzykowski, T., Carroll, K., Michiels, S., Sargent, D. J., Miller, L. L., Elfring, G. L., Pignon, J.-P. & Piedbois, P. (2007). Progression-free survival is a surrogate for survival in advanced colorectal cancer. *Journal of Clinical Oncology* 25: 5218–24.

Byar, D. B., Simon, R. M., Friedewald, W. T., Schlesselman, J. J., DeMets, D. L., Ellenberg, J. H., Gail, M. H. & Ware, J. H. (1976). Randomized clinical trials: Perspectives on some recent ideas. *New England Journal of Medicine* 295: 74–80.

Bypass budget for FY 1986 asks big increase for NCI. (1984). *The Cancer Letter* 10 (20): 1–2.

Byrd, W. M. & Clayton, L. A. (2002). *An American Health Dilemma*, Vol. 2, *Race, Medicine, and Health Care in the United States, 1900–2000*. New York: Routledge.

Cairns, J. & Boyle, P. (with a response by Frei, E.). (1983). Cancer chemotherapy. *Science* 220: 252, 254, 256.

Callon, M., Lascoumes, P. & Barthe, Y. (2009). *Acting in an Uncertain World: An Essay on Technical Democracy*. Cambridge, MA: MIT Press.

Callon, M. & Rabeharisoa, V. (2008). The growing engagement of emergent concerned groups in political and economic life: Lessons from the French Association of Neuromuscular Disease patients. *Science, Technology & Human Values* 33: 230–61.

Cambrosio, A., Cottereau, P., Popowycz, S., Mogoutov, A. & Vichnevskaia, T. (2010). Analysis of heterogeneous networks: The ReseauLu Project. In C. Brossaud & B. Reber, eds., *Digital Cognitive Technologies: Epistemology and Knowledge Society*. London: Wiley: 137–52.

Cambrosio, A., Keating, P., Mercier, S., Lewison, G. & Mogoutov, A. (2006). Mapping the emergence and development of translational cancer research. *European Journal of Cancer* 42: 3150–58.

Campbell, H. (1964). Survival experience of patients from five countries with cancer of the tongue, around 1950–1954. *National Cancer Institute Monograph* 15: 199–207. Washington, DC: Superintendent of Documents, U.S. Government Printing Office.

Canellos, G. P., DeVita, V. T., Gold, G. L., Chabner, B. A., Schein, P. S. & Young, R. C. (1974). Cyclical combination chemotherapy for advanced breast carcinoma. *British Medical Journal* 1: 218–20.

Canellos, G. P., Pocock, S. J., Taylor, S. G., III, Sears, M. E., Klaasen, D. J. & Band, P. R. (1976). Combination chemotherapy for metastatic breast carcinoma: Prospective comparison of multi-drug therapy with 1-phenylalanine mustard. *Cancer* 38: 1882–86.

Cantor, D. (1993). Cancer. In W. F. Bynum & Roy Porter, eds., *Companion Encyclopedia of the History of Medicine, Vol. 1.* London: Routledge: 537–61.

———. (2008). Radium and the origins of the National Cancer Institute. In C. Hannaway, ed., *Biomedicine in the Twentieth Century: Practices, Policies, and Politics.* Amsterdam: IOS Press.

Capdeville, R., Buchdunger, E., Zimmermann, J. & Matter, A. (2002). Glivec (STI571, imatinib), a rationally developed, targeted anticancer drug. *Nature Reviews Drug Discovery* 1: 493–502.

Capell, K. (2009). Novartis: Radically remaking its drug business. CEO Dan Vasella's growth mantra for Novartis is follow the science, not the financials. *BusinessWeek* (June 22): 30–35.

Carbone, P. (1979). Carbone says investigator initiated clinical research gets more for less (Overview of the Cooperative Group Program presented to the NCI Div. of Cancer Treatment Board of Scientific Counselors). *The Cancer Letter* 15 (5): 3–6.

Carbone, P. P., Bono, V., Frei, E., III & Brindley, C. O. (1963). Clinical studies with vincristine. *Blood* 21: 640–47.

Carbone, P. P., Krant, M. J., Miller, S. P., Hall, T. C., Shnider, B. I., Colsky, J., Horton, J., Hosley, H., Miller, J. M. & Frei, E., III. (1965). The feasibility of using randomization schemes early in the clinical trials of new chemotherapeutic agents: Hydroxyurea (NSC-32065). *Clinical Pharmacology and Therapeutics* 6: 17–24.

Carpenter, D. (2010). *Reputation and Power: Organizational Image and Pharmaceutical Regulation at the FDA.* Princeton, NJ: Princeton University Press.

Carpenter, G. & Cohen, S. (1979). Epidermal growth factor. *Annual Review of Biochemistry* 48: 193–216.

Carrese, L. M. & Baker, C. G. (1967). The convergence technique: A method for the planning and programming of research efforts. *Management Science* 13 (8): B420–38.

Carter, S. K. (1972). Single and combination nonhormonal chemotherapy in breast cancer. *Cancer* 30: 1543–55.

———. (1978a). The analysis of adjuvant trials. *Cancer Treatment Reports* 5: 1–5.

——. (1978b). The clinical evaluation of analogues. I. The overall problem. *Cancer Chemotherapy and Pharmacology* 1 (1978): 69–72.

——. (1979). Editorial: Clinical research and experimental design. *Cancer Immunology and Immunotherapy* 6: 1–4.

——. (1995). Cancer chemotherapy 1973–present: A personal perspective. *Journal of Cancer Research and Clinical Oncology* 121: 501–4.

Carter, S. K. & Livingston, R. B. (1975). Cyclophosphamide in solid tumors. *Cancer Treatment Reviews* 2: 295–322.

Carter, S. K. & Mathé, G. (1978). An overview of the 1978 international meeting on comparative clinical trials. *Biomedicine* 28 (special issue): 6–13.

Carter, S. K. & Soper, W. T. (1974). Integration of chemotherapy into combined modality treatment of solid tumors. 1. The overall strategy. *Cancer Treatment Reviews* 1: 1–13.

Cauhan, C. & Eppard, W. (2004). Systematic education and utilization of volunteer patient advocates for cancer clinical trial information dissemination at NCCTG community sites. *Journal of Clinical Oncology* 22 (14S): 1031.

CCIRC members list 10 most important factors in assessing groups' science. (1978). *The Cancer Letter* 4 (9): 2.

CCIRC starts preparing for clinical research review: To include group achievements, problems. (1978). *The Cancer Letter* 4 (9): 1–4.

CCNSC (Cancer Chemotherapy National Service Center). (1959). Cancer Chemotherapy National Service Center specifications for screening chemical agents and natural products against animal tumors. *Cancer Chemotherapy Reports* 1 (1): 42–98.

——. (1959–60a). Program notes. *Cancer Chemotherapy* 1 (2): 43–47.

——. (1959–60b). Program notes. *Cancer Chemotherapy Reports* 1 (6): 115–18.

CCOP recompetition approved, "institutionalized"; board okays concept for eight minority CCOPs. (1989). *The Cancer Letter* 15 (5): 3–5.

CCOP review completed, enough good applications in funding range to get viable programs started. (1983). *The Cancer Letter* 9 (7): 1–2.

CCOP success acknowledged. (1986). *The Cancer Letter* 12 (15): 1–3.

CCOPs at six months. (1984). *The Cancer Letter* 10 (3): 3.

Central dogma reversed. (1970). *Nature* 226: 1198–99.

Chabner, B. A. (1990). In defense of cell-line screening. *Journal of the National Cancer Institute* 82: 1083–85.

——. (2007). Phase II cancer trials: Out of control? *Clinical Cancer Research* 13: 2307–8.

Chabner, B. A. & Roberts, T. G. (2005). Chemotherapy and the war on cancer. *Nature Reviews Cancer* 5: 65–72.

Chadarevian, S. de. (2002). *Designs for Life: Molecular Biology after World War II*. Cambridge: Cambridge University Press.

——. (2003). The making of an icon. *Science* 300: 255–57.

Chalmers, T. C., Block, J. B. & Lee, S. (1972). Controlled studies in clinical can-cer research. *New England Journal of Medicine* 287: 75–78.

Changes in clinical trials mechanism inevitable, but NCI says it will heed critics of proposals. (1986). *The Cancer Letter* 12 (16): 1–4.

Christie, D. A. & Tansey, E. M. (2003). *Leukaemia*. Wellcome Witness Seminar. London: Wellcome Trust Centre.

Clark, B. R. (1977). *Academic Power in Italy: Bureaucracy and Oligarchy in a National University System*. Chicago: University of Chicago Press.

Clayton, L. A. & Byrd, W. M. (1993). The African-American cancer crisis, Part I: The problem. *Journal of Health Care for the Poor and Underserved* 4: 83–101.

Cohen, G. I. (2003). Clinical research by community oncologists. *CA: A Cancer Journal for Clinicians* 53: 73–81.

Coleman, M. P., Quaresma, M., Berrino, F., Lutz, J. M., De Angelis, R., Capo-caccia, R., Baili, P., Rachet, B., Gatta, G., Hakulinen, T., Micheli, A., Sant, M., Weir, H. K., Elwood, J. M., Tsukuma, H., Koifman, S., E., Silva, G. A., Francisci, S., Santaquilani, M., Verdecchia, A., Storm, H. H. & Young, J. L. (2008). Cancer survival in five continents: A worldwide population-based study (CONCORD). *The Lancet Oncology* 9: 730–56.

Collyar, D. (2005). How have patient advocates in the United States benefited cancer research? *Nature Reviews Cancer* 5: 73–78.

———. (2008). The value of clinical trials from a patient perspective. *The Breast Journal* 6: 310–14.

Committee on Immunologically Compromised Rodents. Institute of Laboratory Animal Resources. Commission on Life Sciences. National Research Coun-cil. (1989). *Immunodeficient Rodents: A Guide to Their Immunobiology, Husbandry, and Use*, Washington, DC: National Academies Press.

Cooper, R. G. (1969). Combination chemotherapy in hormone resistant breast cancer. *Proceedings of the American Association for Cancer Research* 10: 15.

Cooper, R. G., Holland, J. F. & Glidewell, O. (1979). Adjuvant chemotherapy of breast cancer. *Cancer* 44: 793–98.

Cooperative group reorganization to be offered by DeVita at clinical trials re-view next March. (1978). *The Cancer Letter* 4 (42): 1–2.

Cooperative groups advised to plan for future of less funding from NCI. (1995). *The Cancer Letter* 21 (24): 1–3.

Copeland, E. M. (1999). Presidential address: Surgical oncology: A specialty in evolution. *Annals of Surgical Oncology* 6: 424–32.

Cornfield, J. (1966). Sequential trials, sequential analysis, and the likelihood principle. *American Statistician* 20: 18–23.

Cox, D. R. (1958). *Planning of Experiments*. New York: John Wiley.

———. (1972). Regression models and life tables (with discussion). *Journal of the Royal Statistical Society B* 34: 187–220.

Cox, J. D. (1995). Evolution and accomplishments of the Radiation Therapy On-cology Group. *International Journal of Radiation Oncology, Biology and Physics* 33: 747–54.

Craft, B. S., Kurzrock, R., Lei, X., Herbst, R., Lippman, S., Fu, S. & Karp, D. D. (2009). The changing face of Phase I cancer clinical trials. *Cancer* 115: 1592–97.

Cragg, G. M., Simon, J. E., Jato, J. G. & Snader, K. M. (1996). Drug discovery and development at the National Cancer Institute: Potential for new phar-maceutical crops. In J. Janick, ed., *Progress in New Crops*. Arlington, VA: ASHS Press, 554–60.

Creager, A. (2002). *The Life of a Virus: TMV as an Experimental Model, 1930–1965*. Chicago: University of Chicago Press.

———. (2009). Phosphorus-32 in the phage group: Radioisotopes as historical tracers of molecular biology. *Studies in History and Philosophy of Biological and Biomedical Sciences* 40: 29–42.

Croce, C. M. (2008). Molecular origins of cancer: Oncogenes and cancer. *New England Journal of Medicine* 358: 502–11.

CTEP clinical trial proposals reshaped by discussion. (1986). *The Cancer Let-ter* 12 (19): 4–5.

CTEP presents options for drastic changes in groups. (1986). *The Cancer Let-ter* 12 (15): 3–4.

Culliton, B. J. (1976). Breast cancer: Reports of new therapy are greatly exagger-ated. *Science* 191: 1029–31.

Curreri, A. R. (1962). Nitrogen mustard as an adjuvant to pulmonary resection in the treatment of carcinoma of the lung. *Cancer Chemotherapy Reports* 16: 123–24.

Curreri, A. R., Ansfield, F. J., McIver, F. A., Waisman, H. A. & Heidelberger, C. (1958). Clinical studies with 5-fluorouracil. *Cancer Research* 18: 478–84.

Custer, R. P. & Bernhard, W. G. (1948). The interrelationship of Hodgkin's dis-ease and other lymphatic tumors. *American Journal of the Medical Sciences* 216: 625–42.

Cutler, S. J., ed. (1964). *International Symposium on End Results of Cancer Therapy. National Cancer Institute Monograph 15*. Washington, DC: Super-intendent of Documents, U.S. Government Printing Office.

Cutler, S. J., Greenhouse, S. W., Cornfield, J. & Schneiderman, M. A. (1966). The role of hypothesis testing in clinical trials. *Journal of Chronic Diseases* 19: 857–82.

Cutler, S. J. & Latourette, H. B. (1959). A national cooperative program for the evaluation of end results in cancer. *Journal of the National Cancer Institute* 22: 633–46.

Cvitkovic-Bertheault, F. (2000). 40 years of progress in chemotherapy. *Pathologie-biologie* 48: 812–18.

Dagher, R. N. & Pazdur, R. (2006). The role of the FDA in the regulation of anti-cancer drugs. *Current Cancer Therapy Reviews* 2: 327–29.

Darling, D. A. (1976). The birth, growth, and blossoming of sequential analysis. In D. B. Owen, W. G. Cochran, H. O. Hartley & J. Neyman, eds., *On the History of Statistics and Probability*. New York: Marcel Dekker.

Daston, L. & Galison, P. (2007). *Objectivity*. New York: Zone Books.

Davies, O. L. (1958). The design of screening tests in the pharmaceutical industry. *Bulletin de l'institut international de statistique* 36: 226–41.

Davis, H. L., Durant, J. R. & Holland, J. F. (1980). Interrelationships: The groups, the NCI and other governmental agencies. In B. Hoogstraten, ed., *Cancer Research: Impact of the Cooperative Groups*. New York: Masson: 371–90.

DCT board approves major clinical trials changes, regional groups, new effort in surgical oncology. (1980). *The Cancer Letter* 6 (40): 1–7.

DCT board committee to recommend limited start on *in vitro* screening. (1985). *The Cancer Letter* 11 (3): 2–8.

DCT board to study "modification" of present Cooperative Group system. (1986). *The Cancer Letter* 12 (24): 4–6.

de Klein, A., Geurts van Kessel, A. H., Grosveld, G., Bartram, C. R., Spurr, N. K., Heisterkamp, N., Groffen, J. & Stephenson, J. R. (1982). A cellular oncogene is translocated to the Philadelphia chromosome in chronic myelocytic leukemia. *Nature* 300: 765–67.

de Vries, E. G. E., Mulder, N. H. & Sleijfer, D. Th. (1999). Medical oncology in the next century. *Netherlands Journal of Medicine* 55: 310–15.

Decades of Progress: 1983–2003, Community Clinical Oncology Program. (2003). Washington, DC: National Cancer Institute, U.S. Department of Health and Human Services.

DeGroot, M. H. (1988). A conversation with George A. Barnard. *Statistical Science* 3: 196–212.

Del Regato, J. A. (1960). Suggestions for incorporating studies, using radiotherapy, into the Clinical Studies of the Cancer Chemotherapy National Cancer Center: Report to the Clinical Studies Panel of the National Cancer Institute, February 1960. *Cancer Chemotherapy Reports* 7: 47–49.

——. (1984). The American Board of Radiology: Its 50th anniversary. *American Journal of Roentgenology* 144: 197–200.

——. (1993). *Radiological Oncologists: The Unfolding of a Medical Specialty*. Reston, VA: Radiology Centennial.

——. (1996). The unfolding of American radiotherapy. *International Journal of Radiation Oncology, Biology and Physics* 35: 5–14.

Descombes, V. (2001). *The Mind's Provisions: A Critique of Cognitivism*. Princeton, NJ: Princeton University Press.

DeVita dissatisfied with cooperative groups on accrual. (1987). *The Cancer Letter* 13 (45): 1–3.

DeVita, V. T. (1984). On special initiatives, critics and the National Cancer Institute. *Cancer Treatment Reports* 68: 1–4.

——. (2003). A selective history of the therapy of Hodgkin's disease. *British Journal of Haematology* 122: 718–27.

DeVita, V. T., Jr. & Chu, E. (2008). A history of cancer chemotherapy. *Cancer Research*, 68: 8643–53.

DeVita, V. T., Oliverio, V. T., Muggia, F. M., Wiernik, P. W., Ziegler, J., Goldin, A., Rubin, D., Henney, J. & Schepartz, S. (1979). The Drug Development and Clinical Trials Program of the Division of Cancer Treatment, National Cancer Institute. *Cancer Clinical Trials* 2: 195–216.

Diamond, L. K. (1985). This week's citation classic. *Current Contents* 14 (April 8): 14.

Dilts, D. M., Sandler, A. B., Baker, M., Cheng, S. K., George, S. L., Karas, K. S., McGuire, S., Menon, G. S., Reusch, J., Sawyer, D., Scoggins, M., Wu, A., Zhou, K. & Schilsky, R. L. (2006). Processes to activate Phase III clinical trials in a cooperative oncology group: The case of cancer and leukemia group B. *Journal of Clinical Oncology* 24: 4553–57.

Dilts, D. M., Sandler, A., Cheng, S., Crites, J., Ferranti, L., Wu, A., Gray, R., MacDonald, J., Marinucci, D. & Comis, R. (2008). Development of clinical trials in a cooperative group setting: The Eastern Cooperative Oncology Group. *Clinical Cancer Research* 14: 3427–33.

DiMasi, J. A. & Grabowski, H. G. (2007). Economics of new oncology drug development. *Journal of Clinical Oncology* 25: 209–16.

Dodier, N. (1995). The conventional foundations of action: Elements of a sociological pragmatics. *Réseaux: The French Journal of Communication* 3 (2): 147–66.

Drazen, J. M. (2003). Controlling research trials. *New England Journal of Medicine* 348: 1377–80.

Druker, B. J. (2002a). Imatinib and chronic myeloid leukemia: Validating the promise of molecularly targeted therapy. *European Journal of Cancer* 38 (Supplement 5): S70–S76.

——. (2002b). STI571 (Gleevec™) as a paradigm for cancer therapy. *Trends in Molecular Medicine* 8: S14–S18.

——. (2007). Targeting hope when nobody believes in you: Dr. Brian Druker's OHSU commencement speech. *The Oregonian* (7 June), available at http://blog.oregonlive.com/oregonianextra/2007/06/dr_brian_drukers_ohsu_commence.html.

——. (2009). Perspectives on the development of imatinib and the future of cancer research. *Nature Medicine* 15: 1149–52.

Druker, B. J. & Lydon, N. B. (2000). Lessons learned from the development of an ABL tyrosine kinase inhibitor for chronic myelogenous leukemia. *Journal of Clinical Investigation* 105: 3–7.

Druker, B. J., Mamon, H. J. & Roberts, T. M. (1989). Oncogenes, growth factors, and signal transduction. *New England Journal of Medicine* 321: 1383–91.

Druker, B. J., Talpaz, M., Resta, D. J., Peng, B., Buchdunger, E., Ford, J. M., Lydon, N. B., Kantarjian, H., Capdeville, R., Ohno-Jones, S. & Sawyer, C. (2001). Efficacy and safety of a specific inhibitor of the BCR-ABL tyrosine kinase in chronic myeloid leukemia. *New England Journal of Medicine* 344: 1031–37.

Druker, B. J., Talpaz, M., Resta, D., Peng, B., Buchdunger, E., Ford, J. & Sawyers, C. L. (1999). Clinical efficacy and safety of an ABL-specific tyrosine kinase inhibitor as targeted therapy for chronic myelogenous leukemia. *Blood* 94 (Supplement 1): 368a.

Druker, B. J., Tamura, S., Buchdunger, E., Ohno, S., Segal, G. M., Fanning, S., Zimmerman, J. & Lydon, N. B. (1996). Effects of a selective inhibitor of the ABL tyrosine kinase on the growth of BCR-ABL positive cells. *Nature Medicine* 2: 561–66.

Duffin, J. (2002a). Poisoning the spindle: Serendipity and discovery of the anti-tumor properties of the vinca alkaloids (Part 1). *Pharmacy in History* 44: 64–76.

——. (2002b). Poisoning the spindle: Serendipity and discovery of the anti-tumor properties of the vinca alkaloids (Part 2). *Pharmacy in History* 44: 105–19.

Dukes, C. E. & Bussey, H. J. (1958). The spread of rectal cancer and its effect on prognosis. *British Journal of Cancer* 12: 309–20.

Eckhouse, S., Lewison, G. & Sullivan, R. (2008). Trends in the global funding and activity of cancer research. *Molecular Oncology* 2: 20–32.

Edler, L. & Burkholder, I. (2006). Overview of Phase I trials. In J. Crowley & D. P. Ankerst, eds., *Handbook of Statistics in Clinical Oncology*, 2nd ed. Boca Raton, FL: CRC Press, 3–30.

Eisenhauer, E. A. (1998). Phase I and II trials of novel anti-cancer agents: Endpoints, efficacy, and existentialism. *Annals of Oncology* 9: 1047–52.

El-Maraghi, R. H. & Eisenhauer, E. A. (2008). Review of Phase II trial designs used in studies of molecular targeted agents: Outcomes and predictors of success in Phase III. *Journal of Clinical Oncology* 26: 1346–54.

Ellenberg, S., Fleming, T. & DeMets, D. L. (2002). *Data Monitoring Committees in Clinical Trials: A Practical Perspective*. Hoboken, NJ: John Wiley.

Ellenberg, S. S. & George, S. L. (2004). Should statisticians reporting to data monitoring committees be independent of the trial sponsor and leadership? *Statistics in Medicine* 23: 1503–5.

Ellis, H. (2002). *A History of Surgery*. Cambridge: Cambridge University Press.

Endicott, K. M. (1957). The Chemotherapy Program. *Journal of the National Cancer Institute* 19: 275–93.

——. (1958). The National Cancer Chemotherapy Program. *Journal of Chronic Diseases* 8 (1): 171–77.

———. (1959). The Cancer Chemotherapy Program of the United States. *Acta Unio Internationalis Contra Cancrum* 15 (Supplement 1): 103–8.

Enrick, N. L. (1946). The U.S. Army Quartermaster Corps' use of sequential sampling inspection. *Industrial Quality Control* 2: 12–15.

Epstein, S. (1996). *Impure Science: AIDS, Activism, and the Politics of Knowledge*. Berkeley: University of California Press.

———. (2007). *Inclusion: The Politics of Difference in Medical Research*. Chicago: University of Chicago Press.

ESMO/ASCO Task Force on Global Curriculum in Medical Oncology. (2004). Recommendations for a global core curriculum in medical oncology. *Journal of Clinical Oncology* 22: 4616–25.

Ezra Greenspan: A trailblazer in combination chemotherapy. (2005). *Cancer Investigation* 1: 1–2.

Faber, K. (1923). *Nosography in Modern Internal Medicine*. New York: Paul B. Hoeber.

Faguet, G. B. (2005). *The War on Cancer: An Anatomy of Failure, a Blueprint for the Future*. Dordrecht: Springer.

Farber, S. (1960). The future of the Cancer Chemotherapy National Program. In B. H. Morrison III, ed., *Conference on Experimental Clinical Cancer Chemotherapy. November 11 and 12, 1959*. Washington, DC: U.S. Department of Health, Education, and Welfare, 293–302.

Farber, S., Toch, R., Manning Sears, E. & Pinkel, D. (1956). Advances in chemotherapy in man. *Advances in Cancer Research* 4: 1–71.

Faul, F., Erdfelder, E., Lang, A.-G. & Buchner, A. (2007). G*Power 3: A flexible statistical power analysis program for the social, behavioral, and biomedical sciences. *Behavior Research Methods* 39: 175–91.

FDA examining endpoints for approval of cancer therapies, division director says. (2003). *The Cancer Letter* 29 (41): 1–5.

FDA to patients: Drop dead. (2002). *Wall Street Journal*, September 24, A18.

FDA. (2001). Food and Drug Administration. Guidance for clinical trial sponsors. On the establishment and operation of clinical trial data monitoring committees. Draft Guidance, available at http://www.fda.gov/Cber/gdlns/clindatmon.pdf.

FDA. (2006). Food and Drug Administration. Guidance for industry, investigators, and reviewers. Exploratory IND studies. January 2006, available at http://www.fda.gov/downloads/Drugs/GuidanceComplianceRegulatoryInformation/Guidances/UCM078933.pdf.

Feds short on money long on regs: Yarbro. (1983). *The Cancer Letter* 9 (12): 1–4.

Feedback: Adjuvant chemotherapy for breast cancer. (1985). *CA: A Cancer Journal for Clinicians* 35: 184–91.

Feinstein, A. R., Sosin, D. M. & Wells, C. K. (1985). The Will Rogers phenomenon: Stage migration and new diagnostic techniques as a source of misleading statistics for survival in cancer. *New England Journal of Medicine* 312: 1604–8.

Fenner, F. & Gibbs, A., eds. (1988). *Portraits of Viruses: A History of Virology.* New York: Karger.

Fiebig, H. H. & Berger, D. F. (1991). Preclinical Phase II trials. In E. Boven & B. Winograd, eds., *The Nude Mouse in Oncology Research.* Boca Raton, FL: CRC Press, 317–26.

Fisher, B. (1979). Presentation to the NCI Division of Cancer Treatment Board of Scientific Counselors. *The Cancer Letter* 15 (15): 6–7.

———. (1991). A biological perspective of breast cancer: Contributions of the National Surgical Adjuvant Breast and Bowel Project clinical trials. *CA: A Cancer Journal for Physicians* 41: 97–111.

Fisher, B., Carbone, P., Economou, S. G., Frelick, R., Glass, A., Lerner, H., Redmond, C., Zelen, M., Band, P., Katrych, D. L., Wolmark, N. & Fisher, E. R. (1975). 1-Phenylalanine mustard (L-PAM) in the management of primary breast cancer. A report of early findings. *New England Journal of Medicine* 292: 117–22.

Fisher, B., Glass, A., Redmond, C., Fisher, E. R., Barton, B., Such, E., Carbone, P., Economou, S., Foster, R., Frelick, R., Lerner, H., Levitt, M., Margolese, R., MacFarlane, J., Plotkin, D., Shibata, H. & Volk, H. (1977). L-phenylalanine mustard (L-PAM) in the management of primary breast cancer. An update of earlier findings and a comparison with those utilizing L-PAM plus 5-fluorouracil (5-FU). *Cancer* 39 (6 Supplement): 2883–903.

Fisher, J. A. (2009). *Medical Research for Hire: The Political Economy of Pharmaceutical Clinical Trials.* Piscataway, NJ: Rutgers University Press.

Flamant, R. & Fohanno, C. (1982). Introduction. In R. Flamant, T. B. Grage & C. Fohanno, eds., *Evaluation of Methods of Treatment and Diagnostic Procedures in Cancer* (UICC Technical Report Series, vol. 70). Geneva: Union for International Cancer Control, 1–3.

Foucault, M. (1963). *Naissance de la Clinique.* Paris: Presses Universitaires de France.

———. (1994a). *Dits et écrits, 1954–1988,* vol. 3, *1976–1979.* Paris: Gallimard.

———. (1994b). *Dits et écrits, 1954–1988,* vol. 4, *1980–1988.* Paris: Gallimard.

Foye, L. V. (1970). The Cooperative Clinical Research Program of the National Cancer Institute. *American Journal of Roentgenology, Radium Therapy, and Nuclear Medicine* 108: 14–18.

Frei, Freireich share one GM award. (1983). *The Cancer Letter* 9: 5.

Frei, E., III. (1960). Remarks made during the "Panel Meeting: Investigation of therapy in solid tumors." In B. H. Morrison III, ed., *Conference on Experimental Clinical Cancer Chemotherapy. November 11 and 12, 1959.* Washington, DC: U.S. Department of Health, Education, and Welfare, 277–92.

——. (1997). Confrontation, passion, and personalization. *Clinical Cancer Research* 3: 2554–62.

Frei, E. & Freireich, E. J. (1965). Progress and perspectives in the chemotherapy of acute leukemia. *Advances in Chemotherapy* 2: 269–98.

Frei, E., III, Holland, J. F., Schneiderman, M. A., Pinkel, D., Selkirk, G., Freireich, E. J., Silver, R. T., Gold, G. L. & Regelson, W. (1958). A comparative study of two regimens of combination chemotherapy in acute leukemia. *Blood* 13: 1126–48.

Freireich, E. J. (1990). Is there any mileage left in the randomized clinical trial? *Cancer Investigations* 8: 231–32.

——. (1997). Who took the clinical out of clinical research?—Mouse *versus* man: Seventh David A. Karnofsky Memorial Lecture, 1976. *Clinical Cancer Research* 3: 2711–22.

Freireich, E. J. & Gehan, E. A. (1974). Historical controls reincarnated: An examination of techniques for clinical therapeutic research. *Cancer Chemotherapy Reports* 58: 623–26.

——. (1979). The limitations of the randomized clinical trial. *Methods in Cancer Research* 17: 277–310.

Freireich, E. J., Gehan, E., Frei, E., III, Schroeder, L. R., Wolman, I. J., Anbari, R., Burgert, E. O., Mills, S. D., Pinkel, D., Selawry, O. S., Moon, J. H., Gendel, B. R., Spurr, C. L., Storrs, R., Haurani, F., Hoogstraten, B. & Lee, S. (Acute Leukemia Group B). (1963). The effect of 6-mercaptopurine on the duration of steroid induced remissions in acute leukemia: A model for evaluation of other potentially useful therapies. *Blood* 21: 699–716.

Freireich, E. J., Judson, G. & Levin, R. H. (1965). Separation and collection of leukocytes. *Cancer Research* 25: 1516–20.

Freireich, E. J., Karon, M. & Frei, E., III. (1964). Quadruple combination therapy (VAMP) for acute lymphocytic leukemia of childhood. *Proceedings of the American Association for Cancer Research* 5: 20.

Freireich, E. J., Schmidt, P. J., Schneiderman, M. A. & Frei, E. (1959). A comparative study of the effect of transfusion of fresh and preserved whole blood on bleeding in patients with acute leukemia. *New England Journal of Medicine* 260: 6–11.

Fricker, J. (2007). Translational cancer research in Europe. *Molecular Oncology* 1: 8–10.

Fujimura, J. H. (1996). *Crafting Science: A Sociohistory of the Quest for the Genetics of Cancer.* Cambridge, MA: Harvard University Press.

Fujimura, J. H. & Chou, D. Y. (1994). Dissent in science: Styles of scientific practice and the controversy over the cause of AIDS. *Social Science & Medicine* 28: 1017–26.

Gail, M. H. (1997). A conversation with Nathan Mantel. *Statistical Science* 12: 88–97.

Galambos, L. & Sturchio, J. L. (1998). Pharmaceutical firms and the transition to biotechnology: A study in strategic innovation. *Business History Review* 72: 250–78.

Galison, P. (1997). *Image and Logic: A Material Culture of Microphysics*. Chicago: University of Chicago Press.

Gaudillière, J.-P. (1998). The molecularization of cancer etiology in the postwar United States: Instruments, politics, and management. In S. de Chadarevian and H. Kamminga, eds., *Molecularizing Biology and Medicine: New Practices and Alliances, 1910s–1970s*. Amsterdam: Harwood Academic, 139–70.

Gaydos, L., Freireich, E. J. & Mantel, N. (1962). The quantitative relation between platelet count and hemorrhage in patients with acute leukemia. *New England Journal of Medicine* 266: 905–9.

Gazdar, A. F. (2009). Personalized medicine and inhibition of EGFR signaling in lung cancer. *New England Journal of Medicine* 361: 1018–20.

GECA (Groupe Européen de Chimiothérapie Anticancéreuse). (1966a). Protocole pour les essais cliniques de traitement des cancers mammaires humains en phase avancée. *European Journal of Cancer* 2: 201–3.

——. (1966b). Protocole pour les essais cliniques de traitement des leucémies et hématosarcomes. *European Journal of Cancer* 2: 85–88.

Gehan, E. A. (1961). The determination of the number of patients required in a preliminary and a follow-up trial of a new chemotherapeutic agent. *Journal of Chronic Diseases* 13: 346–53.

——. (1965). A generalized Wilcoxon test for comparing arbitrarily singly-censored data. *Biometrika* 52: 203–23.

——. (1980). Knowledge acquisition from historical data: Application to cancer clinical trials. *Bulletin du Cancer* 67: 437–45.

——. (1982). Progress of therapy in acute leukemia 1948–1981: Randomized *versus* nonrandomized clinical trials. *Controlled Clinical Trials* 3: 199–207.

——. (1984). The evaluation of therapies: Historical control studies. *Statistics in Medicine* 3: 315–24.

Gehan, E. A. & Freireich, E. J. (1974). Non-randomized controls in clinical cancer trials. *New England Journal of Medicine* 290: 198–203.

——. (1981). Cancer clinical trials: A rational basis for use of historical controls. *Seminars in Oncology* 8: 430–36.

Gehan, E. A. & Schneiderman, M. A. (1990). Historical and methodological developments in clinical trials at the National Cancer Institute. *Statistics in Medicine* 9: 871–80.

Gellhorn, A. (1959). Invited remarks on the current status of research in clinical cancer chemotherapy. *Cancer Chemotherapy Reports* 5: 1–12.

Gellhorn, A. & Hirschberg, E., eds. (1955). Investigation of diverse systems for cancer chemotherapy screening. *Cancer Research Supplement* 3.

George, S. L. (1976). The data center of the European Organization for Research on Treatment of Cancer. *European Journal of Cancer* 12: 245–46.

———. (1980). Sequential trials in cancer research. *Cancer Treatment Reports* 64: 393–97.

George, S. L. & Desu, M. M. (1974). Planning the size and duration of a clinical trial studying the time to some critical event. *Journal of Chronic Diseases* 27: 15–24.

Geran, G. I., Greenberg, N. H., MacDonald, M. M., Schumacker, A. & Abbott, B. J. (1962). Protocols for screening chemical agents and natural products against animal tumors and other biological systems. *Cancer Chemotherapy Reports* 25: 1–66.

Gerlis, L. (1989). Good clinical practice in clinical research. *The Lancet* 333: 1008–9.

Geurts van Kessel, A. H., ten Brinke, H., Boere, W. A., den Boer, W. C., de Groot, P. G., Hagemeijer, A., Khan, P. M. & Pearson, P. L. (1981). Characterization of the Philadelphia chromosome by gene mapping. *Cytogenetics and Cell Genetics* 30: 83–91.

Gilbert, G. N. & Mulkay, M. (1984). *Opening Pandora's Box*. Cambridge: Cambridge University Press.

Gilbert, R. (1939). Radiotherapy in Hodgkin's disease (Malignant Granulomatosis): Anatomic and clinical foundations; governing principles; results. *American Journal of Roentgenology & Radiation Therapy* 41: 198–241.

Gilchrist, K. W., Harrington, D. P., Wolf, B. C. & Neiman, R. S. (1988). Statistical and empirical evaluation of histopathologic reviews for quality assurance in the Eastern Cooperative Oncology Group. *Cancer* 62: 861–68.

Goldberg, I. S., McMahan, C. A., Escher, G. C., Volk, H., Ansfield, F. J. & Olson, K. B. (1973). Secondary chemotherapy of advanced breast cancer. *Cancer* 31: 660–63.

Goldberg, K. B. (2010a). Former group chair responds to Niederhuber: Change but don't dissolve the cooperative groups. *The Cancer Letter* 36 (24): 1, 3–4.

———. (2010b). IOM panel: NCI clinical trials groups at "breaking point," more funds needed. *The Cancer Letter* 36 (14): 1–5.

———. (2010c). Lame duck NCI director: Cancer centers should run cancer clinical trials system. *The Cancer Letter* 36 (24): 1–5.

Goldberg, P. (2005). An "insurgency" targets randomized trials, demands access to investigational drugs. *The Cancer Letter* 31 (31): 1–16.

———. (2010a). Academic oncologists losing control over Phase I trials as for-profits expand. *The Cancer Letter* 36 (11): 1–5.

———. (2010b). FDA to reorganize oncology office into four divisions by disease type. *The Cancer Letter* 36 (30): 1–4.

———. (2010c). Three cooperative groups to combine statistical center back end operations. *The Cancer Letter* 36 (24): 1–3.

Goldin, A., Greenspan, E. M. & Schoenbach, E. B. (1952). Studies on the mechanism of chemotherapeutic agents in cancer. VI Synergistic (additive) action of drugs on a transplantable leukemia in mice. *Cancer* 5: 153–60.

Goldin, A., Muggia, F. M. & Rozencweig, M. (1979a). International activities of the Division of Cancer Treatment, National Cancer Institute. *Cancer Clinical Trials* 2 (1): 29–35.

Goldin, A., Schepartz, S. A., Venditti, J. M. & DeVita, V. T. (1979b). Historical development and current strategy of the National Cancer Institute Drug Development Program. *Methods in Cancer Research* 16: 165–245.

Goldin, A., Venditti, J. M., Muggia, F. M., Rozencweig, M. & DeVita, V. T. (1978). New animal models in cancer chemotherapy. In B. W. Fox, ed., *Advances in Medical Oncology, Research and Education*, Vol. V, *Basis for Cancer Therapy 1*. New York: Pergamon Press, 113–22.

Goldstein, G. & Webb, R. F. (1983). Adjuvant chemotherapy in breast cancer: No news. *The Lancet* 1: 816.

Gollin, F. F., Ansfield, F. J., Curreri, A. R., Heidelberger, C. & Vermund, H. (1962). Combined chemotherapy and irradiation in inoperable bronchogenic carcinoma. *Cancer* 15: 1209–17.

Gonzalez-Crussi, F. (1999). Anaplasia and Wilms Tumor, or what's in a name? *Medical and Pediatric Oncology* 32: 324–25.

Goodman, A. (1963). The initial clinical trial of nitrogen mustard. *American Journal of Surgery* 105: 574–78.

Goodman, J. & Walsh, V. (2001). *The Story of Taxol: Nature and Politics in the Pursuit of an Anti-Cancer Drug*. Cambridge: Cambridge University Press.

Goodman, L. S. & Gilman, A. (1941). *The Pharmacological Basis of Therapeutics*. New York: Macmillan.

Goodman, S. N. (1999). Toward evidence-based medical statistics. 1: The P Value fallacy. *Annals of Internal Medicine* 130: 995–1004.

Goulart, B., Roberts, T. G., Clack, J. & Chabner, B. A. (2008). Reply to the letter to the editor from Banerji. *Clinical Cancer Research* 14: 2513.

Goustin, A. S., Leof, E. B., Shipley, G. D. & Moses, H. L. (1986). Growth factors and cancer. *Cancer Research* 45: 1015–29.

Grage, T. B. & Zelen, M. (1982). The controlled randomized clinical trial in the evaluation of cancer treatment: The dilemma and alternative trial designs. In R. Flamant, T. B. Grage & C. Fohanno, eds., *Evaluation of Methods of Treatment and Diagnostic Procedures in Cancer* (UICC Technical Report Series, vol. 70). Geneva: Union for International Cancer Control, 23–46.

Graziani, Y., Erikson, E. & Erikson, R. L. (1983). The effect of quercetin on the phosphorylation activity of Rous sarcoma virus transforming gene product in vitro and in vivo. *European Journal of Biochemistry* 135: 583–98.

Green, S. (2006). Overview of Phase II trials. In J. Crowley & D. P. Ankerst, eds., *Handbook of Statistics in Clinical Oncology*, 2nd ed. Boca Raton, FL: CRC Press, 199–229.

Green, S., Benedetti, J. & Crowley, J. (1997). *Clinical Trials in Oncology*. New York: Chapman and Hall.

Green, S. J., Fleming, T. R. & O'Fallon J. R. (1987). Policies for study monitoring and interim reporting of results. *Journal of Clinical Oncology* 5: 1477–84.

Greenberg, B. G. (1959). Conduct of cooperative field and clinical trials. *American Statistician* 13: 13–28.

———. (1961). Maintaining interest among members of cooperative clinical study groups. *Cancer Chemotherapy Reports* 14: 189–90.

Greenberg, B. G. & Grizzle, J. E. (1969). Effective statistical consultation for clinical study groups. *Cancer Chemotherapy Reports* 53: 1–2.

Greenberg Report. ([1967] 1988). Organization, review, and administration of cooperative studies: A report from the Heart Special Project Committee to the National Advisory Heart Council, May 1967. *Controlled Clinical Trials* 9: 137–48.

Greene, J. (2007). *Prescribing by Numbers: Drugs and the Definition of Disease*. Baltimore: Johns Hopkins University Press.

Greenhouse, S. W. (1997). Some reflections on the beginnings and development of statistics in "Your Father's NIH." *Statistical Science* 12: 82–87.

Greenspan, E. (1968). Combination antimetabolite and alkylating agent chemotherapy in advanced breast carcinoma. *New York State Journal of Medicine* 68: 780–86.

Grever, M. C. B. (2001). Cancer drug discovery and development. In V. DeVita and S. A. Rosenberg, eds., *Cancer: Principles and Practice of Oncology*. Philadelphia: Lippincott-Raven, 328–39.

Group chairmen balk at DCT board's recommendations for clinical trials. (1980). *The Cancer Letter* 6 (7): 4–6.

Group chairmen unhappy over exclusion from funding at recommended levels. (1983). *The Cancer Letter* 9 (46): 4–7.

Group chairmen unhappy over H & N contracts, NCI protocol reviews. (1978). *The Cancer Letter* 4 (2): 6–8.

Gschwind, A., Fischer, O. M. & Ullrich, A. (2004). The discovery of receptor tyrosine kinases: Targets for cancer therapy. *Nature Reviews Cancer* 4: 361–70.

Gutierrez, M. & Collyar, D. (2008). Patient perspectives on Phase 0 clinical trials. *Clinical Cancer Research* 14: 3689–97.

Gutierrez, M. E, Kummar, S. & Giaccone, G. (2009). Next generation oncology drug development: Opportunities and challenges. *Nature Reviews Clinical Oncology* 6: 259–65.

Hacking, I. (1992a). The self-vindication of the laboratory sciences. In A. Pickering, ed., *Science as Practice and Culture*. Chicago: University of Chicago Press, 29–64.

———. (1992b). Statistical language, statistical truth and statistical reason: The self-authentication of a style of scientific reasoning. In E. McMullin, ed., *The Social Dimensions of Science*. Notre Dame, IN: University of Notre Dame Press, 130–57.

———. (2004). Inaugural lecture: Chair of philosophy and history of scientific concepts at the Collège de France, 16 January 2001. *Economy and Society* 31: 1–14.

Hallock, Y. F. & Cragg, G. M. (2003). National Cooperative Drug Discovery Groups (NCDDGs): A successful model for public private partnerships in cancer drug discovery. *Pharmaceutical Biology* 41: 78–91.

Hampton, T. (2006). Targeted cancer therapies lagging: Better trial design could boost success rate. *JAMA* 296: 1951–52.

Hanks, S. K., Quinn, A. M. & Hunter, T. (1988). The protein kinase family: Conserved features and deduced phylogeny of the catalytic domains. *Science* 241: 42–51.

Hansemann, D. von. (1892). Über die Anaplasie der Geschwulstzellen. *Virchows Archiv* 129: 436–39.

———. (1903). *Studien über die Spezifität, den Altruismus und die Anaplasie der Zellen*. Berlin: Hirschwald.

Hansen, H. H., Bajorin, D. F., Muss, H. B., Purkalne, G., Schrijvers, D. & Stahel, R. (2004). Recommendations for a global core curriculum in medical oncology. *Annals of Oncology* 15: 1603–12.

Harper, P. S. (2006). *First Years of Human Chromosomes: The Beginning of Human Cytogenetics*. Bloxham, UK: Scion Publishing.

Harris, H. (1995). *The Cells of the Body: A History of Somatic Cell Genetics*. Cold Spring Harbor, NY: Cold Spring Harbor Laboratory Press.

Harrison, C. (2009). Glivec patent denied in India. *Nature Reviews Drug Discovery* 8: 606–7.

Hayter, C. R. R. (1998). The clinic as laboratory: The case of radiation therapy, 1896–1920. *Bulletin of the History of Medicine* 72: 663–88.

———. (2005). *An Element of Hope: Radium and the Response to Cancer in Canada, 1900–1940*. Montreal & Kingston: McGill-Queens University Press.

Heidelberger, C. & Ansfield, F. J. (1963). Experimental and clinical use of fluorinated pyrimidines in cancer chemotherapy. *Cancer Research* 23: 1226–43.

Heidelberger, C., Chaudhuri, N. K., Danneberg, P., Mooren, D., Griesbach, L., Duschinsky, R., Schnitzer, J., Pleven, E. & Scheiner, J. (1957). Fluorinated pyrimidines, a new class of tumor-inhibitory compounds. *Nature* 179: 663–66.

Heisterkamp, N., Groffen, J., Stephenson, J. R., Spurr, N. K., Goodfellow, P. N., Solomon, E., Carritt, B. & Bodmer, W. F. (1982). Chromosomal localization of human cellular homologues of two viral oncogenes. *Nature* 299: 747–49.

Henschke, U. K., Leffall, L. D., Mason, C. H., Reinhold, A. W., Schneider, R. L.

& White, J. E. (1973). Alarming increase of the cancer mortality in the U.S. black population (1950–1967). *Cancer* 31: 763–68.

Herzog, H. & Oliveto, E. P. (1992). A history of significant steroid discoveries and developments originating at the Schering Corporation (USA) since 1948. *Steroids* 57: 617–23.

Hester, J. P., Kellogg, R. M., Mulzet, A. P. & Freireich, E. J. (1985). Continuous-flow techniques for platelet concentrate collection: A step toward standardization and yield predictability. *Journal of Clinical Apheresis* 2: 224–30.

Heusler, K. & Kalvoda, J. (1996). Between basic and applied research: Ciba's involvement in steroids in the 1950s and 1960s. *Steroids* 8: 492–503.

Heyn, R. M., Brubaker, C. A., Burchenal, J. H., Cramblett, H. G. & Wolff, J. A. (1960). The comparison of 6-mercaptopurine with the combination of 6-mercaptopurine and azaserine in the treatment of acute leukemia in children: Results of a cooperative study. *Blood* 15: 350–59.

Higgins, G. A. (1984). Adjuvant therapy for carcinoma of the colon and rectum. *International Advances in Surgical Oncology* 7: 77–111.

Higgins, G. A., Serlin, O., Hughes, F. & Dwight, R. W. (1962). The veterans administration surgical adjuvant group—Interim report. *Cancer Chemotherapy Reports* 16: 141–51.

Higgins, G. A. & White, G. E. (1968). Cancer chemotherapy and surgery. *Surgical Clinical of North America* 48 (4): 839–50.

Hill, A. B. (1990). Suspended judgment: Memories of the British streptomycin trial in tuberculosis, the first randomized clinical trial. *Controlled Clinical Trials* 11: 77–79.

Hillier, S. G. (2007). Diamonds are forever: The cortisone legacy. *Journal of Endocrinology* 195: 1–6.

History of the Cancer Chemotherapy Program. (1966). *Cancer Chemotherapy Reports* 50: 349–96.

Hoest, H. & Nissen-Meyer, R. (1960). A preliminary clinical study of cyclophosphamide. *Cancer Chemotherapy Reports* 9: 47–50.

Holden, W. D. & Dixon, W. J. (1962). A study of the use of triethylenethiophosphoramide as an adjuvant in the treatment of colorectal cancer. *Cancer Chemotherapy Reports* 16: 129–34.

Holland, J. F. (1968). Clinical studies of unmaintained remissions in acute lymphocytic leukemia. In (no ed.), *The Proliferation and Spread of Neoplastic Cells: A Collection of Papers Presented at the Twenty-First Annual Symposium on Fundamental Cancer Research, 1967.* Baltimore: Williams and Wilkins, 453–62.

———. (1976a). Breast cancer and chemotherapy. *Science* 192: 1062–63.

———. (1976b). Major advance in breast cancer chemotherapy. *New England Journal of Medicine* 294: 440.

———. (1995). CALGB—The Early Years. *CALGB Newsletter* 4: 6.

Holland, J. F., Frei, E., III, Kufe, D. W., Bast, R. C., Jr., Pollock, R. E. & Weichsel-baum, R. R. (2000). Principles of multidisciplinary management. In R. C. Bast, D. W. Kufe, R. E. Pollock, R. R. Weichselbaum, J. F. Holland & E. Frei III, eds., *Cancer Medicine*, 5th ed. Hamilton, ON: Decker, 986–91.

Hollander, S. & Gordon, M. (1990). The plateau in cytotoxic drug discovery. *Journal of Medicine* 21: 143–63.

Hollingshead, M. G., Alley, M. C., Kaur, G., Pacula-Cox, C. M. & Stinson, S. F. (2004). NCI specialized procedures in preclinical drug evaluation. In B. A. Teicher & P. A. Andrews, eds., *Anticancer Drug Development Guide: Pre-clinical Screening, Clinical Trials, and Approval*, 2nd ed. Totowa, NJ: Hu-mana Press, 153–82.

Holsti, L. R. (1995). The development of clinical radiotherapy since 1896. *Acta Oncologica* 34: 995–1003.

Hoogstraten, B., ed. (1980). *Cancer Research: Impact of the Cooperative Groups.* New York: Masson.

———. (2005). *Cancer Doctor.* New York: iUniverse.

Horiot, J. C., Johansson, K. A., Gonzalez, D. G., van der Schueren, E., van den Bogaert, W. & Notter, G. (1986). Quality assurance control in the EORTC cooperative group of radio therapy. *Radiotherapy and Oncology* 6: 275–84.

Hortobagyi, G. N. (2005). Trastuzumab in the treatment of breast cancer. *New England Journal of Medicine* 353: 1734–36.

Horton, H. V. (1948). A method for obtaining random numbers. *Annals of Mathematical Statistics* 19: 81–85.

Houÿez, F. (2004). Active involvement of patients in drug research, evaluation, and commercialization: European perspective. *Journal of Ambulatory Care Management* 27: 139–45.

Howie, D. (2002). *Interpreting Probability: Controversies and Developments in the Early Twentieth Century.* Cambridge: Cambridge University Press.

Huebner, R. J. & Todaro, G. J. (1969). Oncogenes of RNA tumor viruses as de-terminants of cancer. *Proceedings of the National Academy of Sciences USA* 64: 1087–94.

Hunter, T. & Simon, J. (2007). A not so brief history of the oncogene meeting and its cartoons. *Oncogene* 26: 1260–67.

Hutter, R. V. (1987). At last, worldwide agreement on the staging of cancer. *Archives of Surgery* 122: 1235–39.

IL-1β-targeted antibody approved for rare autoinflammatory disorders. (2009). *Nature Reviews Drug Discovery* 8: 605.

Itri, L., Kisner, D., Moertel, C. & Rockette, H. (1985). Panel discussion: Quality assurance. *Cancer Treatment Reports* 69: 1223–29.

Jackson, H. & Parker, F. (1947). *Hodgkin's Disease and Allied Disorders.* New York: Oxford University Press.

Jackson, J. E. (1960). Bibliography on sequential analysis. *Journal of the American Statistical Association* 55: 561–80.

Javier, R. T. & Butel, J. S. (2008). The history of tumor virology. *Cancer Research* 68: 7693–706.

Johnson, J., Williams, G. & Pazdur, R. (2003). End points and United States Food and Drug Administration approval of oncology drugs. *Journal of Clinical Oncology* 21: 1404–11.

Johnson, N. L. (1961). Sequential analysis: A survey. *Journal of the Royal Statistical Society. Series A (General)* 124: 372–411.

Johnson. R. O., Biselk, H., Andrews, N., Wilson, W., Rochlin, D., Segaloff, A., Krementz, E., Aust, J. & Ansfield, F. (1966). Phase I clinical study of 6V-methyl-pregn-4-ene-3,11,20-trione (NSC-17256). *Cancer Chemotherapy Reports* 50: 671–73.

Jones, S. E. (1978). How to evaluate improvements in cancer treatment: The debate continues. *Biomedicine* 28 (Special issue): 55–57.

Kalish, L. A. & Begg, C. B. (1985). Treatment allocation methods in clinical trials: A review. *Statistics in Medicine* 4: 129–44.

Kantarjian, H., Sawyers, C., Hochhaus, A., Guilhot, F., Schiffer, C., Gambacorti-Passerini, C., Niederwieser, D., Resta, D., Capdeville, R., Zoellner, U., Talpaz, M. & Druker, B. on behalf of the International STI571 CML Study Group. (2002). Hematologic and cytogenetic responses to imatinib mesylate in chronic myelogenous leukemia. *New England Journal of Medicine* 346: 645–52.

Kantarjian, H. M., Cortes, J., La Rosée, P. & Hochhaus, A. (2010). Optimizing therapy for patients with chronic myelogenous leukemia in chronic phase. *Cancer* 116: 1419–30.

Kantarjian, H. M., Talpaz, M., O'Brien, S., Giles, F., Garcia-Manero, G., Faderl, S., Thomas, D., Shan, J., Rios, M. B. & Cortes, J. (2003). Dose escalation of imatinib mesylate can overcome resistance to standard-dose therapy in patients with chronic myelogenous leukemia. *Blood* 101: 473–75.

Kaplan, E. L. (1983). This week's citation classic. *Current Contents* 24: 14.

Kaplan, E. L. & Meier, P. (1958). Nonparametric estimation from incomplete observations. *Journal of the American Statistical Association* 53: 457–81.

Kaplan, H. S. (1962). The radical radiotherapy of regionally localized Hodgkin's disease. *Radiology* 78: 553–61.

Karlberg, J. P. E., Yao, T. J. & Yau, H. K. C. (2009). Industry sponsored oncology clinical trials. *Clinical Trial Magnifier* 2: 402–16.

Karnofsky, D. (1960). Remarks made during the "Panel Meeting: Use of alkylating agents and the future of these agents." In B. H. Morrison III, ed., *Conference on Experimental Clinical Cancer Chemotherapy. November 11 and 12, 1959*. Washington, DC: U.S. Department of Health, Education, and Welfare, 127–48.

——. (1966). The staging of Hodgkin's disease. *Cancer Research* 26: 1090–94.

Kaufmann, M., Pusztai, L. & Biedenkopf Expert Panel Members. (2011). Use of standard markers and incorporation of molecular markers into breast cancer therapy. Consensus recommendations from an international expert panel. *Cancer* 117: 1575–82.

Keating, P. & Cambrosio, A. (2003). *Biomedical Platforms: Realigning the Normal and the Pathological in Late-Twentieth-Century Medicine.* Cambridge, MA: MIT Press.

——. (2009). Who's minding the data? Data managers and data monitoring committees in clinical trials. *Sociology of Health & Illness* 31: 325–42.

Kelley, W. N. & Randolph, M. A., eds. (1994). *Careers in Clinical Research: Obstacles and Opportunities.* Washington, DC: National Academy Press.

Kendall, M. G. & Babington Smith, B. (1938). Randomness and random sampling numbers. *Journal of the Royal Statistical Society* 101 (1): 147–66.

——. (1939a). Second paper on random sampling numbers. *Supplement to the Journal of the Royal Statistical Society* 6: 51–61.

——. (1939b). *Tables of Random Sampling Numbers.* Cambridge: Cambridge University Press.

Kenis, Y. (1977). Dedication. In H. J. Tagnon & M. J. Staquet, eds., *Recent Advances in Cancer Treatment.* New York: Raven Press, v–vi.

Kennedy, B. J. (1970). Oncology in medicine. *CA: A Cancer Journal for Clinicians* 20: 368–71.

——. (1999). Medical oncology: Its origin, evolution, current status, and future. *Cancer* 85: 1–8.

Kennedy, B. J., Calabresi, P., Carbone, P. P., Frei, E., III, Holland, J. F., Owens, A. H., Sleisenger, M. & Beck, J. C. (1973). Training program in medical oncology. *Annals of Internal Medicine* 78: 127–30.

Kent, A. (2007). Should patient groups accept money from drug companies? Yes. *British Medical Journal* 334: 934.

Khleif, S. (2009). Update of FDA's critical path initiative. *Clinical Advances in Hematology & Oncology* 7: 173–74.

Khleif, S. N. & Curt, G. A. (2000). Animal models in developmental therapeutics. In R. C. Bast, T. S. Gansler, J. F. Holland & E. Frei, *Cancer Medicine,* 5th ed. Hamilton, ON: Decker.

Kilpatrick, S. E. (1999). Histologic prognostication in soft tissue sarcomas: Grading versus subtyping or both? A comprehensive review of the literature with proposed practical guidelines. *Annals of Diagnostic Pathology* 3: 48–61.

Klein, J. L. (2000). Economics for a client: The case of statistical quality control and sequential analysis. *History of Political Economy* 32 (Supplement): 27–69.

Kliman, A., Gaydos, L., Schroeder, L. R. & Freireich, E. J. (1961). Repeated plasmapheresis of blood donors as a source of platelets. *Blood* 18: 593–608.

Kola, I. & Landis, J. (2004). Can the pharmaceutical industry reduce attrition rates? *Nature Reviews Drug Discovery* 3: 711–16.

Korn, E. L. (2004). Nontoxicity endpoints in Phase I trial designs for targeted, non-cytotoxic agents. *Journal of the National Cancer Institute* 96: 977–78.

Korn, E. L., Arbuck, S. G., Pluda, J. M., Simon, R., Kaplan, R. S. & Christian M. C. (2001). Clinical trial designs for cytostatic agents: Are new approaches needed? *Journal of Clinical Oncology* 19: 265–72.

Kramer, S. (1973). Radiation therapy and the Cancer Center: The advisory function of the Committee for Radiation Therapy Studies. In J. M. Vaeth, ed., *Frontiers of Radiation Therapy and Radiation, Vol. 8.* Baltimore: University Park Press, 26–28.

——. (1977). The study of patterns of cancer care in radiation therapy. *International Journal of Radiation Oncology, Biology and Physics* 39: 780–87.

Kramer, S., Perez, C., Jenkin, R. D. T. & Cox, J. D. (1980). Radiation therapy. In B. Hoogstraten, ed., *Cancer Research: Impact of the Cooperative Groups.* New York: Masson, 335–40.

Kramer, S., Hanks, G. E. & Diamond, J. J. (1982). Summary results from the Fourth Facilities Master List Survey conducted by Patterns of Care Study. *International Journal of Radiation Oncology, Biology and Physics* 8: 882–88.

Krueger, G. (2004). The formation of the American Society of Clinical Oncology and the development of a medical specialty, 1964–1973. *Perspectives in Biology and Medicine* 47: 537–51.

Krueger, G., Alexander, L., Whippen, D. & Balch, C. (2006). Arnoldus Goudschmit, M.D., Ph.D.: Chemotherapist, visionary, founder of the American Society of Clinical Oncology, 1903–2003. *Journal of Clinical Oncology* 24: 4033–36.

Krul, M. R. L. (1998). The NCI-EORTC symposia on new drugs in cancer therapy: A brief history. *The Oncologist* 3: 269–70.

Kummar, S., Chen, H. X., Wright, J., Holbeck, S., Millin, M. D., Tomaszewski, J., Zweibel, J., Collins, J. & Doroshow, J. H. (2010). Utilizing targeted cancer therapeutic agents in combination: Novel approaches and urgent requirements. *Nature Reviews Drug Discovery* 9: 843–56.

Kusch, M. (2009). Objectivity and historiography. *Isis* 100: 127–31.

——. (2010). Hacking's historical epistemology: A critique of styles of reasoning. *Studies in History and Philosophy of Science* 41: 158–73.

Kushner, R. (1984). Is aggressive adjuvant chemotherapy the Halsted Radical of the 80s? *CA: A Cancer Journal for Clinicians* 34: 345–51.

Lacombe, D., Fumoleau, P., Zwierzina, H., Twelves, C., Hakansson, L., Jayson, G., Lehmann, F. & Verweij, J. (EORTC's New Drug Development Group). (2002). The EORTC and drug development. *European Journal of Cancer* 38: S19–S23.

Laker, T. L. (1989). Assessing the market for cytotoxic drugs. *Medical Marketing & Media* 24: 77–82.

Landecker, H. (2007). *Culturing Life: How Cells Became Technologies.* Cambridge, MA: Harvard University Press.

Lange, N. & MacIntyre, J. (1985). A computerized patient registration and treatment randomization system for multi-institutional clinical trials. *Controlled Clinical Trials* 6: 38–50.

Larry Norton: President-elect of ASCO. (2000). *The Lancet Oncology* 1: 189–92.

Lasagna, L. (1960). Planning clinical chemotherapy studies. In B. H. Morrison III, ed., *Conference on Experimental Clinical Cancer Chemotherapy. November 11 and 12, 1959.* Washington, DC: U.S. Department of Health, Education, and Welfare, 45–50.

Laszlo, J. (1995). *The Cure of Childhood Leukemia: Into the Age of Miracles.* New Brunswick, NJ: Rutgers University Press.

Latour, B. (1987). *Science in Action: How to Follow Scientists and Engineers through Society.* Cambridge, MA: Harvard University Press.

———. (1988). The politics of explanation: An alternative. In S. Woolgar, ed., *Knowledge and Reflexivity: New Frontiers in the Sociology of Knowledge.* London: Sage, 155–76.

———. (2005). *Reassembling the Social: An Introduction to Actor-Network-Theory.* Oxford: Oxford University Press.

Law, L. W. (1952). Effects of combinations of antileukemic agents on an acute lymphocytic leukemia in mice. *Cancer Research* 13: 871–78.

Lebwohl, D. & Canetta, R. (1998). Clinical development of platinum complexes in cancer therapy: An historical perspective and an update. *European Journal of Cancer* 34: 1522–34.

Lee, J. J. & Feng, L. (2005). Randomized Phase II designs in cancer clinical trials: Current status and future directions. *Journal of Clinical Oncology* 23: 4450–57.

Lee, L. E., Jr. (1960). Remarks made during the "Panel Meeting: Investigation of therapy in solid tumors." In B. H. Morrison III, ed., *Conference on Experimental Clinical Cancer Chemotherapy. November 11 and 12, 1959.* Washington, DC: U.S. Department of Health, Education, and Welfare, 71–84.

Lehmann, F., Lacombe, D., Therasse, P. & Eggermont, A. M. M. (2003). Integration of translational research in the European Organization for Research and Treatment of Cancer Research (EORTC) clinical trial cooperative group mechanisms. *Journal of Translational Medicine* 1: 2.

Leiter, J., Bourke, A. R., Fitzgerald, D. B., Schepartz, S. A. & Wodinsky, I. (1962a). Screening data from the Cancer Chemotherapy National Service Center screening laboratories, VII. *Cancer Research* 22 (Supplement): 221–353.

———. (1962b). Screening data from the Cancer Chemotherapy National Service Center screening laboratories, VIII. *Cancer Research* 22 (Supplement): 363–482.

Lejeune, J., Levan, A., Böök, J. A., Chu, E. H. Y., Ford, C. E., Fraccaro, M., Harnden, D. G., Hsu, T. C., Hungerford, D. A., Jacobs, P. A., Makino, S., Puck. T. T., Robinson, A., Tjio, J. H., Catcheside, D. G., Muller, H. J. & Stern, C. (1960). A proposed system of nomenclature of human mitotic chromosomes. *The Lancet* 275: 1063–65.

Lenz, M. (1973). The early workers in clinical radiotherapy of cancer at the Radium Institute of the Curie Foundation, Paris, France. *Cancer* 32: 519–23.

Leong, S. P. L. (2007). *Cancer Clinical Trials: Proactive Strategies*. New York: Springer.

Lerner, B. H. (2001). *The Breast Cancer Wars*. New York: Oxford University Press.

———. (2004). Informed consent: Did cancer patients challenge their physicians in the post–World War II era? *Journal of the History of Medicine and Allied Sciences* 59: 507–21.

Levine, M. & Whelan, T. (2006). Adjuvant chemotherapy for breast cancer—30 years later. *New England Journal of Medicine* 355: 1920–22.

Levinson, A. D., Oppermann, H., Levintow, L., Varmus, H. E. & Bishop, J. M. (1978). Evidence that the transforming gene of avian sarcoma encodes a protein kinase associated with a phosphoprotein. *Cell* 15: 561–72.

Litchfield, J. T. (1960). Sequential analysis, screening, and serendipity. *Journal of Medicinal and Pharmaceutical Chemistry* 2: 469–92.

Littlejohns, P. (2006). Trastuzumab for early breast cancer: Evolution or revolution? *The Lancet Oncology* 7 (1): 22–23.

Longmire, W. P. (1962). The adjuvant chemotherapy gastric study. *Cancer Chemotherapy Reports* 16: 125–28.

Longmire, W. P., Kuzma, J. W. & Dixon, W. J. (1968). The use of triethylenethiophosphoramide [thio-TEPA] as an adjuvant to the surgical treatment of gastric carcinoma. *Annals of Surgery* 167: 293–312.

Lourie, W. I. (1964). Some observations concerning comparability in these data. *National Cancer Institute Monograph* 15: 369–80. Washington, DC: Superintendent of Documents, U.S. Government Printing Office.

Löwy, I. (1995a). Nothing more to be done: Palliative care versus experimental therapy in advanced cancer. *Science in Context* 8: 209–29.

———. (1995b). La standardisation de l'inconnu: Les protocoles thérapeutiques en cancérologie. *Techniques & Culture* 25/26: 73–108.

———. (1998). Immunotherapy of cancer from Coley's toxins to interferons: Molecularization of a Therapeutic Practice. In S. de Chadarevian and H. Kamminga, eds., *Molecularizing Biology and Medicine: New Practices and Alliances, 1910s–1970s*. Amsterdam: Harwood Academic, 249–71.

Löwy, I. & Gaudillière, J.-P. (1998). Disciplining cancer: Mice and the practice of genetic purity. In J.-P. Gaudillière and I. Löwy, eds., *The Invisible Industrialist: Manufactures and the Production of Scientific Knowledge*. London: Macmillan, 209–49.

Lydon, N. (2009). Attacking cancer at its foundation. *Nature Medicine* 15: 1153–57.

Lynch, M. E. (1982). Technical work and critical inquiry: Investigations in a scientific laboratory. *Social Studies of Science* 12: 499–533.

———. (2002). Protocols, practices, and the reproduction of technique in molecular biology. *British Journal of Sociology* 53: 203–20.

Lynch, M., Cole, S. A., McNally, R. & Jordan, K. (2008). *Truth Machine: The Contentious History of DNA Fingerprinting*. Chicago: University of Chicago Press.

Lyttle, C. S., Andersen, R. M., Neymars, K., Schmidt, C., Kohoman, C. H. & Levy, G. S. (1991). National study of internal medicine manpower XVII. Subspecialty fellowships with a special look at hematology and oncology 1988–1989. *Annals of Internal Medicine* 114: 36–42.

MacGregor, A. B. (1966). The search for a chemical cure for cancer. *Medical History* 10: 374–86.

Machin, D. (2004). On the evolution of statistical methods as applied to clinical trials. *Journal of Internal Medicine* 255: 521–28.

Mainland, D. (1960). The clinical trial—Some difficulties and suggestions. *Journal of Chronic Diseases* 11: 484–96.

Maisin, J. H. (1966). *L'Union internationale contre le cancer: De sa fondation à nos jours*. Geneva: UICC.

Majno, G. & Joris, I. (2004). *Cells Tissues and Disease*, 2nd ed. New York: Oxford University Press.

Mantel, N. (1966). Evaluations of survival data and two new rank order statistics arising in its consideration. *Cancer Chemotherapy Reports* 50: 163–70.

Marinus, A. (2002). Quality assurance in EORTC clinical trials. *European Journal of Cancer* 38: S159–61.

Marks, H. M. (1997). *The Progress of Experiment: Science and Therapeutic Reform in the United States, 1900–1990*. Cambridge: Cambridge University Press.

———. (2003). Rigorous uncertainty: Why R. A. Fisher is important. *International Journal of Epidemiology* 32: 932–37.

———. (2004). A conversation with Paul Meier. *Clinical Trials* 1: 131–38.

———. (2006). "Until the sun of science . . . the true Apollo of medicine has risen": Collective investigation in Britain and America, 1880–1910. *Medical History* 50: 147–66.

———. (2009). The science and business of drug pushing. Comments on Jeremy Greene's *Prescribing by Number: Drugs & the Definition of Disease*. 4S Meeting, 31 October 2009, unpublished manuscript.

Marsoni, S. & Wittes, R. (1984). Clinical development of anticancer agents—A National Cancer Institute perspective: A commentary. *Cancer Treatment Reports* 68: 78–85.

Martensen, R. (1997). Medicine, surgery, and gastric cancer in 1900: Debating when to go "radical." *JAMA* 277: 1495.

Maskens, A. P. (1981). Confirmation of the two-step nature of chemical carcinogenesis in the rat colon adenocarcinoma model. *Cancer Research* 41: 1240–45.

Mathé, G. (1973). Guest editorial: Clinical examination of drugs, a scientific and ethical challenge. *Biomedicine* 18: 169–72.

———. (1974). Les essais thérapeutiques cliniques. *Bulletin du Cancer* 61: 191–212.

———. (1978). Towards ecumenism for comparative trials. *Biomedicine* 28 (Special issue): 2–6.

Mauer, J. K., Hoth, D. F., Macfarlane, D., Hammershaimb, L. D. & Wittes, R. D. (1985). Site visit monitoring program of the Clinical Cooperative Groups: Results of the first 3 years. *Cancer Treatment Reports* 69: 1177–87.

McCulloch, E. A. (2003). *The Ontario Cancer Institute: Successes and Reverses at Sherbourne Street*. Montreal & Kingston: McGill-Queen's University Press.

McIntosh, H. (1992). Drug development link between U.S. and Europe yielding results. *Journal of the National Cancer Institute* 84: 13–15.

McIntyre, P. (2007). You need to divorce to become good friends. *Cancer World* (March/April): 32–37.

Medical oncology. (1971). *The Lancet* 2: 419.

Meeting report: First international workshop on chromosomes in leukemia. (1978). *Cancer Research* 38: 867–68.

Meier, P. (1975). Statistics and medical experimentation. *Biometrics* 31: 511–29.

———. (1981). Stratification in the design of a clinical trial. *Controlled Clinical Trials* 1: 355–61.

Meinert, C. L. (1998). Masked monitoring in clinical trials. Blind stupidity? *New England Journal of Medicine* 338: 1381–82.

Mellstedt, H. (2006). The dawn of a golden age of medical oncology. *Hospital Healthcare Europe* (July): 9–10.

Memphis CCOP withdraws, cites lack of patient benefits, excess tests. (1984). *The Cancer Letter* 10 (44): 3–4.

Ménoret, M. (2002). The genesis of the notion of stages in oncology: The French Permanent Cancer Survey (1943–1952). *Social History of Medicine* 15: 291–302.

Mercer, R. D. (1999). The team. *Medical and Pediatric Oncology* 33: 408–9.

Meunier, F. & van Oosterom, A. (2001). Obituary: Professor Henri Tagnon. *European Journal of Cancer* 37: 551–52.

———. (2002). 40 years of the EORTC: The evolution towards a unique network to develop new standards of cancer care. *European Journal of Cancer* 38 (4): S3–S13.

Michaelis, L. C. & Ratain, M. J. (2007). Phase II trials published in 2002: A cross-specialty comparison showing significant design differences between oncology trials and other medical specialties. *Clinical Cancer Research* 13: 2400–2405.

Millar, A. W. & Lynch, K. P. (2003). Rethinking clinical trials for cytostatic drugs. *Nature Reviews Cancer* 3: 540–44.

Miller, J. A. (1994). Brief history of chemical carcinogenesis. *Cancer Letters* 83: 9–14.

Miller, M. (2001). Phase I cancer trials: A collusion of misunderstanding. *Hastings Center Report* 30 (4): 34–43.

MINIMUM Standards. (1959). MINIMUM standards for the commercial production of randombred and inbred laboratory mice. *Cancer Chemotherapy Reports* 1 (1): 99–104.

Minority based CCOP RFA generates wide interest; 44 letters of interest received for eight awards. (1989). *The Cancer Letter* 15 (35): 1–4.

Mintzes, B. (2007). Should patient groups accept money from drug companies? No. *British Medical Journal* 334: 935.

Mirowski, P. & Van Horn, R. (2005). The contract research organization and the commercialization of scientific research. *Social Studies of Science* 35: 503–48.

Mitelman, F. (1983). Catalogue of chromosome aberrations in cancer. *Cytogenetics and Cell Genetics* 36 (1–2): 1–515.

Mok, T. S., Wu, Y. L., Thongprasert, S., Yang, C. H., Chu, D. T., Saijo, N., Sunpaweravong, P., Han, B., Margono, B., Ichinose, Y., Nishiwaki, Y., Ohe, Y., Yang, J. J., Chewaskulyong, B., Jiang, H., Duffield, E. L., Watkins, C. L., Armour, A. A. & Fukuoka M. (2009). Gefitinib or carboplatin-paclitaxel in pulmonary adenocarcinoma. *New England Journal of Medicine* 361: 947–57.

Monks, A., Scudiero, D., Skehan, P., Shoemaker, R., Paull, K., Vistica, D., Hose, C., Langley, J., Cronise, P., Vaigro-Wolff, A., Gray-Goodrich, M., Campbell, H., Mayo, J. & Boyd, M. (1991). Feasibility of a high-flux anticancer drug screen using a diverse panel of culture human tumor cell lines. *Journal of the National Cancer Institute* 83: 757–66.

Moore, G. E. (1962a). Comments: Second conference, experimental clinical cancer chemotherapy. *Cancer Chemotherapy Reports* 16: 121–22.

———. (1962b). The future of cancer chemotherapy. *Cancer Chemotherapy Reports* 16: 589–91.

Moore, G. E., Bross, I. D., Ausman, R., Nadles, S., Jones, R., Slack, N. & Rimm, A. A. (1968). Effects of 5-Fluorouracil (NSC-19893) in 398 patients with cancer. *Cancer Chemotherapy Reports* 52: 641–53.

Morange, M. (1997). From the regulatory vision of cancer to the oncogene paradigm, 1975–1985. *Journal of the History of Biology* 30: 1–29.

Morrison, B. M., III, ed. (1960). *Conference on Experimental Clinical Cancer Chemotherapy. November 11 and 12, 1959.* Washington, DC: U.S. Department of Health, Education, and Welfare.

Morson, B. C. (1985). Cuthbert E. Dukes, OBE, MD, MSc, FRCS, FRCPath, DPH, 1890–1977, *Annals of the Royal College of Surgeons of England* 67: 354.

Mortenson, L. E. (2002). Cancer coalitions and cancer politics. *Seminars in Oncology Nursing* 18: 297–304.

Moss, W. T. (1959). *Therapeutic Radiology: Rationale, Technique, Results.* St. Louis: C. V. Mosby.

Motulsky, H. (1995). *Intuitive Biostatistics.* New York: Oxford University Press.

Mueller, J. M. (2007). Taking TRIPS to India—Novartis, patent law, and access to medicines. *New England Journal of Medicine* 356: 541–43.

Muggia, F. (2005). Ezra Greenspan, M.D.: Pioneer of medical oncology. *The Oncologist* 10: 170–71.

Muss, H. B. (2005). Byrl James Kennedy, MD. *Journal of Clinical Oncology* 23: 3297–98.

Nascimento, A. G. (1999). Grading of sarcomas at the Mayo Clinic 75 years after Broders. *Revista Española de Patología*, 32: 424–26.

Nass, S. J. & Stillman, B., eds. (2003). *Large-Scale Biomedical Science: Exploring Strategies for Future Research.* Washington, DC: National Academies Press.

Nathan, D. G. (2002). Careers in translational clinical research: Historical perspectives, future challenges. *JAMA* 287: 2424–27.

———. (2007). *The Cancer Treatment Revolution.* Hoboken, NJ: John Wiley.

Nathan, D. & Nenz, E. J. (2001). Comprehensive cancer centres and the war on cancer. *Nature Reviews Cancer* 1: 240–45.

National Cancer Institute. (2003). *The Nation's Investment in Cancer Research: A Plan and Budget Proposal for Fiscal Year 2003.* NIH Publication. No. 02–4373. Bethesda, MD: National Cancer Institute.

NCAB votes down effort to delay CCOP RFA, directs subcommittee to look again at program components. (1982). *The Cancer Letter* 8 (7): 1–5.

NCI. (1998). U.S. Department of Health and Human Services, National Institutes of Health and National Cancer Institute. *Understanding Breast Cancer Treatment: A Guide for Patients.* Washington, DC: U.S. Government Printing Office, NIH Publication No. 98–4251.

NCI funds cancer trials support unit. (1999). *Journal of the National Cancer Institute* 91: 1718.

NCI overhauls clinical trials system. (1999). *Journal of the National Cancer Institute* 91: 312.

NDDO splits from EORTC. (1998). *Annals of Oncology* 9: 1145–46.

New funding mechanisms—cooperative agreement, consortium grant—proposed for clinical groups. (1979). *The Cancer Letter* 5 (14): 1–4.

New optimism at the new drugs conference. (1998). *Annals of Oncology* 9: 789.

New surgery grant applications to be reviewed by ad hoc study sections; 54 submitted. (1979). *The Cancer Letter* 5 (25): 1–3.

Newell, G. R. & Sugano, H. (1977). United States–Japan cooperative cancer research program. *Journal of the National Cancer Institute* 58: 455–56.

News from E.O.R.T.C. (1979). *European Journal of Cancer* 15: 1290.

Noble, R. L. (1990). The discovery of the vinca alkaloids—chemotherapeutic agents against cancer. *Biochemistry Cell Biology* 68: 1344–51.

Noer, R. J. (1961). Breast adjuvant chemotherapy: Effectiveness of thio-TEPA as adjuvant to radical mastectomy for breast cancer. *Annals of Surgery* 154: 629–47.

Nowell, P. C. & Hungerford, D. A. (1960). A minute chromosome in human granulocytic leukemia. *Science* 132: 1197–200.

Nowell, P., Dalla-Favera, R., Finan, J., Erikson, J. & Croce, C. (1983). Chromosome translocations, immunoglobulin genes and neoplasia. In J. Rowley & J. E. Ultmann, eds., *Chromosomes and Cancer: From Molecules to Man*, 5th Bristol-Myers Symposium on Cancer Research, University of Chicago. New York: Academic Press, 165–82.

Nowell, P. C., Jackson, L., Weiss, A. & Kurzrock, R. (1988). Historical communication: Philadelphia-positive chronic myelogenous leukemia followed for 27 years. *Cancer Genetics and Cytogenetics* 34: 57–61.

NPCCR (National Program of Cancer Chemotherapy Research). (1959–60). The National Program of Cancer Chemotherapy Research: Information statement. *Cancer Chemotherapy Reports* 1 (1): 3–34.

O'Brien, S. G., Guilhot, F., Larson, R. A., Gathmann, I., Baccarani, M., Cervantes, F., Cornelissen, J. J., Fischer, T., Hochhaus, A., Hughes, T., Lechner, K., Nielsen, J., Rousselot, P., Reiffers, J., Saglio, G., Shepherd, J., Simonsson, B., Gratwohl, A., Goldman, J. M., Kantarjian, H., Taylor, K., Verhoef, G., Bolton, A. E., Capdeville, R. & Druker, B. J., on behalf of the IRIS Investigators. (2003). Imatinib compared with interferon and low-dose cytarabine for newly diagnosed chronic-phase chronic myeloid leukemia. *New England Journal of Medicine* 348: 994–1004.

Oda, K., Matsuoka, Y., Funahashi, A. & Kitano, H. (2005). A comprehensive pathway map of epidermal growth factor receptor signaling. *Molecular Systems Biology* 1: 1–17.

Olby, R. (1994). *The Path to the Double Helix: The Discovery of DNA*. London: Dover.

Oldham, R. K. & Smalley, R. V. (1983). Immunotherapy: The old and the new. *Journal of Biological Response Modifiers* 2: 1–37.

Oliveira, A. M. & Nascimento, A. G. (2001). Grading in soft tissue tumors: Principles and problems. *Skeletal Radiology* 30: 543–59.

Olsen, J. S. (2002). *Bathsheba's Breast: Women, Cancer, and History.* Baltimore: Johns Hopkins University Press.

Osgood, E. E. & Seaman, A. J. (1952). Treatment of chronic leukemia. *JAMA* 150: 1372–79.

O'Shea, J. S. (2004). Specialization in surgical oncology: Historical perspectives. *Annals of Surgical Oncology* 11: 462–64.

Oudshoorn, N. (1994). *Beyond the Natural Body: An Archaeology of Sex Hormones.* London: Routledge.

Ovejera, A. A., Timm, F. M., Houchens, D. P. & Barker, A. D. (1973). Nude mouse facility designed for large-scale drug evaluation and other cancer research activities. In A. Spiegel, ed., *The Laboratory Animal in Drug Testing.* Stuttgart: Fischer, 39–51.

Owen, J. B., Coia, L. R. & Hanks, G. E. (1997). The structure of radiation oncology in the United States in 1994. *International Journal of Radiation Oncology, Biology and Physics* 39: 179–85.

Paik, S., Shak, S., Tang, G., Kim, C., Baker, J., Cronin, M., Baehner, F. L., Walker, M. G., Watson, D., Park, T., Hiller, W., Fisher, E. R., Wickerham, D. L., Bryant, J. & Wolmark, N. (2004). A multigene assay to predict recurrence of tamoxifen-treated, node-negative breast cancer. *New England Journal of Medicine* 351: 2817–26.

Parascandola, M. (2004). Skepticism, statistical methods, and the cigarette: A historical analysis of a methodological debate. *Perspectives in Biology and Medicine* 47: 244–61.

Participants sought for national cooperative drug discovery groups. (1982). *NIH Guide for Grants and Contracts* 11 (November 13): 6–7.

Pasetto, L. M., D'Andrea, M. R., Brandes, A. A., Rossi, E. & Monfardini, S. (2006). The development of platinum compounds and their possible combination. *Critical Reviews in Oncology/Hematology* 60: 59–75.

Pater, J. L., Eisenhauer, E., Shelley, W. & Willan, A. R. (1986). Testing hypotheses in clinical trials: The experience of the National Cancer Institute of Canada Clinical Trials Group. *Cancer Treatment Reports* 70: 1133–36.

Patlak, M. & Nass, S. (2008). *Improving the Quality of Cancer Clinical Trials.* Washington, DC: National Academies Press.

Patlak, M., Nass, S. & Micheel, C. (2009). *Multi-Center Phase III Clinical Trials and NCI Cooperative Groups (Workshop Summary).* Washington, DC: National Academies Press.

Patterson, J. T. (1987). *The Dread Disease: Cancer and Modern American Culture.* Cambridge, MA: Harvard University Press.

Paul, J. (1981). Sir George Beatson and the Royal Beatson Memorial Hospital. *Medical History* 25: 200–201.

Pazdur, R. (2005). The FDA's new oncology office. *Clinical Advances in Hematology & Oncology* 3: 612–13.

Pearce, H. L. & Miller, M. A. (2005). The evolution of cancer research and drug discovery at Lilly Research Laboratories. *Advances in Enzyme Regulation* 45: 229–55.

Pearson, E. S. (1966). The Neyman-Pearson story: 1926–1934. In F. N. David, ed., *Research Papers in Statistics*. New York: John Wiley, 1–24.

Pearson, E. S. & Neyman, J. (1933). On the problem of the most efficient tests of statistical hypotheses. *Philosophical Transactions of the Royal Society of London, Series A* 231: 289–337.

Peters, M. V. (1950). A study of survivals in Hodgkin's disease treated radiologically. *American Journal of Roentgenology* 63: 299–311.

Peters, M. V. & Middlemiss, K. C. H. (1958). A study of Hodgkin's disease treated by irradiation. *American Journal of Roentgenology* 79: 114–21.

Peto, R. (1982). Size of the therapeutic benefit that can be expected in a trial, and its implications in the design and organization of the trial. In R. Flamant, T. B. Grage & C. Fohanno, eds., *Evaluation of Methods of Treatment and Diagnostic Procedures in Cancer* (UICC Technical Report Series, vol. 70). Geneva: Union for International Cancer Control, 49–65.

Peto, R., Pike, M. C., Armitage, P., Breslow, N. E., Cox, D. R., Howard, S. V., Mantel, N., McPherson, K., Peto, J. & Smith, P. G. (1976). Design and analysis of randomized clinical trials requiring prolonged observation of each patient. I. Introduction and design. *British Journal of Cancer* 34: 585–612.

———. (1977). Design and analysis of randomized clinical trials requiring prolonged observation of each patient. II. Analysis and examples. *British Journal of Cancer* 35: 1–35.

Petryna, A. (2009). *When Experiments Travel: Clinical Trials on the Global Search for Human Subjects*. Princeton, NJ: Princeton University Press.

Picard, A. (2005). Be skeptical about the Herceptin hype. *Globe and Mail*, August 4, A15.

Picard, J.-F. (1994). De la médecine expérimentale (1865) à l'INSERM (1964). In C. Debru, J. Gayon & J.-F. Picard, eds., *Les sciences biologiques et médicales en France, 1920–1950*. Paris: CNRS, 329–44.

Piccart-Gebhart, M., Procter, M., Leyland-Jones, B., Goldhirsch, A., Untch, M., Smith, I., Gianni, L., Baselga, J, Bell, R., Jackisch, C., Cameron, D., Dowsett, M., Barrios, C. H., Steger, G., Huang, C.-S., Andersson, M., Inbar, M., Lichinitser, M., Lang, I., Nitz, U., Iwata, H., Thomssen, C., Lohrisch, C., Suter, T. M., Rüschoff, J., Süto, T., Greatorex, V., Ward, C., Straehle, C., McFadden, E., Dolci, S. & Gelber, R., on behalf of the Herceptin Adjuvant (HERA) Trial Study Team. (2005). Trastuzumab after adjuvant chemotherapy in HER2-positive breast cancer. *New England Journal of Medicine* 353: 1659–72.

Piccart, M. J. (2001). The BIG-HERA reunion and the ingredients for a successful collaboration. *B.I.G.-AISBL Newsletter* 3 (2): 1–2.

Pickering, A., ed. (1992). *Science as Practice and Culture*. Chicago: University of Chicago Press.

Pickstone, J. V. (2007). Contested cumulations: Configurations of cancer treatments through the twentieth century. *Bulletin of the History of Medicine* 81: 164–96.

Pil, P. & Lippard, S. J. (2002). Cisplatin and related drugs. In J. R. Berino, ed., *Encyclopedia of Cancer*, vol. 1. San Diego: Academic Press, 525–43.

Pinell, P. (2000). Cancer. In R. Cooter & J. Pickstone, eds., *Routledge Companion to Medicine in the Twentieth Century*. London: Routledge, 671–86.

———. (2002). *The Fight against Cancer: France 1890–1940*. London: Routledge.

Pinkel, D. (2006). Historical Perspective. In Ching-Hon Pui, ed., *Childhood Leukemias*, 2nd ed. Cambridge: Cambridge University Press, 3–20.

Pocock, S. J. (1975). Sequential treatment assignment with balancing for prognostic factors in the controlled clinical trial. *Biometrics* 31: 103–15.

———. (1977). Group sequential methods in the design and analysis of clinical trials. *Biometrika* 64: 191–99.

———. (1979). Allocation of patients to treatment in clinical trials. *Biometrics* 35: 183–97.

———. (1982). Interim analyses for randomized clinical trials: The group sequential approach. *Biometrics* 38: 153–62.

Pocock, S. J. & Lagakos, S. W. (1982). Practical experience of randomization in cancer trials: An international survey. In R. Flamant, T. B. Grage & C. Fohanno, eds., *Evaluation of Methods of Treatment and Diagnostic Procedures in Cancer* (UICC Technical Report Series, vol. 70). Geneva: Union for International Cancer Control, 5–21.

Pocock, S. J. & Simon, R. (1975). Sequential treatment assignment with balancing for prognostic factors in the controlled clinical trial. *Biometrics* 31: 103–15.

Portmann, U. V. (1943). Clinical and pathological criteria as a basis for classifying cases of primary cancer of the breast. *Cleveland Clinic Quarterly* 10: 41–47.

Postan, R. N. (1999). A new look at the original cases of Hodgkin's disease. *Cancer Treatment Reviews* 25: 151–55.

Pritchard, K. I. (2010). Optimizing the delivery of targeted research: An opportunity for comparative effectiveness research. *Journal of Clinical Oncology* 28: 1089–91.

Private dollars increasingly fuel scientific research. (1991). *Journal of the National Cancer Institute* 83: 1526.

Quirke, V. (2009). Inside the drug industry: The cancer research programmes of three companies. Paper presented at the workshop "How cancer changed: Expanding the boundaries of medical interventions." Paris, 2–3 April 2009.

Rabeharisoa, V. (2006). From representation to mediation: The shaping of collective mobilization on muscular dystrophy in France. *Social Science & Medicine* 62: 564–76.

Rabinow, P. (2000). Epoch, presents, events. In M. Lock, A. Young & A. Cambrosio, eds., *Living and Working with the New Medical Technologies: Intersections of Inquiry*. Cambridge: Cambridge University Press, 31–46.

Rader, K. (2004). *Making Mice: Standardizing Animals for American Biomedical Research, 1900–1955*. Princeton, NJ: Princeton University Press.

Radvin, R. G., Lewison, E. F., Slack, N. H., Gardner, B., State, D. & Fisher, B. (1970). Results of a clinical trial concerning the worth of prophylactic oophorectomy for breast carcinoma. *Surgery, Gynecology and Obstetrics* 131: 1055–64.

Rand Corporation (1955). *A Million Random Digits with 100,000 Normal Deviates*. Glencoe, IL: Free Press.

Ratain, M. J. & Glassman, R. (2007). Biomarkers in Phase I oncology trials: Signal, noise, or expensive distraction? *Clinical Cancer Research* 13: 6545–48.

Ravdin, I. S. (1960). The cooperative clinical program. In B. H. Morrison III, ed., *Conference on Experimental Clinical Cancer Chemotherapy. November 11 and 12, 1959*. Washington, DC: U.S. Department of Health, Education, and Welfare, 1–7.

Ray, T. (2009a). AstraZeneca, DxS to commercialize companion Dx for Iressa in Europe; US still uncharted. *Pharmacogenomics Reporter*. Available at http://www.genomeweb.com/print/921536?emc=el&m=458162&l=2&v=c4e8 b24249.

———. (2009b). Conference discussion questions Big Pharma's true motivations for pursuing personalized medicine. *Pharmacogenomics Reporter* 2 (December); available at http://www.genomeweb.com/print/928514?page=show.

———. (2009c). DxS developing KRAS companion Dx for Bristol-Myers Squibb/ ImClone's Erbitux. *Pharmacogenomics Reporter*. Available at http://www .genomeweb.com/print/923176.

Reynolds, T. (2000). How does a drug get to Phase III trials? *Journal of the National Cancer Institute* 92: 1555.

Rheinberger, H.-J. (1997). *Toward a History of Epistemic Things: Synthesizing Proteins in the Test Tube*. Stanford, CA: Stanford University Press.

Ribatti, D. (2007). The contribution of Gianni Bonadonna to the history of chemotherapy. *Cancer Chemotherapy and Pharmacology* 60: 309–12.

Ricciardi, M. (1997). *Constitutive Rules and Institutions*. Paper presented at the joint meeting of the Irish Philosophical Club and the Royal Institute of Philosophy, Ballymanscanlon, February 1997, available at http://www.unipv.it/ deontica/opere/riccia/rules.doc.

Rice, M. (2004a). European patients find a voice. *European Journal of Cancer* 40: 1285.

———. (2004b). Formation of pan-European groups buoys cause of patient advocates. *Journal of the National Cancer Institute* 96: 1498–99.

Riechelmann, R. Sasse, E., Borghesi, G., Miranda, V., Fede, A., Saad, L., Oliveira, V., Barros, E., Campos, M. P., Del Giglio, A. & Saad, E. D. (2009). Randomized Phase II trials (RP2T): Selection design or poor man's Phase III? *Journal of Clinical Oncology* 27 (Supplement): 15S.

Rigal, C. (2003). *Contribution à l'histoire de la recherche médicale: Autour des travaux de Jean Bernard et de ses collaborateurs sur la leucémie aiguë, 1940–1970.* Doctoral thesis (Epistémologie, histoire des sciences et des techniques). Université Paris 7: Denis Diderot.

———. (2008). Neo-clinicians, clinical trials, and the reorganization of medical research in Paris hospitals after the Second World War: The trajectory of Jean Bernard. *Medical History* 52: 511–34.

Rimm, A. A. & Bortin, M. (1978). Clinical trials as a religion. *Biomedicine* 28 (Special issue): 60–63.

Roberts, T. G., Lynch, T. J. & Chabner, B. A. (2003). The Phase III trial in the era of targeted therapy: Unraveling the "go or no go" decision. *Journal of Clinical Oncology* 21: 3683–95.

Roberts, T. M., Kaplan, D., Morgan, W., Keller, T., Mamon, H., Piwnica-Worms, H., Druker, B., Cohen, B., Schaffhausen, B., Whitman, M., Cantley, L. Rapp, U. & Morrison, D. (1988). Tyrosine phosphorylation in signal transduction. *Cold Spring Harbor Symposia on Quantitative Biology* 53: 161–71.

Robinson, W. T. (2008). Innovative early development regulatory approaches: expIND, expCTA, microdosing. *Clinical Pharmacology and Therapeutics* 83: 358–60.

Romond, E. H., Perez, E. A., Bryant, J., Suman, V. J., Geyer, C. E., Davidson, N. E., Tan-Chiu, E., Martino, S., Paik, S., Kaufman, P. A., Swain, S. M., Pisansky, T. M., Fehrenbacher, L., Kutteh, L. A., Vogel, V. G., Visscher, D. W., Yothers, G., Jenkins, R. B., Brown, A. M., Dakhil, S. R., Mamounas, E. P., Lingle, W. L., Klein, P. M., Ingle, J. N. & Wolmark, N. (2005). Trastuzumab plus adjuvant chemotherapy for operable HER2-positive breast cancer. *New England Journal of Medicine* 353: 1673–84.

Rosenberg, S. A. (1999). Development of the concept of Hodgkin's disease as a curable illness: The American experience. In P. M. Mauch, J. O. Armitage, V. Diehl, R. T. Hoppe & L. M. Weiss, eds., *Hodgkin's Disease*. Philadelphia: Lippincott, Williams and Wilkins, 47–57.

Rosner, G. L. & Tsiatis, A. A. (1989). The impact that group sequential tests would have made on ECOG clinical trials. *Statistics in Medicine* 8: 505–16.

Rouse, J. (2002). *How Scientific Practices Matter: Reclaiming Philosophical Naturalism.* Chicago: University of Chicago Press.

Rowley, J. D. (1973). A new consistent chromosomal abnormality in chronic my-elogenous leukaemia identified by quinacrine fluorescence and Giemsa stain-ing. *Nature* 243: 290–93.

———. (1983). Introduction to Symposium. In J. D. Rowley & J. E. Ultmann, eds., *Chromosomes and Cancer: From Molecules to Man*, 5th Bristol-Myers Sympo-sium on Cancer Research, University of Chicago. New York: Academic Press.

Rubin, P. (1985). The emergence of radiation oncology as a distinct medical spe-cialty. *International Journal of Radiation Oncology, Biology and Physics* 11: 1247–70.

Rygaard, J. (1973). *Thymus and Self: Immunobiology of the Mouse Mutant Nude*. London: Wiley.

———. (1978). The nude mouse—Mouse or test tube. In D. P. Houchens & A. A. Ovejera, eds., *Proceedings of the Symposium on the Use of Athymic (Nude) Mice in Cancer Research*. New York: Fischer, 1–7.

Salsberg, D. S. (1992). *The Use of Restricted Significance Tests in Clinical Trials*. New York: Springer-Verlag.

Santoni, P. & Foster, T. (2007). Zero sum game. *Oncology Business Review* (No-vember): 2, 14–17; available at http://www.imshealth.com/imshealth/Global/Content/Document/Value-based%20Medicine%20TL/zero.pdf.

Saul, H. (2008). A new chapter begins for EORTC. *European Journal of Can-cer* 44: 329–33.

Sawyers, C. L. (2001). Cancer treatment in the STI571 era: What will change? *Journal of Clinical Oncology* 19: 13s–16s.

———. (2009). Shifting paradigms: The seeds of oncogene addiction. *Nature Med-icine* 15: 1158–61.

Schepartz, S. A. (1964). Progress in the development of new antineoplastic anti-biotics. *Antimicrobial Agents and Chemotherapy* 10: 500–504.

Schilsky, R. (1997). Message from the group chair. *Cancer and Leukemia Group B Newsletter* 6 (1): 2; available at http://www.calgb.org/Public/publications/calgabs/1997/spring1997.pdf.

———. (2002). End point in cancer clinical trials and the drug approval process. *Clinical Cancer Research* 8: 935–38.

Schilsky, R. L., McIntyre, O. R., Holland, J. F. & Frei, E., III. (2006). A concise history of the Cancer and Leukemia Group B. *Clinical Cancer Research* 12 (Supplement): 3553s–55s.

Schneiderman, M. A. (1961). Statistical problems in the screening search for an-ticancer drugs by the National Cancer Institute of the United States. In H. De Jonge, ed., *Quantitative Methods in Pharmacology, Proceedings of a Sympo-sium Held in Leyden on May 10–13, 1960*. Amsterdam: North Holland Pub-lishing Company, 232–47.

———. (1962). The clinical excursion into 5-fluorouracil. *Journal of Chronic Dis-eases* 15: 283–95.

Schoenfeld, D. A. & Tsiatis, A. A. (1987). A modified log rank test for highly stratified data. *Biometrika* 74: 167–75.

Schultz, M. D. (1975). The supervoltage story. *American Journal of Roentgenology* 124: 541–59.

Schwartsmann, G., Ratain, M. J., Cragg, G. M., Wong, J. E., Saijo, N., Parkinson, D. R., Fujiwara, R., Pazdur, R., Newman, D. J., Dagher, R. & Di Leone, L. (2002). Anticancer drug discovery and development throughout the world. *Journal of Clinical Oncology* 20 (September 15 Supplement): 47s–59s.

Schwartz, D. & Lellouch, J. (1967). Explanatory and pragmatic attitudes in therapeutic trials. *Journal of Chronic Diseases* 20: 637–48.

Schwartz, D., Flamant, R. & Lellouch, J. (1969). *L'essai thérapeutique chez l'homme*, 1st ed. Paris: Flammarion.

——. (1981). *L'essai thérapeutique chez l'homme*. 2nd ed. Paris: Flammarion.

Schwartz, D., Flamant, R., Lellouch, J. & Roquette, C. (1960). *Les essais thérapeutiques cliniques*. Paris: Masson.

Schwarz, H. A. (1961). Presurgical intra-arterial use of nitrogen mustard in operations for cancer. *Cancer Chemotherapy Reports* 11: 1–3.

Scoggins, J. F. & Ramsey, S. D. (2010). A national cancer clinical trials system for the 21st century: Reinvigorating the NCI cooperative group program. *Journal of the National Cancer Institute* 102: 1371.

Searle, J. R. (1995). *The Construction of Social Reality*. New York: Free Press.

Sessoms, S. M., Coghill, R. D. & Waalkes, T. P. (1960). Review of the Cancer Chemotherapy National Service Center Program. *Cancer Chemotherapy Reports* 7: 25–46.

Shah, N. P. & Sawyers, C. L. (2003). Mechanisms of resistance to STI571 in Philadelphia chromosome-associated leukemias. *Oncogene* 22: 7389–95.

Shampo, M. A. (2001). Dukes and Broders: Pathologic classifications of cancer of the rectum. *Journal of Pelvic Surgery* 7: 5–7.

Sharifi, N. & Steinman, R. A. (2002). Targeted chemotherapy: Chronic myelogenous leukemia as a model. *Journal of Molecular Medicine* 80: 219–32.

Shaw, M. T. (1977). Medical oncology: An adolescent or a mature specialty? *The Lancet*: 135–36.

Sheer, D., Hiorns, L. R., Swallow, D. M., Povey, S., Goodfellow, P. N., Heisterkamp, N., Groffer, J., Stephenson, J. R. & Solomon, E. (1983). Genetic mapping of the 15;17 chromosome translocation associated with acute promyelocytic leukemia. In J. Rowley & J. E. Ultmann, eds., *Chromosomes and Cancer: From Molecules to Man*, 5th Bristol-Myers Symposium on Cancer Research, University of Chicago. New York: Academic Press, 183–93.

Shewhart, W. A. (1926). Quality control charts. *Bell System Technical Journal* 5: 593–606.

Shimkin, M. B. (1977). *Contrary to Nature: Being an Illustrated Commentary on Some Persons and Events of Historical Importance in the Devel-*

opment of Knowledge concerning Cancer. Washington, DC: U.S. Dept. of Health, Education, and Welfare, Public Health Service, National Institutes of Health.

——. (1978). Lost colony: The Laboratory of Experimental Oncology. *Journal of the National Cancer Institute* 60: 479–88.

Shrum, W., Genuth, J. & Chompalov, I. (2007). *Structures of Scientific Collaboration.* Cambridge, MA: MIT Press.

Silverstein, A. M. (2002). *Paul Ehrlich's Receptor Immunology: The Magnificent Obsession.* San Diego, CA: Academic Press.

Simon, R. (1989). Optimal two-stage designs for Phase II clinical trials. *Controlled Clinical Trials* 10: 1–10.

——. (1993). Discussion. *Statistics in Medicine* 12: 521–25.

——. (1997). A Conversation with Marvin A. Schneiderman. *Statistical Science* 12: 98–102.

——. (1999). The role of statisticians in intervention trials. *Statistical Methods in Medical Research* 8: 281–86.

——. (2000). Comment. *Statistical Science* 15: 103–8.

Simon, S. D. (2006). *Statistical Evidence in Medical Trials.* Oxford: Oxford University Press.

Simone, J. (2004). ASCO annual meeting: Too big? Too commercial? *Oncology Times* 26 (14): 4, 13, 15.

Sinha, G. (2007). United Kingdom becomes the cancer clinical trials recruitment capital of the world. *Journal of the National Cancer Institute* 99: 420–22.

Skipper, H. E., Schabel, F. M., Jr. & Wilcox, W. S. (1964). Experimental evaluation of potential anticancer agents XIII. On the criteria and kinetics associated with "curability" of experimental leukemia. *Cancer Chemotherapy Reports* 35: 1–111.

Slamon, D. J., Clark, G. M., Wong, S. G., Levin, W. J., Ullrich, A. & McGuire, W. L. (1987). Human breast cancer: Correlation of relapse and survival with amplification of the HER-2/neu oncogene. *Science* 235: 177–81.

Sleijfer, S. & Wierner, E. (2008). Dose selection in Phase I studies: Why we should always go for the top. *Journal of Clinical Oncology* 26: 1576–78.

Sneed, T. B., Kantarjian, H. M., Talpaz, M., O'Brien, S., Rios, M. B., Bekele, B. N., Zhou, X., Resta, D., Wierda, W., Faderl, S., Giles, F. & Cortes, J. E. (2003). The significance of myelosuppression during therapy with imatinib mesylate in patients with chronic myelogenous leukemia in chronic phase. *Cancer* 100: 116–21.

Sobin, L. H. (2001). TNM: Principles, history, and relation to other prognostic factors. *Cancer* 91 (Supplement): 1589–92.

Sobin, L. H., Gospodarowicz, M. K., Wittekind, Ch., eds. (2010). *UICC International Union Against Cancer—TNM Classification of Malignant Tumours,* 7th ed. Chichester, UK: Wiley-Blackwell.

Sonpavde, G., Galsky, M. D., Hutson, T. & Von Hoff, D. D. (2009). Patient selection for Phase II trials. *American Journal of Clinical Oncology* 32: 216–19.

Sox, H. C. (2010). Comparative effectiveness research: A progress report. *Annals of Internal Medicine* 153: 469–72.

Soto, A. M. & Sonnenschein, C. (2004). The somatic mutation theory of cancer: Growing problems with the paradigm? *Bioessays* 26: 1097–107.

Staquet, M. J., ed. (1972). *The Design of Clinical Trials in Cancer Therapy; Proceedings of a Course in Clinical Pharmacology, Held at the Institut Jules Bordet in Brussels from 2 to 5 May 1972.* Brussels: Editions scientifiques européennes.

———. (1976). The practice of cooperative clinical trials. *European Journal of Cancer* 12: 241–43.

Staquet, M., Sylvester, R. & Jasmin, C. (1980). Guidelines for the preparation of EORTC cancer clinical trials protocols. *European Journal of Cancer* 16: 871–75.

Stehlin, D., Varmus, H. E., Bishop, J. M. & Vogt, P. K. (1976). DNA related to the transforming gene(s) of avian sarcoma viruses is present in normal avian DNA. *Nature* 260: 170–73.

Steinle, S. (1996). Europa Donna unites Europe against breast cancer. *Journal of the National Cancer Institute* 88: 1614–15.

Sterne, J. A. C. & Smith, G. D. (2001). Sifting the evidence: What's wrong with significance tests. *BMJ* 322: 226–31.

Strasser, B. J. (2006). A world in one dimension: Linus Pauling, Francis Crick, and the central dogma of molecular biology. *History and Philosophy of the Life Sciences* 28: 491–512.

Strom, S. & Fleischer-Black, M. (2003). Drug maker's vow to donate cancer medicine falls short. *New York Times*, June 5, A1, C7.

Sykes, M. P., Karnofsky, D. A., Philips, F. S. & Burchenal, J. H. (1953). Clinical studies on triethylenephosphoramide compounds with nitrogen mustard-like activity. *Cancer* 6: 142–48.

Sylvester, R. J., Machin, D. & Staquet, M. J. (1978). A comparison of the alternative methods of calculating survival curves arising from clinical trials. *Biomedicine* 28 (September Special issue): 49–53.

Sylvester, R. J., Pinedo, H. M., De Pauw, M., Staquet, M. J., Buyse, M. E., Renard, J. & Bonadonna, G. (1981). Quality of institutional participation in multicenter clinical trials. *New England Journal of Medicine* 305: 852–55.

Sylvester, R., Van Glabbeke, M., Collette, L., Suciu, S., Baron, B., Legrand, C., Gorlia, T., Sollins, G., Coens, C., Declerck, L. & Therasse, P. (2002). Statistical methodology of Phase III cancer clinical trials: Advances and future perspectives. *European Journal of Cancer* 38: S162–68.

Tagnon, H. J. (1964). Introduction à la clinique cancérologique (Leçon inaugurale faite le 6 février 1964). *La Revue Médicale de Bruxelles* 20 (5): 281–87.

———. (1976). E.O.R.T.C. news. Definition and function of the E.O.R.T.C. *European Journal of Cancer* 12: 239–40.

———. (1977). European oncology revisited. In H. J. Tagnon & M. J. Staquet, eds., *Recent Advances in Cancer Treatment*. New York: Raven Press, 149–53.

Talpaz, M., Silver, R. T., Druker, B. Goldman, J. M., Gambacorti-Passerini, C., Guilhot, F., Schiffer, C. A., Fischer, T., Deininger, M. W. N., Lennard, A. L., Hochhaus, A., Ottmann, O. G., Gratwohl, A., Baccarani, M., Strone, R., Tura, S., Mahon, F.-X., Fernandes-Reese, S., Gathmann, I., Capdeville, R., Kantarjian, H. M. & Sawyers, C. L. (2001). Gleevec/Glivec (imatinib mesylate, STI-571) in patients with chronic myeloid leukemia in myeloid blast crisis: Updated results of a Phase II study. *Blood* 98: 845a.

Talpaz, M., Shah, N. P., Kantarjian, H., Donato, N., Nicoll, J., Paquette, R., Cortes, J., O'Brien, S., Nicaise, C., Bleickardt, E., Blackwood-Chirchir, M. A., Iyer, V., Chen, T. T., Huang, F., Decillis, A. P. & Sawyers, C. L. (2006). Dasatinib in imatinib-resistant Philadelphia chromosome-positive leukemias. *New England Journal of Medicine* 354: 2531–41.

Tan, A. R. & Swain, S. M. (2002). Ongoing adjuvant trials with trastuzumab in breast cancer. *Seminars in Oncology* 30 (5 Supplement 16): 54–64.

The technical history of radiology: A monograph prepared on the 75th anniversary of the Radiological Society of North America. (1989). *Radiographics* 9 (6): 1095–283.

Temin, H. M. & Mizutani, S. (1970). RNA-dependent DNA polymerase in virions of Rous carcinoma virus. *Nature* 226: 1211–13.

Temkin, O. (1963). The scientific approach to disease: Specific entity and individual sickness. In A. C. Crombie, ed., *Scientific Change*. London: Heinemann Educational, 629–47.

Therasse, P. (2002). Quality assurance within EORTC: Past, present and future. *European Journal of Cancer* 38: S151.

Therasse, P., Arbuck, S. G., Eisenhauer, E. A., Wanders, J., Kaplan, R. S., Rubinstein, L., Verweij, J., van Glabbeke, M., van Oosterom, A. T., Christian, M. C. & Gwyther, S. G. (2000). New guidelines to evaluate the response to treatment in solid tumors. *Journal of the National Cancer Institute* 92: 205–16.

Timmermans, S. & Berg, M. (2003). *The Gold Standard: The Challenge of Evidence-Based Medicine and Standardization in Health Care*. Philadelphia: Temple University Press.

Timothy, F. E. (1980). The reach to recovery program in America and Europe. *Cancer* 46: 1059–60.

Tippett, L. H. C. (1927). *Random Sampling Numbers*. Issue 15 of Tracts for Computers. Cambridge: Cambridge University Press.

Tormey, D. C. (1975). Combined chemotherapy and surgery in breast cancer: A review. *Cancer* 36: 881–92.

Translation of "new biology" into clinical use identified by DeVita as NCI's major challenge. (1985). *The Cancer Letter* 11 (33): 1–4.

Trusheim, M. R., Berndt, E. R. & Douglas, F. L. (2007). Stratified medicine: Strategic and economic implications of combining drugs and clinical biomarkers. *Nature Reviews—Drug Discovery* 6: 287–93.

Tubiana, M. (1991). *La lumière dans l'ombre: Le cancer hier et demain*. Paris: Odile Jacob.

———. (1995). *Histoire de la pensée médicale: Les chemins d'Esculape*. Paris: Flammarion.

UICC. (1958). Union for International Cancer Control. Committee on Clinical Stage Classification and Applied Statistics. *Clinical Stage Classification and Presentation of Results, Malignant Tumors of the Breast and Larynx*. Paris.

Ultmann, J. E., Hirschberg, E. & Gellhorn, A. (1962). Clinical cancer therapy in a single institution. *Journal of Chronic Diseases* 15: 259–63.

Uy, R. & Wold, F. (1977). Posttranslational covalent modification of proteins. *Science* 198: 890–96.

Van Glabbeke, M., Steward, W., & Armand, J. P. (2001). Non-randomised Phase II trials of drug combinations: Often meaningless, sometimes misleading. Are there alternative strategies? *European Journal of Cancer* 38: 635–38.

van Oosterom, A. T., Judson, I., Verweij, J., Stroobants, S., Donato di Paola, E., Dimitrijevic, S., Martens, M., Webb, A., Sciot, R., Van Glabbeke, M., Silberman, S. & Nielsen, O. S. (2001). Safety and efficacy of Imatinib (STI571) in metastatic gastrointestinal stromal tumours: A Phase I study. *The Lancet* 358: 1421–23.

Vantongelen, K., Rotmensz, N. & van der Schueren, E. (1989). Quality control of validity of data collected in clinical trials. *European Journal of Cancer and Clinical Oncology* 25: 1241–47.

Vantongelen, K., Steward, W., Blackedge, G., Verweij, J. & van Oosterom, A. (1991). EORTC joint ventures in quality control: Treatment-related variables and data acquisition in chemotherapy trials. *European Journal of Cancer* 27: 201–7.

Varmus, H. (1985). Viruses, genes, and cancer: The discovery of cellular oncogenes and the their role in neoplasia. *Cancer* 55: 2324–28.

———. (1989). Retroviruses and oncogenes I. Nobel Lecture. December 8, 1989.

———. (2009). *The Art and Politics of Science*. New York: W. W. Norton.

Vasella, D. (2003). *Magic Cancer Bullet: How a Tiny Orange Pill Is Rewriting Medical History*. New York: HarperBusiness.

Venditti, J. M., Wesley, R. A. & Plowman, J. (1984). Current NCI preclinical antitumor screening in vivo: Results of tumor panel screening, 1976–1982, and future directions. *Advances in Pharmacology and Chemotherapy* 20: 1–20.

Veronesi, U., Bonadonna, G. & Valagussa, P. (2008). Lessons from the initial adjuvant cyclophosphamide, methotrexate, and fluorouracil studies in operable breast cancer. *Journal of Clinical Oncology* 26: 342–44.

Verweij, J., Casali, P. G., Zalcberg, J., LeCesne, A., Reichardt, P., Blay, J. Y., Issels, R., van Oosterom, A., Hogendoorn, P. C., Van Glabbeke, M., Bertulli, R. & Judson I. (2004). Progression-free survival in gastrointestinal stromal tumours with high-dose imatinib: Randomised trial. *The Lancet* 364: 1127–34.

Veyne, P. (2008). *Foucault: Sa pensée, sa personne.* Paris: Albin Michel.

Visco, F. (2007). The National Breast Cancer Coalition: Setting the standard for advocate collaboration in clinical trials. In S. P. L. Leong, ed., *Cancer Clinical Trials: Proactive Strategies.* New York: Springer, 143–56.

Vogelstein, B., Fearon, E. R., Hamilton, S. R., Kern, S. E., Preisinger, A. C., Leppert, M., Nakamura, Y., White, R., Smits, A. M. & Bos, J. L. (1988). Genetic alterations during colorectal-tumor development. *New England Journal of Medicine* 319: 525–32.

Vogt, P. (2009). A humble chicken virus that changed biology and medicine. *The Lancet Oncology* 10: 96.

Wagener, D. J. Th., Vermorken, J. B., Hansen, H. H. & Hossfeld, D. K. (1998). The ESMO programme of certification and training for medical oncology. *Annals of Oncology* 9: 585–87.

Wagner, R. P. (1999). Rudolph Virchow and the genetic basis of somatic ecology. *Genetics* 151: 917–20.

Wagstaff, A. (2004). Bob Pinedo: Bringing two worlds together. *Cancer World* (September–October): 28–33.

Wailoo, K. (2005). *The Strange Career of Race and Cancer.* New York: Oxford University Press.

Wald, A. (1945). Sequential tests of statistical hypotheses. *Annals of Mathematical Statistics* 16: 117–86.

——. (1947). *Sequential Analysis.* New York: Wiley.

Wallhauser, A. (1933). Hodgkin's disease. *Archives of Pathology*: 522–62.

Wallis, W. A. (1980). The Statistical Research Group, 1942–1945. *Journal of the American Statistical Association* 75: 320–30.

Walter, W. A. & King, T. J. (1977). The Cancer Centers program. *Journal of the National Cancer Institute* 59 (Supplement): 645–49.

Wasserman, T. H., Slavik, M. & Carter, S. K. (1974). Review of CCNU in clinical cancer therapy. *Cancer Treatment Reviews* 1 (1974): 131–51.

Watson, J. D., Baker, T. A., Bell, S. P., Gann, A., Levine, M. & Losick, R. (2008). *Molecular Biology of the Gene,* 6th ed. Cold Spring Harbor, NY: Cold Spring Harbor Laboratory Press.

Weinberg, R. A. (1996). *Racing to the Beginning of the Road: The Search for the Origins of Cancer.* New York: W. H. Freeman.

Weiss, L. (2000a). The morphologic documentation of clinical progression, invasion metastasis—staging. *Cancer and Metastasis Reviews* 19: 303–13.

——. (2000b). The prediction of invasion, metastasis and progression by grading. *Cancer and Metastasis Reviews* 19: 315–25.

Weiss, R. B. (1992). The anthracyclines: Will we ever find a better Doxorubicin? *Seminars in Oncology* 19: 670–86.

Welch, H. G., Schwartz, L. M. & Woloshin, S. (2000). Are increasing 5-year survival rates evidence of success against cancer? *JAMA* 283: 2975–78.

Whalen, J. (2009). FDA approves a novel Novartis drug. *Wall Street Journal,* June 19, B1–B2.

Whang-Peng, J., Canellos, G. P., Carbone, P. P. & Tjio, J. H. (1968). Clinical implications of cytogenetic variants in chronic myelocytic leukemia. *Blood* 32: 755–66.

White House to appoint Richard Klausner, molecular biologist, as NCI's 11th director. (1995). *The Cancer Letter* 21 (30): 1–4.

Witte, O. N., Dasgupta, A. & Baltimore, D. (1980). Abelson Murine leukemia virus is phosphorylated *in vitro* to form phosphotyrosine. *Nature* 283: 826–31.

Witte, O. N., Rosenberg, N., Paskind, M., Shields, A. & Baltimore, D. (1978). Identification of an Abelson Murine leukemia virus-encoded protein present in transformed fibroblast and lymphoid cells. *Proceedings of the National Academy of Sciences* 75: 2488–92.

Wittes, group chairmen agree on payment for case, other methods to increase patient accrual. (1987). *The Cancer Letter* 13 (34): 1–7.

Wittes, R. E., Friedman, M. A. & Simon, R. (1986). Some thoughts on the future of clinical trials in cancer. *Cancer Treatment Reports* 70: 241–50.

Wittes, R. E. & Yoder, O. (1998). One community. *Annals of Oncology* 9: 251–54.

Wolfowitz, J. (1952). Abraham Wald, 1902–1950. *Annals of Mathematical Statistics* 23: 1–13, 29–33.

Wong, A. J., Ruppert, J. M., Eggleston, J., Hamilton, S. R., Baylin, S. B. & Vogelstein, B. (1986). Gene amplification of c-myc and n-myc in small cell carcinoma of the lung. *Science* 233: 461–64.

Wooldridge, D. E. (Wooldridge Committee). (1965). *Biomedical Science and Its Administration: A Study of the National Institutes of Health.* Washington, DC: The White House.

Working group to review clinical trials programs. (1996). *The Cancer Letter* 22 (15): 4–6.

Workman, P. & Collins, I. (2008). Modern cancer drug discovery: Integrating targets technologies and treatments. In S. Neidle, ed., *Cancer Drug Discovery and Design*. New York: Academic Press, 3–38.

Yaish, P., Gazit, A., Gilon, C. & Levitzki, A. (1988). Blocking of EGF-dependent cell proliferation by EGF receptor kinase inhibitors. *Science* 242: 933–35.

Yoder, O. C. (1997). Clinical development straddling the Atlantic. Interview with Dr. Omar C. Yoder. *Cancer Centers around the World* 1: 9.

Yu, L. R. & Veenstra, T. D. (2007). AACR-FDA-NCI cancer biomarkers collaborative. *Expert Review of Molecular Diagnosis* 7: 507–9.

Yule, G. U. (1938). A test of Tippett's random sampling numbers. *Journal of the Royal Statistical Society* 101: 167–72.

Yuspa, S. H., Hennings, H, & Saffiotti, U. (1976). Cutaneous chemical carcinogenesis: Past, present, and future. *Journal of Investigative Dermatology* 67: 199–208.

Zelen, M. (1974). The randomization and stratification of patients to clinical trials. *Journal of Chronic Diseases* 27: 365–75.

Zerhouni, E. A. (2003). The NIH roadmap. *Science* 302: 63–64, 72.

———. (2005). US biomedical research: Basic, translational, and clinical sciences. *JAMA* 294: 1352–58.

Ziegler, K. (1911). *Die Hodgkinsche Krankheit*. Jena: G Fischer.

Zubrod, C. G. (1960a). Remarks made during the "Panel Meeting: Newer techniques and some problems in cooperative group studies." In B. H. Morrison III, ed., *Conference on Experimental Clinical Cancer Chemotherapy. November 11 and 12, 1959*. Washington, DC: U.S. Department of Health, Education, and Welfare, 277–92.

———. (1960b). Remarks made during the "Panel Meeting: Use of alkylating agents and the future of these agents." In B. H. Morrison III, ed., *Conference on Experimental Clinical Cancer Chemotherapy. November 11 and 12, 1959*. Washington, DC: U.S. Department of Health, Education, and Welfare, 127–48.

———. (1979). Historical milestones in curative chemotherapy. *Seminars in Oncology* 6: 490–505.

———. (1980). The cure of cancer by chemotherapy: Reflections of how it happened. The fifth Myron Karon Memorial Lecture. *Medical and Pediatric Oncology* 8: 107–14.

———. (1984). Origins and development of chemotherapy research at the National Cancer Institute. *Cancer Treatment Reports* 68: 9–19.

Zubrod, C. G., Schepartz, S., Leiter, J., Endicott, K. M., Carrese, L. M. & Baker, C. G. (1966). The chemotherapy program of the National Cancer Institute: History, analysis, and plans. *Cancer Chemotherapy Reports* 50: 349–540.

Zubrod, C. G., Schneiderman, M., Frei, E., III, Brindley, C., Oviedo, R., Gorman, J., Jones, Jr., R., Jonsson, U., Colsky, J., Chalmers, T., Ferguson, B.,

Dederick, M., Holland, J., Selawry, O., Regelson, W., Lasagna, L. & Owens, A. H., Jr. (1969). Appraisal of methods for the study of chemotherapy of cancer in man: Comparative therapeutic trial of nitrogen mustard and triethylene thiophosphoramide. *Journal of Chronic Disease* 11: 7–33.

Zwieten, M. J. van. (1984). General discussion, including a brief review of animal models in breast cancer research. In M. J. van Zwieten, ed., *The Rat as Animal Model in Breast Cancer Research: A Histopathologcal Study of Radiation and Hormone-Induced Rat Mammary Tumors*. The Hague: Martinus Nijhoff, 199–224.

Index